A FIRST COURSE IN
GRAPH THEORY

GARY CHARTRAND

and

PING ZHANG

Western Michigan University

DOVER PUBLICATIONS, INC.
Mineola, New York

Bibliographical Note

This Dover edition, first published in 2012, is a revised and corrected republication of *Introduction to Graph Theory,* originally published in 2005 by McGraw-Hill Higher Education, Boston.

Library of Congress Cataloging-in-Publication Data

Chartrand, Gary.
    A first course in graph theory / Gary Chartrand and Ping Zhang.
       p. cm.
    Previous edition published as: Introduction to graph theory. Boston : McGraw-Hill Higher Education, c2005
    Includes bibliographical references and index.
    ISBN-13: 978-0-486-48368-9
    ISBN-10: 0-486-48368-1
    1. Graph theory. I. Zhang, Ping, 1957– II. Chartrand, Gary. Introduction to graph theory. III. Title.

QA166.C455 2012
511'.5—dc23

2011038125

Manufactured in the United States by RR Donnelley
48368107   2016
www.doverpublications.com

Dedicated to the memory of the many mathematicians whose contributions, linked in a variety of ways, have led to the development of graph theory.

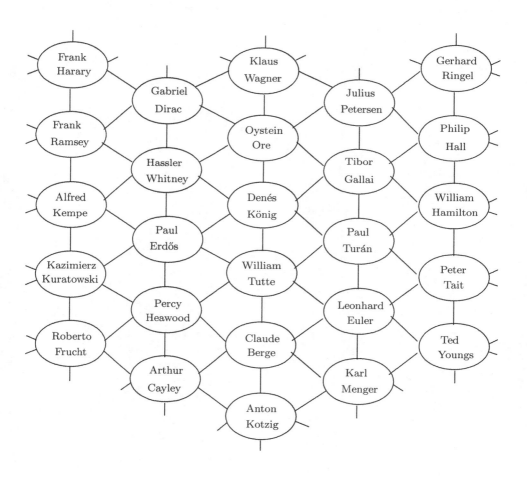

From Königsberg to König's book,
So runs the graphic tale.
And still it grows more colorful ...

— — — Blanche Descartes (1969)

CONTENTS

13. Domination

PREFACE

Perhaps it's not so surprising that when we (the authors) were learning mathematics, we thought that we were being taught some well-known facts – facts that had been around forever. It wasn't until later that we started to understand that these facts (the word "theorem" was beginning to become part of our vocabulary) had not been around forever and that *people* had actually discovered these facts. Indeed, *names* of people were becoming part of the discussion.

Mathematics has existed for many centuries. In the ancient past, certain cultures developed their own mathematics. This was certainly the case with Egypt, Babylonia, Greece, China, India and Japan. In recent centuries, there has become only one international mathematics. It has become more organized and has been divided into more clearly defined areas (even though there is significant overlap). While this was occurring, explanations (proofs) as to why mathematical statements are true were becoming more structured and clearly written.

The goal of this book is to introduce undergraduates to the mathematical area called *graph theory*, which only came into existence during the first half of the 18th century. This area didn't start to develop into an organized branch of mathematics until the second half of the 19th century and there wasn't even a book on the subject until the first half of the 20th century. Since the second half of the 20th century, however, the subject has exploded.

It is our intent to describe some of the major topics of this subject to you and to inform you of some of the people who helped develop and shape this area. In the beginning, most of these people were just like you – students who enjoyed mathematics but with a great sense of curiosity. As with everything else (though not as often talked about), mathematics has its non-serious side and we've described some of this as well. Even the most brilliant mathematicians don't know everything and we've presented some topics that have not been well-studied and in which the answers (and even the questions) are not known. This will give you the chance to do some creative thinking of your own. In fact, maybe the next person who will have an influence on this subject is you.

Part of what makes graph theory interesting is that graphs can be used to model situations that occur within certain kinds of problems. These problems can then be studied (and possibly solved) with the aid of graphs. Because of this, graph models occur frequently throughout this textbook. However, graph theory is an area of mathematics and consequently concerns the study of mathematical ideas – of concepts and their connections with each other. The topics and results we have included were chosen because we feel they are interesting, important and/or are representative of the subject.

As we said, this text has been written for undergraduates. Keeping this in mind, we have included a proof of a theorem if we believe it is appropriate, the proof technique is informative and if the proof is not excessively long. We would like to think that the material in this text will be useful and interesting for mathematics students as well as for other students whose areas of interest include graphs. This text is also appropriate for self-study.

We have included three appendixes. In Appendix 1, we review some important facts about sets and logic. Appendix 2 is devoted to equivalence relations and functions while Appendix 3 describes methods of proof. We understand how frustrating it is for students (or anyone!) who try to read a proof that is not reader-friendly and which leaves too many details for the reader to supply. Consequently, we have endeavored to give clear, well-written proofs.

Although this can very well be said about any area of mathematics or indeed about any scholarly activity, we feel that appreciation of graph theory is enhanced by being familiar with many of the people, past and present, who were or are responsible for its development. Consequently, we have included several remarks that we find interesting about some of the "people of graph theory." Since we believe that these people are part of the story that is graph theory, we have discussed them within the text and not simply as footnotes. We often fail to recognize that mathematics is a living subject. Graph theory was created by *people* and is a subject that is still evolving.

There are several sections that have been designated as "Excursion." These can be omitted with no negative effect if this text is being used for a course. In some cases, an Excursion is an area of graph theory we find interesting but which the instructor may choose not to discuss due to lack of time or because it's not one of his or her favorites. In other cases, an Excursion brings up a sidelight of graph theory that perhaps has little, if any, mathematical content but which we simply believe is interesting.

There are also sections that we have designated as "Exploration." These sections contain topics with which students can experiment and use their imagination. These give students opportunities to practice asking questions. In any case, we believe that this might be fun for some students.

As far as using this text for a course, we consider the first three chapters as introductory. Much of this could be covered quite quickly. Students could read these chapters on their own. It isn't necessary to cover connectivity and Menger's Theorem if the instructor chooses not to do so. Sections 8.3, 9.2, 10.3 and 11.2 could easily be omitted, while material from Chapters 12 and 13 can be covered according to the instructor's interest.

Solutions or hints for the odd-numbered exercises in the regular sections of the text, references, an index of mathematical terms, an index of people and a list of symbols are provided at the end of the text.

It was because of discussions we had with Robert Ross that we decided to write "An Introduction to Graph Theory." We thank him for this and for his encouragement. We especially thank John Grafton, Senior Reprint Editor at Dover Publications, whose encouragement led us to revise the book, with its new title "A First Course in Graph Theory." We are most grateful to the reviewers of the original edition who gave us many valuable suggestions: Jay Bagga, Ball State University; Richard Borie, University of Alabama; Anthony Evans, Wright

State University; Mark Ginn, Appalachian State University; Mark Goldberg, Rensselaer Polytechnic Institute; Arthur Hobbs, Texas A&M University; Garth Isaak, Lehigh University; Daphne Liu, California State University, Los Angeles; Alan Mills, Tennessee Technological University; Dan Pritikin, Miami University; John Reay, Western Washington University; Yue Zhao, University of Central Florida.

Gary Chartrand and Ping Zhang
May 2011

Chapter 1

Introduction

1.1   Graphs and Graph Models

A major publishing company has ten editors (referred to by $1, 2, \ldots, 10$) in the scientific, technical and computing areas. These ten editors have a standard meeting time during the first Friday of every month and have divided themselves into seven committees to meet later in the day to discuss specific topics of interest to the company, namely, advertising, securing reviewers, contacting new potential authors, finances, used and rented copies, electronic editions and competing textbooks. This leads us to our first example.

Example 1.1   The ten editors have decided on the seven committees: $c_1 = \{1, 2, 3\}$, $c_2 = \{1, 3, 4, 5\}$, $c_3 = \{2, 5, 6, 7\}$, $c_4 = \{4, 7, 8, 9\}$, $c_5 = \{2, 6, 7\}$, $c_6 = \{8, 9, 10\}$, $c_7 = \{1, 3, 9, 10\}$. They have set aside three time periods for the seven committees to meet on those Fridays when all ten editors are present. Some pairs of committees cannot meet during the same period because one or two of the editors are on both committees. This situation can be modeled visually as shown in Figure 1.1.

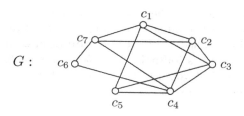

Figure 1.1: A graph

In this figure, there are seven small circles, representing the seven committees and a straight line segment is drawn between two circles if the committees they represent have at least one committee member in common. In other words, a straight line segment between two small circles (committees) tells us that these two committees should not be scheduled to meet at the same time. This gives us a picture or a "model" of the committees and the overlapping nature of their membership.   ◇

What we have drawn in Figure 1.1 is called a graph. Formally, a **graph** G consists of a finite nonempty set V of objects called **vertices** (the singular is **vertex**) and a set E of 2-element subsets of V called **edges**. The sets V and E are the **vertex set** and **edge set** of G, respectively. So a graph G is a pair (actually an *ordered* pair) of two sets V and E. For this reason, some write $G = (V, E)$. At times, it is useful to write $V(G)$ and $E(G)$ rather than V and E to emphasize that these are the vertex and edge sets of a particular graph G. Although G is the common symbol to use for a graph, we also use F and H, as well as G', G'' and G_1, G_2, etc. Vertices are sometimes called **points** or **nodes** and edges are sometimes called **lines**. Indeed, there are some who use the term **simple graph** for what we call a graph. Two graphs G and H are **equal** if $V(G) = V(H)$ and $E(G) = E(H)$, in which case we write $G = H$.

It is common to represent a graph by a diagram in the plane (as we did in Figure 1.1) where the vertices are represented by points (actually small circles – open or solid) and whose edges are indicated by the presence of a line segment or curve between the two points in the plane corresponding to the appropriate vertices. The diagram itself is then also referred to as a graph. For the graph G of Figure 1.1 then, the vertex set of G is $V(G) = \{c_1, c_2, \cdots, c_7\}$ and the edge set of G is

$$
\begin{aligned}
E(G) \quad &= \quad \{\{c_1, c_2\}, \{c_1, c_3\}, \{c_1, c_5\}, \{c_1, c_7\}, \{c_2, c_3\}, \{c_2, c_4\}, \{c_2, c_7\}, \\
&\qquad \{c_3, c_4\}, \{c_3, c_5\}, \{c_4, c_5\}, \{c_4, c_6\}, \{c_4, c_7\}, \{c_6, c_7\}\}.
\end{aligned}
$$

Let's consider another situation. Have you ever encountered this sequence of integers before?

$$1,\ 1,\ 2,\ 3,\ 5,\ 8,\ 13,\ 21,\ 34,\ 55,\ \ldots$$

Every integer in the sequence is the sum of the two integers immediately preceding it (except for the first two integers of course). These numbers are well known in mathematics and are called the **Fibonacci numbers**. In fact, these integers occur so often that there is a journal (*The Fibonacci Quarterly*, frequently published *five* times a year!) devoted to the study of their properties. Our second example concerns these numbers.

Example 1.2 Consider the set $S = \{2, 3, 5, 8, 13, 21\}$ of six specific Fibonacci numbers. There are some pairs of distinct integers belonging to S whose sum or difference (in absolute value) also belongs to S, namely, $\{2, 3\}$, $\{2, 5\}$, $\{3, 5\}$, $\{3, 8\}$, $\{5, 8\}$, $\{5, 13\}$, $\{8, 13\}$, $\{8, 21\}$ and $\{13, 21\}$. There is a more visual way of identifying these pairs, namely by the graph H of Figure 1.2. In this case, $V(H) = \{2, 3, 5, 8, 13, 21\}$ and

$$E(H) = \{\{2,3\}, \{2,5\}, \{3,5\}, \{3,8\}, \{5,8\}, \{5,13\}, \{8,13\}, \{8,21\}, \{13,21\}\}. \ \diamondsuit$$

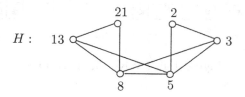

Figure 1.2: Another graph

When dealing with graphs, it is customary and simpler to represent an edge $\{u, v\}$ by uv (or vu). If uv is an edge of G, then u and v are said to be **adjacent** in G. The number of vertices in G is often called the **order** of G, while the number of edges is its **size**. Since the vertex set of every graph is nonempty, the order of every graph is at least 1. A graph with exactly one vertex is called a **trivial graph**, implying that the order of a **nontrivial graph** is at least 2. The graph G of Figure 1.1 has order 7 and size 13, while the graph H of Figure 1.2 has order 6 and size 9. We often use n and m for the order and size, respectively, of a graph. So, for the graph G of Figure 1.1, $n = 7$ and $m = 13$; while for the graph H of Figure 1.2, $n = 6$ and $m = 9$.

A graph G with $V(G) = \{u, v, w, x, y\}$ and $E(G) = \{uv, uw, vw, vx, wx, xy\}$ is shown in Figure 1.3(a). There are occasions when we are interested in the structure of a graph and not in what the vertices are called. In this case, a graph is drawn without labeling its vertices. For this reason, the graph G of Figure 1.3(a) is a **labeled graph** and Figure 1.3(b) represents an **unlabeled graph**.

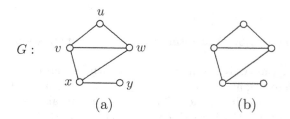

(a)   (b)

Figure 1.3: A labeled graph and an unlabeled graph

Let us now turn to yet another situation.

Example 1.3 Suppose that we have two coins, one silver and one gold, placed on two of the four squares of a 2×2 checkerboard. There are twelve such configurations, shown in Figure 1.4, where the shaded coin is the gold coin.

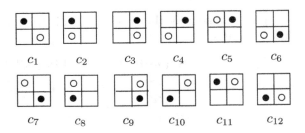

Figure 1.4: Twelve configurations

A configuration can be transformed into other configurations according to certain rules. Specifically, we say that the configuration c_i can be transformed into the configuration c_j $(1 \leq i,j \leq 12, i \neq j)$ if c_j can be obtained from c_i by performing exactly one of the following two steps:

(1) moving one of the coins in c_i horizontally or vertically to an unoccupied square;

(2) interchanging the two coins in c_i.

Necessarily, if c_i can be transformed into c_j, then c_j can be transformed into c_i. For example, c_2 can be transformed (i) into c_1 by shifting the silver coin in c_2 to the right, (ii) into c_4 by shifting the gold coin to the right or (iii) into c_8 by interchanging the two coins (see Figure 1.5).

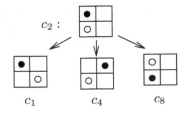

Figure 1.5: Transformations of the configuration c_2

Now consider the twelve configurations shown in Figure 1.4. Some pairs c_i, c_j of these configurations, where $1 \leq i,j \leq 12$, $i \neq j$, can be transformed into each other and some pairs cannot. This situation can also be represented by a graph, say by a graph F where $V(F) = \{c_1, c_2, \ldots, c_{12}\}$ and $c_i c_j$ is an edge of F if c_i and c_j can be transformed into each other. This graph F is shown in Figure 1.6. $\diamond$

Let's look at a somewhat related example.

Example 1.4. Suppose that we have a collection of 3-letter English words, say

ACT, AIM, ARC, ARM, ART, CAR, CAT, OAR, OAT, RAT, TAR.

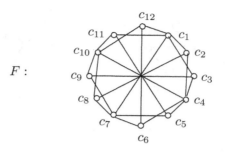

$F:$

Figure 1.6: Modeling transformations of twelve configurations

We say that a word W_1 can be transformed into a word W_2 if W_2 can be obtained from W_1 by performing exactly one of the following two steps:

(1) interchanging two letters of W_1;

(2) replacing a letter in W_1 by another letter.

Therefore, if W_1 can be transformed into W_2, then W_2 can be transformed into W_1. This situation can be modeled by a graph G, where the given words are the vertices of G and two vertices are adjacent in G if the corresponding words can be transformed into each other. This graph is called **the word graph of the set of words.** For the 11 words above, its word graph G is shown in Figure 1.7.

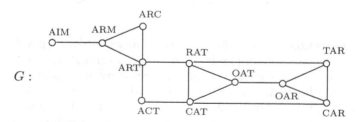

$G:$

Figure 1.7: The word graph of a set of 11 words

In this case, a graph G is called **a word graph** if G is the word graph of some set S of 3-letter words. For example, the (unlabeled) graph G of Figure 1.8(a) is a word graph because it is the word graph of the set $S = \{$BAT, BIT, BUT, BAD, BAR, CAT, HAT$\}$, as shown in Figure 1.8(b). (This idea is related to the concept of "isomorphic graphs," which will be discussed in Chapter 3.) ◇

We conclude this section with one last example.

Example 1.5 Figure 1.9 shows the traffic lanes at the intersection of two busy streets. When a vehicle approaches this intersection, it could be in one of the nine lanes: L1, L2, ..., L9.

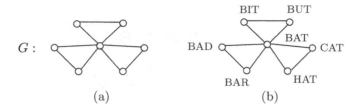

Figure 1.8: A word graph

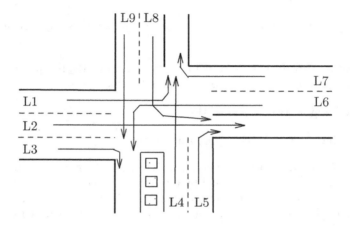

Figure 1.9: Traffic lanes at street intersections

This intersection has a traffic light that informs drivers in vehicles in the various lanes when they are permitted to proceed through the intersection. To be sure, there are pairs of lanes containing vehicles that should not enter the intersection at the same time, such as L1 and L7. However, there would be no difficulty for vehicles in L1 and L5 to drive through this intersection at the same time. This situation can be represented by the graph G of Figure 1.10, where $V(G) = \{L1, L2, \ldots, L9\}$ and two vertices (lanes) are joined by an edge if vehicles in these two lanes cannot safely enter the intersection at the same time, as there would be a possibility of an accident.   ◇

What we have just seen is how five different situations can be represented by graphs. Actually, in each case, there is a set involved: (1) a set of committees, (2) a set of integers, (3) a set of configurations consisting of two coins on a 2×2 checkerboard, (4) a set of 3-letter words, (5) a set of traffic lanes at a street intersection. Certain pairs of elements in each set are related in some manner: (1) two committees have a member in common, (2) the sum or difference (in absolute value) of two integers in the set also belongs to the set, (3) two configurations can be transformed into each other according to some rule, (4) two 3-letter words can be transformed into each other by certain movements of letters, (5) cars in

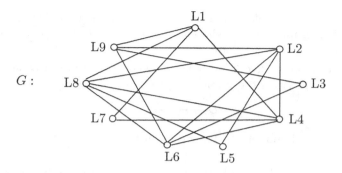

Figure 1.10: The graph G in Example 1.5

certain pairs of traffic lanes cannot enter the intersection at the same time. In each case, a graph G is defined whose vertices are the elements of the set and two vertices of G are adjacent if they are related as described above. The graph G then *models* the given situation. Often questions concerning the situations described above arise and can be analyzed by studying the graphs that model them. We will encounter such questions throughout the text and discuss how graphs can be used to help us answer the questions.

Exercises for Section 1.1

1.1 What is a logical question to ask in Example 1.1? Answer this question.

1.2 Create an example of your own similar to Example 1.1 with nine editors and eight committees and then draw the corresponding graph.

1.3 Let $S = \{2, 3, 4, 7, 11, 13\}$. Draw the graph G whose vertex set is S and such that $ij \in E(G)$ for $i, j \in S$ if $i + j \in S$ or $|i - j| \in S$.

1.4 Let $S = \{-6, -3, 0, 3, 6\}$. Draw the graph G whose vertex set is S and such that $ij \in E(G)$ for $i, j \in S$ if $i + j \in S$ or $|i - j| \in S$.

1.5 Create your own set S of integers and draw the graph G whose vertex set is S and such that $ij \in E(G)$ if i and j are related by some rule imposed on i and j.

1.6 Consider the twelve configurations $c_1, c_2, \ldots, c_{12}$ in Figure 1.4. For every two configurations c_i and c_j, where $1 \le i, j \le 12$, $i \ne j$, it may be possible to obtain c_j from c_i by first shifting one of the coins in c_i horizontally or vertically *and* then interchanging the two coins. Model this by a graph F such that $V(F) = \{c_1, c_2, \ldots, c_{12}\}$ and $c_i c_j$ is an edge of F if c_i and c_j can be transformed into each other by this 2-step process.

1.7 Following Example 1.4,

(a) give an example of ten 3-letter words, none of which are mentioned in Example 1.4 and whose corresponding word graph has at least six edges. Draw this graph.

(b) give a set of five 3-letter words whose word graph is shown in Figure 1.11 (with the vertices appropriately labeled).

Figure 1.11: The graph in Exercise 1.7(b)

(c) give a set of five 3-letter words whose word graph is shown in Figure 1.12 (with the vertices appropriately labeled).

Figure 1.12: The graph in Exercise 1.7(c)

1.8 Let S be a finite set of 3-letter and/or 4-letter words. In this case, the word graph $G(S)$ of S is that graph whose vertex set is S and such that two vertices (words) w_1 and w_2 are adjacent if either (1) or (2) below occurs:

(1) one of the words can be obtained from the other by replacing one letter by another letter,

(2) w_1 is a 3-letter word and w_2 is a 4-letter word and w_2 can be obtained from w_1 by the insertion of a single letter (anywhere, including the beginning or the end) into w_1.

(a) Find six sets $S_1, S_2, \ldots, S_6$ of 3-letter and/or 4-letter words so that for each integer i ($1 \le i \le 6$) the graph G_i of Figure 1.13 is the word graph of S_i.

(b) For another graph H (of your choice), determine whether H is a word graph of some set.

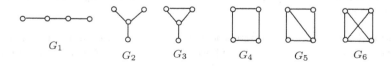

Figure 1.13: The graphs for Exercise 1.8(a)

1.9 Define a word graph differently from the word graphs defined in Example 1.4 and Exercise 1.8 and illustrate your definition.

1.10 Figure 1.14 illustrates the traffic lanes at the intersection of two streets. When a vehicle approaches this intersection, it could be in one of the seven lanes: L1, L2, ..., L7. Draw a graph G that models this situation, where $V(G) = \{$L1, L2, ..., L7$\}$ and where two vertices are joined by an edge if vehicles in these two lanes cannot safely enter this intersection at the same time.

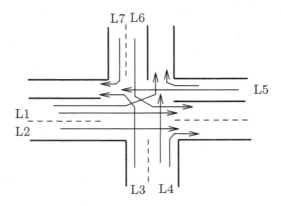

Figure 1.14: Traffic lanes at a street intersection in Exercise 1.10

1.2 Connected Graphs

In order to analyze certain situations that can be modeled by graphs, we must have a better understanding of graphs. As with all areas of mathematics, there is a certain amount of terminology with which we must first be familiar in order to discuss graphs and their properties. Becoming aware of this fundamental terminology is our current goal. First, let's review some concepts and introduce others. Recall that a graph G consists of a finite nonempty set V of vertices and a set E of 2-element subsets of V called edges. If $e = uv$ is an edge of G, then the adjacent vertices u and v are said to be **joined** by the edge e. The vertices u and v are referred to as **neighbors** of each other. In this case, the vertex u and the edge e (as well as v and e) are said to be **incident** with each other. Distinct edges incident with a common vertex are **adjacent edges**.

As we mentioned earlier, although graphs are defined in terms of sets, it is customary and convenient to represent graphs by (and, in fact, to consider them as) diagrams. A graph G with vertex set $V = \{u, v, w, x, y\}$ and edge set $E = \{uv, vw, vx, vy, wy, xy\}$ is shown in Figure 1.15. Since this graph has five vertices and six edges, its order is 5 and its size is 6. In this graph G, the vertices u and v are adjacent, while u and w are not adjacent. The vertex v is incident with the edge vw but not with the edge wy. The edges uv and vw are adjacent, but uv and xy are not adjacent.

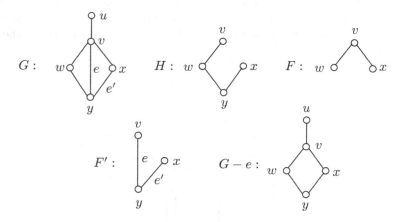

Figure 1.15: A graph G and some of its subgraphs

A graph H is called a **subgraph** of a graph G, written $H \subseteq G$, if $V(H) \subseteq V(G)$ and $E(H) \subseteq E(G)$. We also say that G contains H as a subgraph. If $H \subseteq G$ and either $V(H)$ is a proper subset of $V(G)$ or $E(H)$ is a proper subset of $E(G)$, then H is a **proper subgraph** of G. So the graph H of Figure 1.15 is a subgraph of the graph G shown in that figure; indeed, H is a proper subgraph of G. If a subgraph of a graph G has the same vertex set as G, then it is a **spanning subgraph** of G.

A subgraph F of a graph G is called an **induced subgraph** of G if whenever u and v are vertices of F and uv is an edge of G, then uv is an edge of F as well. Therefore, the graph H of Figure 1.15 is not an induced subgraph of the graph G of Figure 1.15 since, for example, $v, x \in V(H)$ and $vx \in E(G)$ but $vx \notin E(H)$. On the other hand, the graph F of Figure 1.15 is an induced subgraph of G. If S is a nonempty set of vertices of a graph G, then the **subgraph of G induced by** S is the induced subgraph with vertex set S. This induced subgraph is denoted by $G[S]$. For a nonempty set X of edges, the **subgraph $G[X]$ induced by** X has edge set X and consists of all vertices that are incident with at least one edge in X. This subgraph is called an **edge-induced subgraph** of G. Sometimes $\langle S \rangle_G$ and $\langle X \rangle_G$ are used for $G[S]$ and $G[X]$, respectively. The graph F' of Figure 1.15 is an edge-induced subgraph of G in that figure; indeed, $F' = G[X']$, where $X' = \{e, e'\}$.

Any proper subgraph of a graph G can be obtained by removing vertices and edges from G. For an edge e of G, we write $G-e$ for the spanning subgraph of G whose edge set consists of all edges of G except e. More generally, if X is a set of edges of G, then $G-X$ is the spanning subgraph of G with $E(G-X) = E(G)-X$. For the graph G of Figure 1.15 and $e = vy$, the subgraph $G - e$ is shown. If $X = \{e_1, e_2, \ldots, e_k\}$, then we also write $G - X$ as $G - e_1 - e_2 - \cdots - e_k$.

For a vertex v of a nontrivial graph G, the subgraph $G - v$ consists of all vertices of G except v and all edges of G except those incident with v. For a proper subset U of $V(G)$, the subgraph $G - U$ has vertex set $V(G) - U$ and its edge set consists of all edges of G joining two vertices in $V(G) - U$. Necessarily, $G - U$ is an induced subgraph of G. For $U = \{u, y\}$ in the graph G of Figure 1.15, $G - U$ is the subgraph F shown in that figure.

If u and v are nonadjacent vertices of a graph G, then $e = uv \notin E(G)$. By $G + e$, we mean the graph with vertex set $V(G)$ and edge set $E(G) \cup \{e\}$. Thus G is a spanning subgraph of $G + e$.

Many of the concepts that occur in graph theory and which we will investigate in detail later concern various ways in which one can "move about" in a graph. In particular, if we think of the vertices of a graph as locations and the edges as roads between certain pairs of locations, then the graph can be considered as modeling some community. There is a variety of kinds of trips that can be taken in the community.

Let's start at some vertex u of a graph G. If we proceed from u to a neighbor of u and then to a neighbor of that vertex and so on, until we finally come to a stop at a vertex v, then we have just described a walk from u to v in G. More formally, a $u - v$ **walk** W in G is a sequence of vertices in G, beginning with u and ending at v such that consecutive vertices in the sequence are adjacent, that is, we can express W as

$$W = (u = v_0, v_1, \ldots, v_k = v), \tag{1.1}$$

where $k \geq 0$ and v_i and v_{i+1} are adjacent for $i = 0, 1, 2, \ldots, k - 1$. Each vertex v_i $(0 \leq i \leq k)$ and each edge $v_i v_{i+1}$ $(0 \leq i \leq k - 1)$ is said to lie on or belong to W. Notice that the definition of the walk W does not require the listed vertices to be distinct; in fact, even u and v are not required to be distinct. However, every two consecutive vertices in W are distinct since they are adjacent. If $u = v$, then the walk W is **closed**; while if $u \neq v$, then W is **open**. As we move from one vertex of W to the next, we are actually encountering or traversing edges of G, possibly traversing some edges of G more than once. The number of edges encountered in a walk (including multiple occurrences of an edge) is called the **length** of the walk. Thus the length of the walk W defined in (1.1) is k.

For the graph G of Figure 1.16,

$$W = (x, y, w, y, v, w) \tag{1.2}$$

is therefore a walk, indeed an $x - w$ walk of length 5 (one less than the number of occurrences of vertices in the walk). A walk of length 0 is a **trivial walk**. So $W = (v)$ is a trivial walk. (By this definition, those people who feel guilty about not exercising need not feel guilty any longer as going for a daily "walk" just became easier.)

Provided we continue to proceed from a vertex to one of its neighbors (and eventually stop), there is essentially no conditions on a walk. However, there will be occasions when we want to place restrictions on certain types of walks.

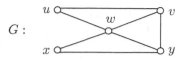

Figure 1.16: Illustrating walks in a graph

Borrowing terminology from the Old West, we define a $u - v$ **trail** in a graph G to be a $u - v$ walk in which no edge is traversed more than once. Thus, the $x - w$ walk W in (1.2) is *not* an $x - w$ trail as the edge wy is repeated. On the other hand,

$$T = (u, w, y, x, w, v) \tag{1.3}$$

is a $u - v$ trail in the graph G of Figure 1.16. Notice that this trail T repeats the vertex w. This is perfectly permissible. Although the definition of a trail stipulates that no edge can be repeated, no such condition is placed on vertices.

A $u - v$ walk in a graph in which no vertices are repeated is a $u - v$ **path**. While the $u - v$ trail T in (1.3) is not a $u - v$ path in the graph G of Figure 1.16 (since the vertex w is repeated),

$$P = (u, w, y, v)$$

is a $u - v$ path. If no vertex in a walk is repeated (thereby producing a path), then no edge is repeated either. Hence every path is a trail.

If a $u - v$ walk in a graph is followed by a $v - w$ walk, then a $u - w$ walk results. In particular, a $u - v$ path followed by a $v - w$ path is a $u - w$ walk W but not necessarily a $u - w$ path, as vertices in W may be repeated. While not every walk is a path, if a graph contains a $u - v$ walk, then it must also contain a $u - v$ path. This is our first theorem.

Theorem 1.6 *If a graph G contains a $u - v$ walk of length ℓ, then G contains a $u - v$ path of length at most ℓ.*

Proof. Among all $u - v$ walks in G, let

$$P = (u = u_0, u_1, \ldots, u_k = v)$$

be a $u - v$ walk of smallest length k. Therefore, $k \leq \ell$. We claim that P is a $u - v$ path. Assume, to the contrary, that this is not the case. Then some vertex of G must be repeated in P, say $u_i = u_j$ for some i and j with $0 \leq i < j \leq k$. If we then delete the vertices $u_{i+1}, u_{i+2}, \ldots, u_j$ from P, we arrive at the $u - v$ walk

$$(u = u_0, u_1, \ldots, u_{i-1}, u_i = u_j, u_{j+1}, \ldots, u_k = v)$$

whose length is less than k, which is impossible. Therefore, as claimed, P is a $u - v$ path of length $k \leq \ell$.   ■

A **circuit** in a graph G is a closed trail of length 3 or more. Hence a circuit begins and ends at the same vertex but repeats no edges. A circuit can be described by choosing any of its vertices as the beginning (and ending) vertex provided the vertices are listed in the same cyclic order. In a circuit, vertices can be repeated, in addition to the first and last. For example, in the graph G of Figure 1.16,

$$C = (y, w, u, v, w, x, y) \text{ or } C = (x, y, w, u, v, w, x) \text{ or } C = (w, x, y, w, u, v, w)$$
(1.4)

is a circuit. A circuit that repeats no vertex, except for the first and last, is a **cycle**. A k-**cycle** is a cycle of length k. A 3-cycle is also referred to as a **triangle**. A cycle of odd length is called an **odd cycle**; while, not surprisingly, a cycle of even length is called an **even cycle**. In the graph G of Figure 1.16, the circuit C in (1.4) is not a cycle, while

$$C' = (x, y, v, w, x)$$

is a cycle, namely a 4-cycle. If a vertex of a cycle is deleted, then a path is obtained. This is not necessarily true for circuits, however.

The vertices and edges of a trail, path, circuit or cycle in a graph G form a subgraph of G, also called a **trail**, **path**, **circuit** or **cycle**. Hence a path, for example, is used to describe both a manner of traversing certain vertices and edges of G and a subgraph consisting of those vertices and edges. The graph G of Figure 1.16 is shown again in Figure 1.17. Thus the subgraphs G_1, G_2, G_3, G_4 of the graph G are a trail, path, circuit and cycle, respectively.

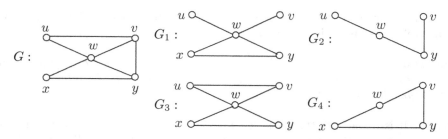

Figure 1.17: Trails, paths, circuits and cycles as subgraphs of a graph

We will have a special interest in graphs G in which it is possible to travel from each vertex of G to any other vertex of G. If G contains a $u - v$ path, then u and v are said to be **connected** and u **is connected to** v (and v is connected to u). So, saying that u and v are connected only means that there is some $u - v$ path in G; it doesn't say that u and v are joined by an edge. Of course, if u *is* joined to v, then u is connected to v as well. A graph G is **connected** if every two vertices of G are connected, that is, if G contains a $u - v$ path for every pair u, v of vertices of G. By Theorem 1.6, G is connected if and only if G contains a $u - v$ walk for every pair u, v of vertices of G. Since every vertex is connected to itself, the trivial graph is connected.

A graph G that is not connected is called **disconnected**. A connected subgraph of G that is not a proper subgraph of any other connected subgraph of G

is a **component** of G. The number of components of a graph G is denoted by $k(G)$. Thus a graph G is connected if and only if $k(G) = 1$. While the graph G of Figure 1.16 is connected, the graph H of Figure 1.18 is disconnected since, for example, there is no $s - w$ path in H. There is no $x - z$ path either. The graph H has three components, namely H_1, H_2 and H_3 and so $k(H) = 3$.

For subgraphs $G_1, G_2, \ldots, G_k$, $k \geq 2$, of a graph G, with mutually disjoint vertex sets, we write $G = G_1 \cup G_2 \cup \ldots \cup G_k$ if every vertex and every edge of G belong to exactly one of these subgraphs. In this case, G is the **union** of the graphs $G_1, G_2, \ldots, G_k$. In particular, we write $G = G_1 \cup G_2 \cup \ldots \cup G_k$ if $G_1, G_2, \ldots, G_k$ are components of G. That is, every graph is the union of its components. Therefore, we can write $H = H_1 \cup H_2 \cup H_3$ for the graphs in Figure 1.18.

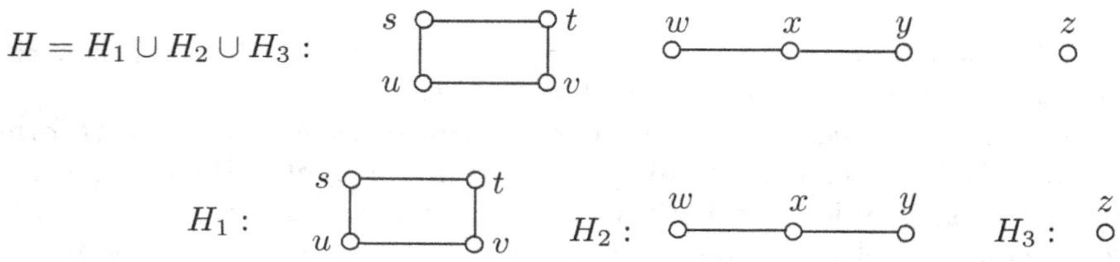

Figure 1.18: A disconnected graph and its components

Components can also be defined by means of an equivalence relation. (Equivalence relations are reviewed in Appendix 2.1.)

Theorem 1.7 *Let R be the relation defined on the vertex set of a graph G by $u\ R\ v$, where $u, v \in V(G)$, if u is connected to v, that is, if G contains a $u - v$ path. Then R is an equivalence relation.*

Proof. It is immediate that R is reflexive and symmetric. It remains therefore only to show that R is transitive. Let $u, v, w \in V(G)$ such that $u\ R\ v$ and $v\ R\ w$. Hence G contains a $u - v$ path P' and a $v - w$ path P''. As we have seen earlier, following P' by P'' produces a $u - w$ walk W. By Theorem 1.6, G contains a $u - w$ path and so $u\ R\ w$.   ∎

The equivalence relation described in Theorem 1.7 produces a partition of the vertex set of every graph G into equivalence classes. The subgraph of G induced by the vertices in an equivalence class is a component of G. Exercise 1.14 asks you to show this. As a consequence, we have the following:

> *Each vertex and each edge of a graph G belong to exactly one component of G. This implies that if G is a disconnected graph and u and v are vertices belonging to different components of G, then $uv \notin E(G)$.*

The following theorem provides a sufficient condition for a graph of order at least 3 to be connected.

Theorem 1.8 *Let G be a graph of order 3 or more. If G contains two distinct vertices u and v such that G − u and G − v are connected, then G itself is connected.*

Proof. Suppose that G contains distinct vertices u and v such that $G − u$ and $G − v$ are connected. To show that G itself is connected, we show that every two vertices of G are connected. Let x and y be two vertices of G. We consider two cases.

Case 1. $\{x, y\} \neq \{u, v\}$*, say* $u \notin \{x, y\}$. Then x and y are vertices in $G − u$. Since $G − u$ is connected, there is an $x − y$ path P in $G − u$. Hence P is in G and x and y are connected in G.

Case 2. $\{x, y\} = \{u, v\}$*, say* $x = u$ *and* $y = v$. We show that u and v are connected in G. Since the order of G is at least 3, there is a vertex w in G such that $w \neq u, v$. Since $G − v$ is connected, $G − v$ contains a $u − w$ path P'. Furthermore, since $G − u$ is connected, $G − u$ contains a $w − v$ path P''. Therefore, P' followed by P'' produces a $u−v$ walk. By Theorem 1.6, G contains a $u − v$ path and so u and v are connected in G. ∎

If G is the disconnected graph consisting of two vertices u and v and no edges, then the subgraphs $G − u$ and $G − v$ are (trivially) connected. Therefore, in Theorem 1.8, it is essential that the order of the graph under consideration be at least 3.

If u and v are vertices in a connected graph G, then there must be a $u − v$ path in G. However, it is quite possible that G contains several $u − v$ paths. For example, in the graph G of Figure 1.16, all of the following are $u − y$ paths:

$$P' = (u, v, y) \qquad P'' = (u, w, v, y) \qquad P''' = (u, v, w, x, y).$$

The length of P' is 2, the length of P'' is 3 and the length of P''' is 4. There is no $u − y$ path of length 1 in this graph since u and y are not adjacent and there are no $u − y$ paths of length 5 or more as G only has five vertices.

Let G be a connected graph of order n and let u and v be two vertices of G. The **distance** between u and v is the smallest length of any $u − v$ path in G and is denoted by $d_G(u, v)$ or simply $d(u, v)$ if the graph G under consideration is clear. Hence if $d(u, v) = k$, then there exists a $u − v$ path

$$P = (u = v_0, v_1, \ldots, v_k = v) \tag{1.5}$$

of length k in G but no $u − v$ path of smaller length exists in G. A $u − v$ path of length $d(u, v)$ is called a $u − v$ **geodesic**. In fact, since the path P in (1.5) is a $u − v$ geodesic, not only is $d(u, v) = d(u, v_k) = k$ but $d(u, v_i) = i$ for every i with $0 \leq i \leq k$. Exercise 1.16 asks you to verify this. If $u = v$, then $d(u, v) = 0$. If $uv \in E(G)$, then $d(u, v) = 1$. In general, $0 \leq d(u, v) \leq n − 1$ for every two vertices u and v (distinct or not) in a connected graph of order n. For the vertices u and y in the graph G of Figure 1.16, $d(u, y) = 2$. If G is disconnected, then there are some pairs x, y of distinct vertices of G such that there is no $x - y$ path in G. In this case, $d(x, y)$ is not defined.

At times, it is useful to visualize the vertices of a connected graph according to their distances from a given vertex. The graph H of Figure 1.19(a) is redrawn in Figure 1.19(b) to indicate those vertices at a given distance from the vertex t. The vertex t (the only vertex whose distance from t is 0) is drawn at the top. The vertices one level down are the neighbors of t. The next level consists of those vertices whose distance from t is 2 and so on. Observe that two adjacent vertices must either belong to the same level or to neighboring levels.

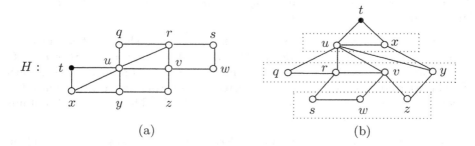

Figure 1.19: Distances from a given vertex

The greatest distance between any two vertices of a connected graph G is called the **diameter** of G and is denoted by $\operatorname{diam}(G)$. The diameter of the graph H of Figure 1.19 is 3. The path $P' = (y, u, r, s)$ is a $y - s$ geodesic whose length is $\operatorname{diam}(H)$.

If G is a connected graph such that $d(u, v) = \operatorname{diam}(G)$ and $w \neq u, v$, then no $u - w$ geodesic can contain v, for otherwise $d(u, w) > d(u, v) = \operatorname{diam}(G)$, which is impossible.

Let's return to Theorem 1.8, where we proved that if a graph G of order 3 contains two distinct vertices u and v such that $G - u$ and $G - v$ are connected, then G is connected. Actually, the converse of this theorem is also true; that is, if G is a connected graph of order at least 3, then G must contain two vertices u and v such that $G - u$ and $G - v$ are both connected. We are now in a position to prove this theorem as well.

Theorem 1.9 *If G is a connected graph of order 3 or more, then G contains two distinct vertices u and v such that $G - u$ and $G - v$ are connected.*

Proof. Let u and v be two vertices of G such that $d(u, v) = \operatorname{diam}(G)$. We claim that $G - u$ and $G - v$ are both connected. Suppose that this is not the case. Then at least one of $G - u$ and $G - v$ is disconnected, say $G - v$ is disconnected. Therefore, $G - v$ contains two vertices x and y that are not connected in $G - v$. However, since G is connected, the vertices u and x are connected in G, as are u and y.

Let P' be an $x - u$ geodesic in G and let P'' be a $u - y$ geodesic in G. Since $d_G(u, v) = \text{diam}(G)$, the vertex v cannot lie on either P' or on P'', so P' and P'' are paths in $G - v$. The path P' followed by P'' produces an $x - y$ walk W in $G - v$. By Theorem 1.6, $G - v$ contains an $x - y$ path and so x and y are connected in $G - v$. This is a contradiction.   ∎

Theorem 1.9 gives a property that every connected graph of order at least 3 must have. That is, Theorem 1.9 provides a *necessary condition* for a graph to be connected. Actually, Theorem 1.9 is true even if the order of G is 2, but we stated Theorem 1.9 as we did so we could combine Theorems 1.8 and 1.9 into a single *necessary and sufficient condition* for a graph to be connected, which we state next.

Theorem 1.10 *Let G be a graph of order 3 or more. Then G is connected if and only if G contains two distinct vertices u and v such that $G - u$ and $G - v$ are connected.*

Exercises for Section 1.2

1.11 Let G be the graph of Figure 1.20, let $X = \{e, f\}$, where $e = ru$ and $f = vw$, and let $U = \{u, w\}$. Draw the subgraphs $G - X$ and $G - U$ of G.

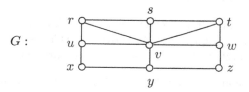

Figure 1.20: The graph G in Exercises 1.11 and 1.12

1.12 For the graph G of Figure 1.20, give an example of each of the following or explain why no such example exists.

(a) An $x - y$ walk of length 6.

(b) A $v - w$ trail that is not a $v - w$ path.

(c) An $r - z$ path of length 2.

(d) An $x - z$ path of length 3.

(e) An $x - t$ path of length $d(x, t)$.

(f) A circuit of length 10.

(g) A cycle of length 8.

(h) A geodesic whose length is $\text{diam}(G)$.

1.13 (a) Give an example of a connected graph G containing three vertices u, v and w such that $d(u, v) = d(u, w) = d(v, w) = \mathrm{diam}(G) = 3$.

(b) Does the question in (a) suggest another question?

1.14 For a graph G, a component of G has been defined as (1) a connected subgraph of G that is not a proper subgraph of any other connected subgraph of G and has been described as (2) a subgraph of G induced by the vertices in an equivalence class resulting from the equivalence relation defined in Theorem 1.7. Show that these two interpretations of components are equivalent.

1.15 Draw all connected graphs of order 5 in which the distance between every two distinct vertices is odd. Explain why you know that you have drawn all such graphs.

1.16 Let $P = (u = v_0, v_1, \cdots, v_k = v)$, $k \geq 1$, be a $u - v$ geodesic in a connected graph G. Prove that $d(u, v_i) = i$ for each integer i with $1 \leq i \leq k$.

1.17 (a) Prove that if P and Q are two longest paths in a connected graph, then P and Q have at least one vertex in common.

(b) Prove or disprove: Let G be connected graph of diameter k. If P and Q are two geodesics of length k in G, then P and Q have at least one vertex in common.

1.18 A graph G of order 12 has vertex set $V(G) = \{c_1, c_2, \ldots, c_{12}\}$ for the twelve configurations in Figure 1.4. A "move" on this checkerboard corresponds to moving a single coin to an unoccupied square, where

(1) the gold coin can only be moved horizontally or diagonally,

(2) the silver coin can only be moved vertically or diagonally.

Two vertices c_i and c_j $(i \neq j)$ are adjacent if it is possible to move c_i to c_j by a single move.

(a) What vertices are adjacent to c_1 in G?

(b) What vertices are adjacent to c_2 in G?

(c) Draw the subgraph of G induced by $\{c_2, c_6, c_9, c_{11}\}$.

(d) Give an example of a $c_1 - c_7$ path in G.

1.19 Theorem 1.10 states that a graph G of order 3 or more is connected if and only if G contains two distinct vertices u and v such that $G - u$ and $G - v$ are connected. Based on this, one might suspect that the following statement is true. *Every connected graph G of order 4 or more contains three distinct vertices u, v and w such that $G - u$, $G - v$ and $G - w$ are connected.* Is it?

1.20   (a) Let u and v be distinct vertices in a connected graph G. There may
           be several connected subgraphs of G containing u and v. What is
           the minimum size of a connected subgraph of G containing u and v?
           Explain your answer.

       (b) Does the question in (a) suggest another question to you?

1.3   Common Classes of Graphs

As we continue to study graphs, we will see that there are certain graphs that are
encountered often and it is useful to be familiar with them. In many instances,
there is special notation reserved for these graphs.

We have already seen that paths and cycles are certain kinds of walks and
subgraphs in graphs. These terms are also used to describe certain kinds of
graphs. If the vertices of a graph G of order n can be labeled (or relabeled)
$v_1, v_2, \cdots, v_n$ so that its edges are $v_1 v_2, v_2 v_3, \cdots, v_{n-1} v_n$, then G is called a **path**;
while if the vertices of a graph G of order $n \geq 3$ can be labeled (or relabeled)
$v_1, v_2, \cdots, v_n$ so that its edges are $v_1 v_2, v_2 v_3, \cdots, v_{n-1} v_n$ and $v_1 v_n$, then G is
called a **cycle**. A graph that is a path of order n is denoted by P_n, while a graph
that is a cycle of order $n \geq 3$ is denoted by C_n. Several paths and cycles are
shown in Figure 1.21.

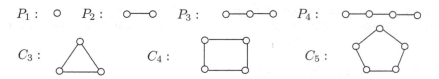

Figure 1.21: Paths and cycles

A graph G is **complete** if every two distinct vertices of G are adjacent. A
complete graph of order n is denoted by K_n. Therefore, K_n has the maximum
possible size for a graph with n vertices. Since every two distinct vertices of K_n
are joined by an edge, the number of pairs of vertices in K_n is $\binom{n}{2}$ and so

$$\text{the size of } K_n \text{ is } \binom{n}{2} = \frac{n(n-1)}{2}. \tag{1.6}$$

Therefore, the complete graph K_3 has three edges, K_4 has six edges and K_5 has
ten edges. The five smallest complete graphs are shown in Figure 1.22. Notice
that P_1 and K_1 represent the same graph, as do P_2 and K_2, as well as C_3 and
K_3. Although there are edges that cross in the drawings of K_4 and K_5, the
points of intersection do not represent vertices.

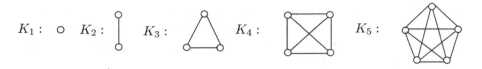

Figure 1.22: Complete graphs

The graphs that are drawn in Figures 1.21 and 1.22 bring up some points that need to be discussed. Although we have attempted to draw these graphs in a manner that makes them easy to visualize, this is certainly not a requirement when drawing a graph, as its vertices can be placed in any convenient location. Figure 1.23 shows a variety of ways to draw the path P_4 and the complete graph K_4.

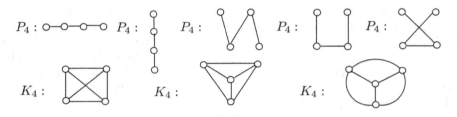

Figure 1.23: The graphs P_4 and K_4

Since the disconnected graph G in Figure 1.24 has two components that are complete graphs of order 4, one that is C_5 and one that is P_3, we write this graph as $G = 2K_4 \cup C_5 \cup P_3$.

Figure 1.24: The graph $G = 2K_4 \cup C_5 \cup P_3$

The **complement** $\overline{G}$ of a graph G is that graph whose vertex set is $V(G)$ and such that for each pair u, v of distinct vertices of G, uv is an edge of $\overline{G}$ if and only if uv is not an edge of G. Observe that if G is a graph of order n and size m, then $\overline{G}$ is a graph of order n and size $\binom{n}{2} - m$. The graph $\overline{K}_n$ then has n vertices and no edges; it is called the **empty graph** of order n. Therefore, empty graphs have empty edge sets. In fact, if G is any graph of order n, then $G - E(G)$ is the empty graph $\overline{K}_n$. By definition, no graph can have an empty vertex set. A graph H and its complement are shown in Figure 1.25. Both of these graphs are connected. Although a graph and its complement need not both be connected, at least one must be connected.

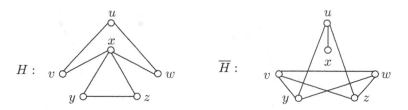

Figure 1.25: A graph and its complement

Theorem 1.11 *If G is a disconnected graph, then $\overline{G}$ is connected.*

Proof. Since G is disconnected, G contains two or more components. Let u and v be two vertices of $\overline{G}$. We show that u and v are connected in $\overline{G}$. If u and v belong to different components of G, then u and v are not adjacent in G and so u and v are adjacent in $\overline{G}$. Hence $\overline{G}$ contains a $u - v$ path of length 1. Suppose next that u and v belong to the same component of G. Let w be a vertex of G that belongs to a different component of G. Then $uw, vw \notin E(G)$, implying that $uw, vw \in E(\overline{G})$ and so (u, w, v) is a $u - v$ path in $\overline{G}$. ∎

We now turn to graphs whose vertex sets can be partitioned in special ways. A graph G is a **bipartite graph** if $V(G)$ can be partitioned into two subsets U and W, called **partite sets**, such that every edge of G joins a vertex of U and a vertex of W. It's not always easy to tell at a glance whether a graph is bipartite. For example, the connected graphs G_1 and G_2 of Figure 1.26 are bipartite, as every edge of G_1 joins a vertex of $U_1 = \{u_1, x_1, y_1\}$ and a vertex of $W_1 = \{v_1, w_1\}$, while every edge of G_2 joins a vertex of $U_2 = \{u_2, w_2, y_2\}$ and a vertex of $W_2 = \{v_2, x_2, z_2\}$. The bipartite nature of these graphs is illustrated in Figure 1.26. By letting $U = U_1 \cup U_2$ and $W = W_1 \cup W_2$, we see that every edge of $G = G_1 \cup G_2$ joins a vertex of U and a vertex of W. This illustrates the observation that a graph is bipartite if and only if each of its components is bipartite.

Certainly not every graph is bipartite. For example, consider the 5-cycle C_5 in Figure 1.27. If C_5 were bipartite, then its vertex set could be partitioned into two sets U and W such that every edge of C_5 joins a vertex of U and a vertex of W. The vertex v_1 must belong to either U or W, say $v_1 \in U$. Since $v_1 v_2$ is an edge of C_5, it follows that $v_2 \in W$. Since $v_2 v_3$ is an edge of C_5, it follows that $v_3 \in U$. Similarly, $v_4 \in W$ and $v_5 \in U$. However, $v_1, v_5 \in U$ and $v_1 v_5$ is an edge of C_5. This is a contradiction. Therefore, C_5 is not bipartite. In fact, no odd cycle is bipartite. Indeed, any graph that contains an odd cycle is not bipartite. The converse is true as well, which may come as a surprise.

Theorem 1.12 *A nontrivial graph G is a bipartite graph if and only if G contains no odd cycles.*

Proof. We have already seen that if a graph contains an odd cycle, then it's

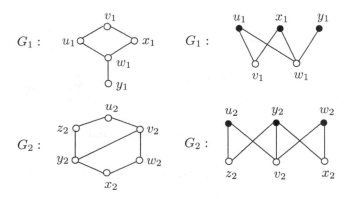

Figure 1.26: Bipartite graphs

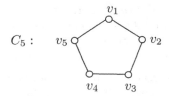

Figure 1.27: A 5-cycle: A graph that is not bipartite

not bipartite. To prove the converse, let G be a nontrivial graph having no odd cycles. We show that G is bipartite. Because of our earlier observation that a graph is bipartite if and only if each of its components is bipartite, we may assume that G is connected. Let u be any vertex of G, let U consist of all vertices of G whose distance from u is even and let W consist of all vertices whose distance from u is odd. Thus $\{U, W\}$ is a partition of $V(G)$. Since $d(u, u) = 0$, it follows that $u \in U$. We claim that every edge of G joins a vertex of U and a vertex of W.

Assume, to the contrary, that there exist two adjacent vertices in U or two adjacent vertices in W. Since these two situations are similar, we will assume that there are vertices v and w in W such that $vw \in E(G)$. Since $d(u, v)$ and $d(u, w)$ are both odd, $d(u, v) = 2s + 1$ and $d(u, w) = 2t + 1$ for nonnegative integers s and t. Let $P' = (u = v_0, v_1, \ldots, v_{2s+1} = v)$ be a $u - v$ geodesic and let $P'' = (u = w_0, w_1, \ldots, w_{2t+1} = w)$ be a $u - w$ geodesic in G. Certainly, P' and P'' have their initial vertex u in common but they may have other vertices in common as well. Among the vertices P' and P'' have in common, let x be the last vertex. Perhaps $x = u$. In any case, $x = v_i$ for some integer $i \geq 0$. Thus $d(u, v_i) = i$. Since x is on P'' and w_i is the only vertex of P'' whose distance from u is i, it follows that $x = w_i$. So $x = v_i = w_i$. However then, $C = (v_i, v_{i+1}, \ldots, v_{2s+1}, w_{2t+1}, w_{2t}, \ldots, w_i = v_i)$ is a cycle of length

$$[(2s+1)-i] + [(2t+1)-i] + 1 = 2s + 2t - 2i + 3 = 2(s+t-i+1)+1$$

and so C is an odd cycle, which is a contradiction. ∎

We know that if G is a bipartite graph, then $V(G)$ can be partitioned into two subsets U and W, called partite sets, such that every edge of G joins a vertex of U and a vertex of W. However, this does not mean that every vertex of U is adjacent to every vertex of W. If this does happen, however, then we call G a **complete bipartite graph**. A complete bipartite graph with $|U| = s$ and $|W| = t$ is denoted by $K_{s,t}$ or $K_{t,s}$. If either $s = 1$ or $t = 1$, then $K_{s,t}$ is a **star**. Several complete bipartite graphs are shown in Figure 1.28, including the star $K_{1,3}$. Observe that $K_{2,2}$ is the same graph as C_4, although it is certainly not drawn the same way that we drew C_4 in Figure 1.21. When two graphs G and H are the same except possibly for the way that they're drawn or their vertices are labeled, then we write $G \cong H$. (The technical term for this is that these graphs are isomorphic. We'll discuss this in Chapter 3.) If the structures of G and H are different, then we write $G \ncong H$.

$K_{1,3}$ $K_{2,2}$ $K_{2,3}$ $K_{3,3}$

Figure 1.28: Complete bipartite graphs

Bipartite graphs belong to a more general class of graphs. A graph G is a k-**partite graph** if $V(G)$ can be partitioned into k subsets $V_1, V_2, \ldots, V_k$ (once again called **partite sets**) such that if uv is an edge of G, then u and v belong to different partite sets. If, in addition, every two vertices in different partite sets are joined by an edge, then G is a **complete k-partite graph**. If $|V_i| = n_i$ for $1 \le i \le k$, then we denote this complete k-partite graph by $K_{n_1, n_2, \ldots, n_k}$. The complete k-partite graphs are also referred to as **complete multipartite graphs**. If $n_i = 1$ for every i ($1 \le i \le k$), then $K_{n_1, n_2, \ldots, n_k}$ is the complete graph K_k. Complete 2-partite graphs are thus complete bipartite graphs. Several complete multipartite graphs are shown in Figure 1.29.

$K_{2,4}$ $K_{1,1,1} = K_3$ $K_{2,2,2}$ $K_{1,2,3}$

Figure 1.29: Complete multipartite graphs

There are several ways to produce a new graph from a given pair of graphs. For two vertex-disjoint graphs G and H, we have already mentioned the union $G \cup H$ of G and H as that (disconnected) graph with vertex set $V(G) \cup V(H)$ and edge set $E(G) \cup E(H)$. The **join** $G + H$ consists of $G \cup H$ and all edges joining a vertex of G and a vertex of H. The join of P_3 and K_2 is shown in Figure 1.30.

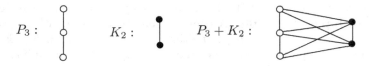

Figure 1.30: The join of two graphs

For two graphs G and H, the **Cartesian product** $G \times H$ has vertex set $V(G \times H) = V(G) \times V(H)$, that is, every vertex of $G \times H$ is an ordered pair (u, v), where $u \in V(G)$ and $v \in V(H)$. The Cartesian product of G and H is often denote by $G \;\square\; H$ as well. Two distinct vertices (u, v) and (x, y) are adjacent in $G \times H$ if either (1) $u = x$ and $vy \in E(H)$ or (2) $v = y$ and $ux \in E(G)$. Figure 1.31 shows the Cartesian product of P_3 and K_2.

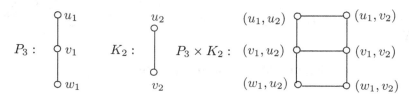

Figure 1.31: The Cartesian product of two graphs

Some additional comments about Cartesian products of graphs are useful. First, the definition of Cartesian product tells us that the order in which the graphs G and H are written is structurally irrelevant, that is, $G \times H$ and $H \times G$ are the same graph, that is, they are isomorphic graphs.

There is an informal way of drawing the graph $G \times H$ (or $H \times G$) that doesn't require us to label the vertices. Replace each vertex x of G by a copy H_x of the graph H. Let u and v be two vertices of G. If u and v are adjacent in G, then we join corresponding vertices of H_u and H_v by an edge. If u and v are not adjacent in G, then we add no edges between H_u and H_v. This is illustrated in Figure 1.32.

Notice that $K_2 \times K_2$ is the 4-cycle. The graph $C_4 \times K_2$ is often denoted by Q_3 and is called the **3-cube**. More generally, we define Q_1 to be K_2 and for $n \geq 2$, define Q_n to be $Q_{n-1} \times K_2$. The graphs Q_n are then called n-**cubes** or **hypercubes**. The n-cube can also be defined as that graph whose vertex set is the set of ordered n-tuples of 0s and 1s (commonly called n-**bit strings**) and where two vertices are adjacent if their ordered n-tuples differ in exactly

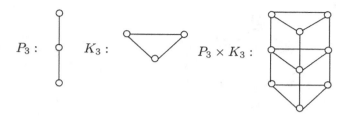

Figure 1.32: The Cartesian product of two graphs

one position (coordinate). The n-cubes for $n = 1, 2, 3$ are shown in Figure 1.33, where their vertices are labeled by n-bit strings.

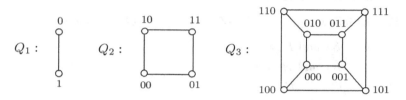

Figure 1.33: The n-cubes for $1 \leq n \leq 3$

Exercises for Section 1.3

1.21 Draw the graph $3P_4 \cup 2C_4 \cup K_4$.

1.22 Let G be a disconnected graph. By Theorem 1.11, $\overline{G}$ is connected. Prove that if u and v are any two vertices of $\overline{G}$, then $d_{\overline{G}}(u, v) = 1$ or $d_{\overline{G}}(u, v) = 2$. Therefore, if G is a disconnected graph, then $\text{diam}(\overline{G}) \leq 2$.

1.23 Consider the following question: For a given positive integer k, does there exist a connected graph G whose complement $\overline{G}$ is also connected and contains four distinct vertices u, v, x, y for which $d_G(u, v) = k = d_{\overline{G}}(x, y)$?

   (a) Show that the answer to this question is yes if $k = 1$ or $k = 2$.

   (b) Find the largest value of k for which the answer to this question is yes.

1.24 Determine whether the graphs G_1 and G_2 of Figure 1.34 are bipartite. If a graph is bipartite, then redraw it indicating the partite sets; if not, then give an explanation as to why the graph is not bipartite.

1.25 Let G be a graph of order 5 or more. Prove that at most one of G and $\overline{G}$ is bipartite.

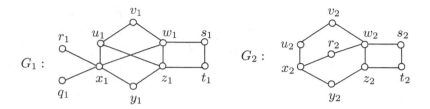

Figure 1.34: Graphs in Exercise 1.24

1.26 Suppose that the vertex set of a graph G is a (finite) set of integers. Two vertices x and y are adjacent if $x + y$ is odd. To which well-known class of graphs is G a member?

1.27 For the following pairs G, H of graphs, draw $G + H$ and $G \times H$.

    (a) $G = K_5$ and $H = K_2$.

    (b) $G = \overline{K}_5$ and $H = \overline{K}_3$.

    (c) $G = C_5$ and $H = K_1$.

1.28 We have seen that for $n \geq 1$, the n-cube Q_n is that graph whose vertex set is the set of n-bit strings, where two vertices of Q_n are adjacent if they differ in exactly one coordinate.

    (a) For $n \geq 2$, define the graph R_n to be that graph whose vertex set is the set of n-bit strings, where two vertices of R_n are adjacent if they differ in exactly two coordinates. Draw R_2 and R_3.

    (b) For $n \geq 3$, define the graph S_n to be that graph whose vertex set is the set of n-bit strings, where two vertices of S_n are adjacent if they differ in exactly three coordinates. Draw S_3 and S_4.

1.4 Multigraphs and Digraphs

There are occasions when graphs may not be an appropriate model for a problem we are investigating. We now describe two variations of graphs that we will encounter from time to time. In a graph, two vertices are either adjacent or they are not, that is, two vertices are joined by one edge or no edges. A **multigraph** M consists of a finite nonempty set V of vertices and a set E of edges, where every two vertices of M are joined by a finite number of edges (possibly zero). If two or more edges join the same pair of (distinct) vertices, then these edges are called **parallel edges**. In a **pseudograph**, not only are parallel edges permitted but an edge is also permitted to join a vertex to itself. Such an edge is called a **loop**. If a loop e joins a vertex v to itself, then e is said to be a loop at v. There can be any finite number of loops at the same vertex in a pseudograph. In Figure 1.35, M_1 and M_2 are multigraphs, M_3 is a pseudograph and M_4 is a graph. In fact, M_4 is a multigraph and all four are pseudographs.

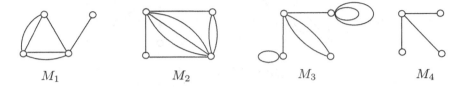

$$M_1 \qquad\qquad M_2 \qquad\qquad M_3 \qquad\qquad M_4$$

Figure 1.35: Multigraphs and pseudographs

If M is a multigraph with vertex set V, then it is no longer appropriate to regard an edge of M as a 2-element subset of V as we must somehow indicate the multiplicity of the edge.

Let's return to Example 1.2 where we considered the set $S = \{2, 3, 5, 8, 13, 21\}$ as well as those pairs of integers of S whose sum or difference (in absolute value) belongs to S. The graph H of Figure 1.2 models this situation. In H there is an edge joining the vertices 3 and 5, indicating that $3 + 5 \in S$ or $|3 - 5| \in S$. In this case, however, *both* $3 + 5 \in S$ and $|3 - 5| \in S$, but there is no way of knowing this from H. The multigraph M of Figure 1.36 supplies this information. However, even in this case, the existence of a single edge between a pair i, j of vertices doesn't tell us whether $i + j \in S$ or $|i - j| \in S$; it only tells us that one of these occurs. Thus the multigraph M of Figure 1.36 is a better model of this situation.

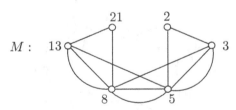

Figure 1.36: A multigraph

A **digraph** (or **directed graph**) D is a finite nonempty set V of objects called **vertices** together with a set E of *ordered pairs* of distinct vertices. The elements of E are called **directed edges** or **arcs**. If (u, v) is a directed edge, then we indicate this in a diagram representing D by drawing a directed line segment or curve from u to v. Then u is said to be **adjacent to** v and v is **adjacent from** u. The vertices u and v are also said to be **incident with** the directed edge (u, v). Arcs (u, v) and (v, u) may both be present in some directed graph. If, in the definition of *digraph*, for each pair u, v of distinct vertices, at most one of (u, v) and (v, u) is a directed edge, then the resulting digraph is an **oriented graph**. Thus an oriented graph D is obtained by assigning a direction to each edge of some *graph* G. The digraph D is also called an **orientation** of G. Figure 1.37 shows two digraphs D_1 and D_2, where D_2 is an oriented graph but D_1 is not.

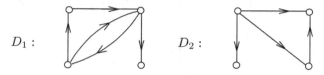

Figure 1.37: Digraphs

Next, we return to Example 1.3, where we considered twelve configurations of two coins (one silver, one gold), which were denoted by $c_1, c_2, \ldots, c_{12}$. Now, we say that c_i can be transformed into c_j if c_j can be obtained by moving one of the coins in c_i to the right or up. Modeling this situation requires a digraph, namely, the digraph D shown in Figure 1.38, which is an oriented graph.

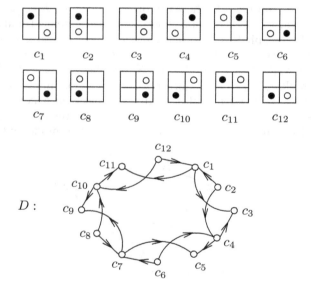

Figure 1.38: Modeling twelve configurations by a digraph

Exercises for Section 1.4

1.29   (a) Let $S = \{2, 3, 4, 7, 11, 13\}$. Construct the multigraph M whose vertex set is S and where ij is an edge for distinct elements i and j in S whenever $i + j \in S$ and ij is an edge whenever $|i - j| \in S$. In other words, i and j are joined by two edges if both $i+j \in S$ and $|i-j| \in S$.

(b) How are the problem and solution in (a) affected if we remove the word "distinct"?

1.30 Consider the twelve configurations c_i, $1 \le i \le 12$, in Figure 1.38. Draw the digraph D, where $V(D) = \{c_1, c_2, \ldots, c_{12}\}$ and where (c_i, c_j) is a directed edge of D if it is possible to obtain c_j by rotating the configuration c_i either $90°$ or $180°$ clockwise about the midpoint of the checkerboard.

1.31 Using the twelve configurations in Figure 1.38, define a transformation different from the one described in Exercise 1.30 which can be modeled by a digraph but not by a graph.

1.32 Let S and A be two finite nonempty sets of integers. Define a digraph D with $V(D) = A$, where (x, y) is an arc of D if $x \ne y$ and $y - x \in S$.

(a) Draw the digraph D for $A = \{0, 1, 2, 3, 4\}$ and $S = \{-2, 1, 2, 4\}$.

(b) What can be said about D if A and S consist only of odd integers?

(c) How can the question in (b) be generalized?

(d) If $|A| = |S| = 5$, how large can the size of D be?

1.33 A digraph D has vertex set $\{-3, 3, 6, 12\}$ and $(i, j) \in D$ if $i \ne j$ and $i \mid j$, that is, j is a multiple of i. Draw the digraph D.

Chapter 2

Degrees

2.1   The Degree of a Vertex

There are many numbers, referred to as **parameters**, associated with a graph G. Knowing the values of certain parameters provides us with information about G but rarely tells us the entire structure of G. (These comments are tied in with the concept of isomorphic graphs, which will be discussed in Chapter 3.) We've already mentioned the best known parameters: the order and the size. There are also numbers associated with each vertex of a graph. We now consider the best known of these.

The **degree of a vertex** v in a graph G is the number of edges incident with v and is denoted by $\deg_G v$ or simply by $\deg v$ if the graph G is clear from the context. Also, $\deg v$ is the number of vertices adjacent to v. Recall that two adjacent vertices are referred to as neighbors of each other. The set $N(v)$ of neighbors of a vertex v is called the **neighborhood** of v. Thus $\deg v = |N(v)|$.

A vertex of degree 0 is referred to as an **isolated vertex** and a vertex of degree 1 is an **end-vertex** (or a **leaf**). The **minimum degree** of G is the minimum degree among the vertices of G and is denoted by $\delta(G)$; the **maximum degree** of G is denoted by $\Delta(G)$. So if G is a graph of order n and v is any vertex of G, then

$$0 \leq \delta(G) \leq \deg v \leq \Delta(G) \leq n - 1.$$

The graph G of Figure 2.1 has order 6 and size 5. Each vertex of G is labeled with its degree. Since G contains an isolated vertex, namely u, it follows that $\delta(G) = 0$. Furthermore, w has the largest degree in G and so $\Delta(G) = \deg w = 4$. Both v and z are end-vertices of G since $\deg v = \deg z = 1$. If we add the degrees of the vertices of G, we obtain $0 + 1 + 1 + 2 + 2 + 4 = 10$, which happens to be twice the size of G. This is not a coincidence as we show in our next theorem, which is often referred to as **The First Theorem of Graph Theory**, so-called because it is likely that anyone studying graph theory for the first time would discover this result as his or her own first theorem on the subject. Although we've already discovered some theorems in Chapter 1, we'll follow the trend and also refer to the following theorem as the First Theorem of Graph Theory.

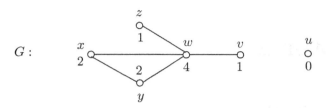

Figure 2.1: A graph G with $\delta(G) = 0$ and $\Delta(G) = 4$

Theorem 2.1 (The First Theorem of Graph Theory) *If G is a graph of size m, then*

$$\sum_{v \in V(G)} \deg v = 2m.$$

Proof. When summing the degrees of the vertices of G, each edge of G is counted twice, once for each of its two incident vertices. ∎

The First Theorem of Graph Theory is useful in solving problems such as the following.

Example 2.2 *A certain graph G has order 14 and size 27. The degree of each vertex of G is 3, 4 or 5. There are six vertices of degree 4. How many vertices of G have degree 3 and how many have degree 5?*

Solution. Let x be the number of vertices of G having degree 3. Since the order of G is 14 and six vertices have degree 4, eight vertices have degree 3 or 5. So there are $8 - x$ vertices of degree 5. Summing the degrees of the vertices of G and applying the First Theorem of Graph Theory, we obtain

$$
\begin{aligned}
3 \cdot x + 4 \cdot 6 + 5 \cdot (8 - x) &= 2 \cdot 27 \\
3x + 24 + 40 - 5x &= 54 \\
-2x &= -10 \\
x &= 5
\end{aligned}
$$

and so $8 - x = 3$. Thus G has five vertices of degree 3 and three vertices of degree 5.   ◇

The method we used to solve the problem in Example 2.2 tells us that there is a unique solution. Perhaps other methods of solving this problem might have occurred to you, such as trying to draw the graph. Consider the graph of Figure 2.2, each of whose vertices is labeled by its degree. This graph has order 14, size 27 and six vertices of degree 4, which are characteristics of the graph G of Example 2.2. We see that the graph of Figure 2.2 has five vertices of degree 3 and three vertices of degree 5, solving the problem. Even though this provides the correct answers to our question, the explanation is not correct; for how do we know that the graph we have just drawn is *the* graph G referred to in the problem and therefore gives us the correct answer?

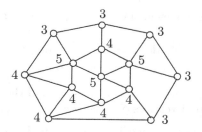

Figure 2.2: A graph of order 14 and size 27

Another possible "solution" might go something like this: We know that we are looking for integers x and y such that $x + y = 8$ and

$$3x + 4 \cdot 6 + 5y = 2 \cdot 27 = 54.$$

By observation, we see that $x = 5$ and $y = 3$ satisfy this equation. Thus we have "solved" the problem. But this "solution" also has its drawbacks. How do we know that this is the *only* solution? (Of course, we could try all possible values of x and y.) The solution that we gave for Example 2.2 shows that there is only one solution for each of x and y and that the solutions do not depend on the graph under consideration (provided it has order 14, size 27 and six vertices of degree 4). Just as when asked to solve $x^2 - x = 3x - 4$ for a real number x, it is not enough to simply note that $x = 2$ is a root. It is required to find *all* roots and even if $x = 2$ is the *only* root, we are obliged to show that this is so.

Suppose that G is a bipartite graph of size m with partite sets $U = \{u_1, u_2, \ldots, u_s\}$ and $W = \{w_1, w_2, \ldots, w_t\}$. Since every edge of G joins a vertex of U and a vertex of W, it follows that adding the degrees of the vertices in U (or in W) gives the number of edges in G, that is,

$$\sum_{i=1}^{s} \deg u_i = \sum_{j=1}^{t} \deg w_j = m. \tag{2.1}$$

A vertex of even degree is called an **even vertex**, while a vertex of odd degree is an **odd vertex**. Returning to the graph G of Figure 2.2, we see that it has six even vertices and eight odd vertices. In particular, the number of odd vertices of G is even. We show that this is the case for every graph.

Corollary 2.3 *Every graph has an even number of odd vertices.*

Proof. Let G be a graph of size m. Divide $V(G)$ into two subsets V_1 and V_2, where V_1 consists of the odd vertices of G and V_2 consists of the even vertices of G. By the First Theorem of Graph Theory,

$$\sum_{v \in V(G)} \deg v = \sum_{v \in V_1} \deg v + \sum_{v \in V_2} \deg v = 2m.$$

The number $\sum_{v \in V_2} \deg v$ is even since it is a sum of even integers. Thus

$$\sum_{v \in V_1} \deg v = 2m - \sum_{v \in V_2} \deg v,$$

which implies that $\sum_{v \in V_1} \deg v$ is even. Since each of the numbers $\deg v$, $v \in V_1$, is odd, the number of odd vertices of G is even. ∎

There is a great deal of information that can be learned about a graph from the degrees of its vertices. For example, if a graph G of order n contains a vertex of degree $n - 1$, then G is connected. In order to see why this is true, suppose that $\deg w = n - 1$. Therefore, w is adjacent to all other vertices of G. To show that G is connected, we need to show that every pair x, y of vertices of G are connected, that is, G contains an $x - y$ path. This is certainly true if one of x and y is w. If neither x nor y is w, then since w is adjacent to both x and y, it follows that (x, w, y) is an $x - y$ path and consequently G contains an $x - y$ path.

This degree condition is certainly not necessary for a graph to be connected. For example, for $n \geq 4$, the path P_n of order n is connected but contains no vertex of degree greater than 2. Next, we present another degree condition that implies that a graph is connected and more.

Theorem 2.4 *Let G be a graph of order n. If*

$$\deg u + \deg v \geq n - 1$$

for every two nonadjacent vertices u and v of G, then G is connected and $\text{diam}(G) \leq 2$.

Proof. We show that every two distinct vertices of G are connected by a path of length at most 2. Let $x, y \in V(G)$. If $xy \in E(G)$, then (x, y) is a path and x and y are certainly connected. Hence we may assume that $xy \notin E(G)$. Therefore, $\deg x + \deg y \geq n - 1$, which implies that there must be a vertex w that is adjacent to both x and y. So (x, w, y) is a path in G, as desired. ∎

Theorem 2.4 implies that if G is a graph of order n such that $\deg v \geq (n-1)/2$ for every vertex v of G, then G must be connected.

Corollary 2.5 *If G is a graph of order n with $\delta(G) \geq (n - 1)/2$, then G is connected.*

Proof. For every two nonadjacent vertices u and v of G,

$$\deg u + \deg v \geq \frac{n-1}{2} + \frac{n-1}{2} = n - 1.$$

By Theorem 2.4, G is connected. ∎

According to Corollary 2.5, if G is a graph of order $n = 7$ and $\delta(G) \geq (7-1)/2 = 3$, then G is connected. Also, if G is a graph of order $n = 8$ and $\delta(G) \geq (8-1)/2 = 3.5$, then G is connected. Of course, in the latter case, this says that if G is a graph of order $n = 8$ and $\delta(G) \geq 4$, then G is connected. For an even integer n, Corollary 2.5 then says that if G is a graph of order n with $\delta(G) \geq n/2$, then G is connected.

Let's return to Theorem 2.4. This theorem tells us then that if the sum of the degrees of any two nonadjacent vertices of a graph G of order n is "large enough," then G is connected. According to Theorem 2.4, $n-1$ is large enough. Obviously, if the sum of the degrees of any two nonadjacent vertices of G is at least n, then G must be connected as well. But what if the sum of the degrees of any two nonadjacent vertices of G is at least $n-2$? Does that also guarantee that G is connected? What we are now discussing is the **sharpness** of the bound in Theorem 2.4. That is, would Theorem 2.4 still be true if we replace $n-1$ by a smaller integer? If not, then Theorem 2.4 cannot be improved and the bound is sharp.

As it turns out, the bound in Theorem 2.4 *is* sharp. For example, suppose that n is even, say $n = 2k$ and consider the graph $G = 2K_k$, that is, G is the disconnected graph with two components each of which is K_k (see Figure 2.3). If u and v are two nonadjacent vertices in this graph G, then u and v must be in different components and each has degree $k-1$. So

$$\deg u + \deg v = (k-1) + (k-1) = 2k - 2 = n - 2.$$

Therefore, if the sum of the degrees of any two nonadjacent vertices of a graph G of order n is at least $n-2$, then there is no guarantee that G is connected.

G :

Figure 2.3: A disconnected graph of order $n = 2k$ such that the sum of the degrees of any two nonadjacent vertices is $n - 2$

Observe also that if G is a disconnected graph of order n, then (since G has at least two components) some component G_1 of G has order n_1 that is at most $n/2$. Every vertex of G_1 has degree at most $n_1 - 1 \leq (n/2) - 1 = (n-2)/2$ and so $\delta(G) \leq (n-2)/2$. (This observation actually provides a proof by contrapositive of Corollary 2.5.) If G has three components, then the order of some component of G is at most $n/3$. More generally, if G has k components, then the order of some component of G is at most n/k.

The concept of degree has counterparts in both multigraphs, pseudographs and digraphs. For a vertex v in a multigraph or pseudograph G, the **degree** $\deg v$ of v in G is the number of edges of G incident with v, where there is a contribution of 2 for each loop at v in a pseudograph. For the pseudograph G of Figure 2.4,

$$\deg u_1 = 4,\ \deg u_2 = 6,\ \deg u_3 = 6,\ \deg u_4 = 4.$$

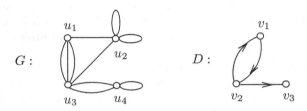

Figure 2.4: Illustrating degrees in a multigraph and a digraph

For a vertex v in a digraph D, the **outdegree** $\operatorname{od} v$ of v is the number of vertices of D to which v is adjacent, while the **indegree** $\operatorname{id} v$ of v is the number of vertices of D from which v is adjacent. For the digraph D of Figure 2.4,

$$\operatorname{od} v_1 = \operatorname{id} v_1 = 1,\ \operatorname{od} v_2 = 2,\ \operatorname{id} v_2 = 1,\ \operatorname{od} v_3 = 0,\ \operatorname{id} v_3 = 1.$$

Exercises for Section 2.1

2.1 Give an example of the following or explain why no such example exists:

   (a) a graph of order 7 whose vertices have degrees 1, 1, 1, 2, 2, 3, 3.

   (b) a graph of order 7 whose vertices have degrees 1, 2, 2, 2, 3, 3, 7.

   (c) a graph of order 4 whose vertices have degrees 1, 3, 3, 3.

2.2 Give an example of the following or explain why no such example exists:

   (a) a graph that has no odd vertices.

   (b) a noncomplete graph, all of whose vertices have degree 3.

   (c) a graph G of order 5 or more with the property that $\deg u \neq \deg v$ for every pair u, v of adjacent vertices of G.

   (d) a noncomplete graph H of order 5 or more with the property that $\deg u \neq \deg v$ for every pair u, v of nonadjacent vertices of H.

2.3 The degree of each vertex of a certain graph of order 12 and size 31 is either 4 or 6. How many vertices of degree 4 are there?

2.4 Give an example of a graph G of order 6 and size 10 such that $\delta(G) = 3$ and $\Delta(G) = 4$.

2.5 The degree of every vertex of a graph G of order 25 and size 62 is 3, 4, 5 or 6. There are two vertices of degree 4 and 11 vertices of degree 6. How many vertices of G have degree 5?

2.6 Prove that if a graph of order $3n$ ($n \geq 1$) has n vertices of each of the degrees $n - 1, n$ and $n + 1$, then n is even.

2.7 (a) Prove that if G is a bipartite graph of size m with partite sets U and W, then $m = \sum_{u \in U} \deg u = \sum_{w \in W} \deg w$.

   (b) Let G be a bipartite graph of order 22 with partite sets U and W, where $|U| = 12$. Suppose that every vertex in U has degree 3, while every vertex of W has degree 2 or 4. How many vertices of G have degree 2?

2.8 Let G be a graph of order n. If $\deg u + \deg v + \deg w \geq n - 1$ for every three pairwise nonadjacent vertices u, v and w of G, must G be connected?

2.9 Show that if G is a disconnected graph containing exactly two odd vertices, then these odd vertices must be in the same component of G.

2.10 We have already seen that if G is a graph of order n such that $\deg u + \deg v \geq n - 2$ for every two nonadjacent vertices u and v of G, then G might be disconnected.

   (a) Show that there exists a connected graph G of order n such that $\deg u + \deg v \geq n - 2$ for every two nonadjacent vertices u and v and for which $\deg x + \deg y = n - 2$ for some pair x, y of nonadjacent vertices of G.

   (b) Let G be a graph of order n. Prove that if $\deg u + \deg v \geq n - 2$ for every pair u, v of nonadjacent vertices of G, then G has at most two components.

   (c) Is the bound in part (b) sharp?

2.11 Corollary 2.5 states that if G is a graph of order n with $\delta(G) \geq (n - 1)/2$, then G is connected. Is the bound $(n - 1)/2$ sharp, that is, in this case, can $(n - 1)/2$ be replaced by $(n - 2)/2$ and obtain the same conclusion?

2.12 Prove that if G is a graph of order n such that $\Delta(G) + \delta(G) \geq n - 1$, then G is connected and $\mathrm{diam}(G) \leq 4$. Show that the bound $n - 1$ is sharp.

2.13 Let G be a graph of order $n \geq 2$.

   (a) Prove that if $\deg v \geq (n-2)/3$ for every vertex v of G, then G contains at most two components.

   (b) Show that the bound in (a) is sharp.

2.14 A graph G has the property that every edge of G joins an odd vertex with an even vertex. Show that G is bipartite and has even size.

2.15 A certain connected graph G has the property that for every two vertices u and v of G, the length of each $u - v$ path is even or the length of each $u - v$ path is odd. Prove that G is bipartite.

2.16 The degree of every vertex of a graph G of order $2n + 1 \geq 5$ is either $n + 1$ or $n + 2$. Prove that G contains at least $n + 1$ vertices of degree $n + 2$ or at least $n + 2$ vertices of degree $n + 1$.

2.17 Let G be a connected graph containing a vertex w such that (1) $\deg w \not\equiv 0 \pmod 3$ and (2) $\deg u + \deg v \equiv 0 \pmod 3$ for every two adjacent vertices u and v of G. Prove that G is bipartite and contains no vertex x such that $\deg x \equiv 0 \pmod 3$.

2.18 Let G be a graph of order 8 with $V(G) = \{v_1, v_2, \cdots, v_8\}$ such that $\deg v_i = i$ for $1 \leq i \leq 7$. What is $\deg v_8$?

2.2   Regular Graphs

We have already mentioned that $0 \leq \delta(G) \leq \Delta(G) \leq n - 1$ for every graph G of order n. If $\delta(G) = \Delta(G)$, then the vertices of G have the same degree and G is called **regular**. If $\deg v = r$ for every vertex v of G, where $0 \leq r \leq n - 1$, then G is r-**regular** or **regular of degree** r. The only regular graphs of order 4 or 5 are shown in Figure 2.5. There is no 1-regular or 3-regular graph of order 5, as no graph contains an odd number of odd vertices by Corollary 2.3.

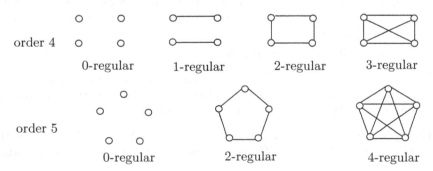

Figure 2.5: Some regular graphs

A 3-regular graph is also referred to as a **cubic graph**. The graphs K_4, $K_{3,3}$ and Q_3 are cubic graphs; however, the best known cubic graph may very well be the **Petersen graph**, shown in Figure 2.6. We will see this graph again. (Indeed, Section 8.5 is devoted to this graph.)

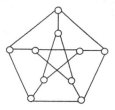

Figure 2.6: The Petersen graph

By Corollary 2.3, there are no r-regular graphs of order n if r and n are both odd. However, provided $0 \leq r \leq n - 1$, there are no other restrictions on the existence of an r-regular graph of order n. In the next proof, we will be considering a graph G with vertex set $V(G) = \{v_1, v_2, \ldots, v_n\}$ and performing arithmetic on the subscripts of the vertices. We follow the standard practice of performing the arithmetic modulo n. For example, if $n = 6$ and $i = 5$, then the vertex v_{i+2} denotes v_1.

Theorem 2.6 *Let r and n be integers with $0 \leq r \leq n - 1$. There exists an r-regular graph of order n if and only if at least one of r and n is even.*

Proof. As we already mentioned, there is no r-regular graph of order n if r and n are both odd. It remains only to verify the converse. So let r and n be integers with $0 \leq r \leq n - 1$ such that at least one of r and n is even. We construct an r-regular graph $H_{r,n}$ of order n. Let $V(H_{r,n}) = \{v_1, v_2, \ldots, v_n\}$. First, assume that r is even. Then $r = 2k \leq n - 1$ for some nonnegative integer $k \leq (n-1)/2$. For each i ($1 \leq i \leq n$), we join v_i to $v_{i+1}, v_{i+2}, \ldots, v_{i+k}$ and to $v_{i-1}, v_{i-2}, \ldots, v_{i-k}$. If we think of arranging the vertices $v_1, v_2, \ldots, v_n$ cyclically, then each vertex v_i is adjacent to the k vertices that immediately follow v_i and the k vertices that immediately precede v_i. Thus $H_{r,n}$ is r-regular. For $r = 4$ and $n = 10$, the graph $H_{4,10}$ is shown in Figure 2.7(a).

Second, assume that r is odd. Then $n = 2\ell$ is even. Also, $r = 2k + 1 \leq n - 1$ for some nonnegative integer $k \leq (n-2)/2$. We join v_i to the $2k$ vertices described above as well as to $v_{i+\ell}$. In this case, we again think of arranging the vertices $v_1, v_2, \ldots, v_n$ cyclically and joining each vertex v_i to the k vertices immediately following it, the k vertices immediately preceding it and the unique vertex "opposite" v_i. Thus $H_{r,n}$ is r-regular. For $r = 5$ and $n = 10$, the graph $H_{5,10}$ is shown in Figure 2.7(b). ∎

The graphs $H_{r,n}$ described above are called **Harary graphs**, named for Frank Harary. We will visit him again. Also, we will visit these graphs again in Section 5.3.

The proof of Theorem 2.6 that we have presented is a *constructive proof*, that is, we actually constructed a graph with the desired properties and didn't

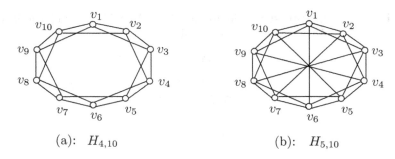

(a): $H_{4,10}$ (b): $H_{5,10}$

Figure 2.7: A 4-regular graph and a 5-regular graph, both of order 10

simply show that *some* graph with these properties exists. Although the proof
of Theorem 2.6 doesn't suggest it, there is little restriction on the subgraphs
that regular graphs can contain. Of course, if H is an r-regular graph, then H
cannot contain any graph G as a subgraph if $\Delta(G) > r$. On the other hand, if
G is a graph with $\Delta(G) \le r$ for some integer r, then G is a subgraph (indeed,
an *induced* subgraph) of some r-regular graph H, as we now see.

Theorem 2.7 *For every graph G and every integer $r \ge \Delta(G)$, there exists an*
r-regular graph H containing G as an induced subgraph.

Proof. If G is r-regular, then we let $H = G$. Thus, we may assume that G is not
an r-regular graph. Suppose that G has order n and $V(G) = \{v_1, v_2, \ldots, v_n\}$.
Let G' be another copy of G with $V(G') = \{v'_1, v'_2, \ldots, v'_n\}$, where each vertex v'_i
in G' corresponds to v_i in G for $1 \le i \le n$. We now construct a graph G_1 from
G and G' by adding the edges $v_i v'_i$ for all vertices v_i $(1 \le i \le n)$ of G for which
$\deg v_i < r$. Then G is an induced subgraph of G_1 and $\delta(G_1) = \delta(G) + 1$. If G_1
is r-regular, then we let $H = G_1$. If not, then we continue this procedure until
we arrive at an r-regular graph G_k, where $k = r - \delta(G)$. The graph G_k is the
desired graph H.   ∎

To illustrate the construction described in Theorem 2.7, consider the graph
G of Figure 2.8, where $\Delta(G) = 4$ and $\delta(G) = 2$. We seek a 4-regular graph
H containing G as an induced subgraph. First, we construct a graph G_1 from
two copies of G by joining all pairs of corresponding vertices in these copies
whose degrees are less than 4. Then $\Delta(G_1) = 4$, $\delta(G_1) = 3$ and G is an induced
subgraph of G_1. We then construct H from two copies of G_1 by joining all
pairs of corresponding vertices in these copies whose degrees are 3. Then H is
4-regular and G is an induced subgraph of H.

Theorem 2.7 appears in the first book entirely devoted to graph theory, published in 1936 and written by Dénes König. Although König stated and proved
the theorem for multigraphs, the proof he presented suggests the proof we gave
for graphs in Theorem 2.7. The proof of Theorem 2.7 does not construct an
r-regular graph of smallest order containing a given graph G with $\Delta(G) \le r$ as

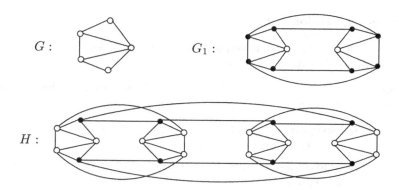

Figure 2.8: A 4-regular graph H containing G as an induced subgraph

an induced subgraph. Paul Erdös and Paul J. Kelly discovered a method for finding the smallest order of such an r-regular graph. We consider an example dealing with this problem.

Example 2.8 *For the graph G of Figure 2.9, find a 5-regular graph H of minimum order containing G as an induced subgraph.*

G :

Figure 2.9: The graph G in Example 2.8

Solution. Since $\delta(G) = 2$ and the order of G is 6, the order of such a graph H must be at least 9. However, there does not exist a 5-regular graph of order 9, so the order of H must be at least 10. Hence to construct a 5-regular graph containing G as an induced subgraph, it is necessary to add at least four vertices u, v, w, x to G (see Figure 2.10(a)). Joining u, v, w, x to the vertices of G and to each other as indicated in Figure 2.10(b) produces a 5-regular graph H containing G as an induced subgraph. Thus the minimum order of such a graph H is 10. $\diamondsuit$

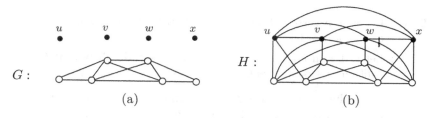

(a)   (b)

Figure 2.10: The graph H in Example 2.8

Exercises for Section 2.2

2.19 Construct an r-regular graph of order 6 and an s-regular graph of order 7 for all possible values of r and s.

2.20 Show that if G is a connected graph that is not regular, then G contains adjacent vertices u and v such that $\deg u \neq \deg v$.

2.21 (a) Find spanning subgraphs G_0, G_1, G_2, G_3 of the Petersen graph, where G_r is r-regular for $0 \leq r \leq 3$.

   (b) Find induced subgraphs F_0, F_1, F_2, F_3 in the Petersen graph, where F_r is r-regular for $0 \leq r \leq 3$.

   (c) How can the problem in (b) be revised so that it would be more interesting (and more challenging)?

2.22 For the graph G of Figure 2.11, construct a 3-regular graph H containing G as an induced subgraph

   (a) using the proof of Theorem 2.7. What is the order of H?

   (b) such that H has the smallest possible order. What is this order?

G :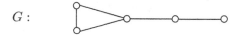

Figure 2.11: The graph G in Exercise 2.22

2.23 For each of the following paths, construct a 3-regular graph H of minimum order containing the path as an induced subgraph: (a) P_5, (b) P_6, (c) P_7.

2.24 What is the minimum order of a 3-regular graph H containing the graph G in Figure 2.12 as an induced subgraph?

G :

Figure 2.12: The graph G in Exercise 2.24

2.25 (a) Let v be a vertex of a graph G. Show that if $G - v$ is 3-regular, then G has odd order.

   (b) Let G be an r-regular graph, where r is odd. Show that G does not contain any component of odd order.

2.26 (a) Show that a graph G is regular if and only if $\overline{G}$ is regular.

   (b) Show that if G and $\overline{G}$ are both r-regular for some nonnegative integer r, then G has odd order.

2.27 Prove that if G is an r-regular bipartite graph with $r \geq 1$ and partite sets U and W, then $|U| = |W|$.

2.28 Investigate the following question: Does there exist a graph G and an integer r with $\delta(G) < r$ and $\Delta(G) \leq r$, such that the r-regular graph H in Theorem 2.7 that contains G as an induced subgraph has the smallest order among all r-regular graphs with this property?

2.29 (a) Prove that if G is a graph of order n, then $\delta(G) + \delta(\overline{G}) \leq n - 1$.

   (b) Prove, for a graph G of order n, that $\delta(G) + \delta(\overline{G}) = n - 1$ if and only if G is regular.

   (c) Prove that a graph G is regular if and only if G contains a vertex v such that $\deg_G v = \delta(G)$ and $\deg_{\overline{G}} v = \delta(\overline{G})$.

   (d) What can we say about a graph G with the property that for every vertex v of G, either $\deg_G v = \delta(G)$ or $\deg_{\overline{G}} v = \delta(\overline{G})$ but not both?

2.30 Beginning with $G = K_1$, use the construction in Theorem 2.7 to produce a 3-regular graph H (containing G as an induced subgraph). What famous graph is H?

2.3   Degree Sequences

Although we've been discussing graphs all of whose vertices have the same degree, it is more typical for the vertices of a graph to have a variety of degrees. If the degrees of the vertices of a graph G are listed in a sequence s, then s is called a **degree sequence** of G. For example, all of the sequences

$$s : 4, 3, 2, 2, 2, 1, 1, 1, 0; \quad s' : 0, 1, 1, 1, 2, 2, 2, 3, 4; \quad s'' : 4, 3, 2, 1, 2, 2, 1, 1, 0 \quad (2.2)$$

are degree sequences of the graph G of Figure 2.13, each of whose vertices is labeled by its degree. The sequence s is non-increasing, s' is non-decreasing and s'' is neither. Determining a degree sequence of a graph is not difficult. There is a converse question that is considerably more intriguing, however.

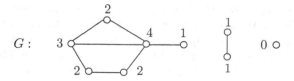

Figure 2.13: A graph with degree sequence $4, 3, 2, 2, 2, 1, 1, 1, 0$

Suppose that we are given a finite sequence s of nonnegative integers. Is s a degree sequence of some graph? A finite sequence of nonnegative integers is called **graphical** if it is a degree sequence of some graph. Of course, all of the sequences in (2.2) are graphical.

Example 2.9 *Which of the following sequences are graphical?*

(1) $s_1 : 3, 3, 2, 2, 1, 1$

(2) $s_2 : 6, 5, 5, 4, 3, 3, 3, 2, 2$

(3) $s_3 : 7, 6, 4, 4, 3, 3, 3$

(4) $s_4 : 3, 3, 3, 1$

Solution.

(1) The sequence s_1 is graphical. Indeed, it is a degree sequence of the graph G_1 of Figure 2.14.

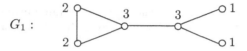

Figure 2.14: A graph with degree sequence $3, 3, 2, 2, 1, 1$

(2) Since s_2 has an odd number of terms that are odd integers, s_2 cannot be a degree sequence of a graph (for otherwise, such a graph would have an odd number of odd vertices, contradicting Corollary 2.3). Therefore, s_2 is not graphical.

(3) The sequence s_3 is also not graphical; for otherwise, s_3 would be a degree sequence of a graph of order 7 and containing a vertex of degree 7. (The degree of any vertex in a graph of order 7 is at most $7 - 1 = 6$.)

(4) The sequence s_4 contains four terms, all of which are at most $4 - 1 = 3$. Also, s_4 contains an even number of terms that are odd integers. Yet, s_4 is not graphical. Assume, to the contrary, that s_4 is graphical. Then there is a graph G_4 of order 4 with $V(G_4) = \{u, v, w, x\}$ such that $\deg u = \deg v = \deg w = 3$ and $\deg x = 1$. This implies that each of u, v and w is adjacent to all other vertices of G_4, including x, but x is adjacent to only one of u, v and w. This is impossible. $\diamond$

The sequence s_4 shows that determining which sequences are graphical is potentially difficult. We present a theorem that will help us to efficiently decide whether a given sequence is graphical. This theorem is due to Václav Havel and S. Louis Hakimi and is often referred to as the Havel-Hakimi Theorem, despite the fact that Havel and Hakimi gave independent proofs and wrote separate papers that include this theorem. To use this theorem, we assume that we are beginning with a non-increasing sequence.

Theorem 2.10 *A non-increasing sequence* $s : d_1, d_2, \ldots, d_n$ $(n \geq 2)$ *of non-negative integers, where* $d_1 \geq 1$, *is graphical if and only if the sequence*

$$s_1 : d_2 - 1, d_3 - 1, \ldots, d_{d_1+1} - 1, d_{d_1+2}, \ldots, d_n$$

is graphical.

Proof. First, assume that s_1 is graphical. Then there is a graph G_1 with $V(G_1) = \{v_2, v_3, \ldots, v_n\}$ such that

$$\deg_{G_1} v_i = \begin{cases} d_i - 1 & \text{if } 2 \leq i \leq d_1 + 1 \\ d_i & \text{if } d_1 + 2 \leq i \leq n. \end{cases}$$

We construct a graph G from G_1 by adding a new vertex v_1 and the d_1 edges $v_1 v_i$ for $2 \leq i \leq d_1 + 1$. Since $\deg_G v_i = d_i$ for $1 \leq i \leq n$, it follows that s is a degree sequence of G and so s is graphical.

Proving the converse is more challenging. Assume that s is graphical. Suppose that a graph H has degree sequence s and contains a vertex u of degree d_1 such that u is adjacent to vertices whose degrees are $d_2, d_3, \ldots, d_{d_1+1}$. Then s_1 is a degree sequence of $H - u$ and the proof of the converse is complete. We show next that there must be a graph H with degree sequence s containing a vertex u of degree d_1 that is adjacent to vertices whose degrees are $d_2, d_3, \ldots, d_{d_1+1}$.

Assume, to the contrary, that there is no graph with degree sequence s containing a vertex of degree d_1 that is adjacent to vertices whose degrees are $d_2, d_3, \ldots, d_{d_1+1}$. Among all graphs with degree sequence s, let G be one with $V(G) = \{v_1, v_2, \ldots, v_n\}$ such that $\deg v_i = d_i$ for $1 \leq i \leq n$ and the sum of the degrees of vertices adjacent to v_1 is as large as possible. Since v_1 is not adjacent to vertices having degrees $d_2, d_3, \ldots, d_{d_1+1}$ (that is, v_1 is not adjacent to vertices with the next d_1 highest degrees), v_1 must be adjacent to a vertex v_s having a smaller degree than a vertex v_r to which v_1 is *not* adjacent. That is, there exist vertices v_r and v_s with $d_r > d_s$ such that v_1 is adjacent to v_s but not to v_r. Since $\deg v_r = d_r > d_s = \deg v_s$, there exists a vertex v_t such that v_t is adjacent to v_r but not to v_s. Consider the graph G' obtained from G by removing the edges $v_1 v_s$ and $v_r v_t$ and adding the edges $v_1 v_r$ and $v_s v_t$ (see Figure 2.15, where a dashed line means no edge).

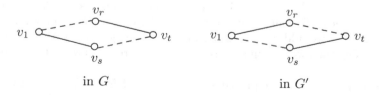

in G in G'

Figure 2.15: Edges in (and not in) G and G' in the proof of Theorem 2.10

Then G and G' have the same vertex set; indeed, s is a degree sequence of both G and G'. However, the sum of the degrees of the vertices adjacent to v_1 in G' is larger than that in G, which produces a contradiction. ∎

We now illustrate Theorem 2.10 with the following two examples.

Example 2.11 *Decide whether the sequence $s : 5, 4, 3, 3, 2, 2, 2, 1, 1, 1$ is graphical.*

Solution. Deleting 5 from s and subtracting 1 from the next five terms, we obtain

$$s_1' : 3, 2, 2, 1, 1, 2, 1, 1, 1.$$

Reordering this sequence (so that a non-increasing sequence results and we can apply Theorem 2.10 again), we have

$$s_1 : 3, 2, 2, 2, 1, 1, 1, 1, 1.$$

By Theorem 2.10, s is graphical if and only if s_1 is. To decide whether s_1 is graphical, we choose to continue this procedure since s_1 is relatively complicated. Deleting 3 from s_1 and subtracting 1 from the next three terms, we obtain

$$s_2' = s_2 : 1, 1, 1, 1, 1, 1, 1, 1.$$

By Theorem 2.10, s_1 is graphical if and only if s_2 is. But s_2 is so simple that we can quickly observe that s_2 is a (actually *the*) degree sequence of the graph $G_2 = 4K_2$ of Figure 2.16. Therefore, s_2 is graphical, so s_1 and s are graphical as well. $\diamond$

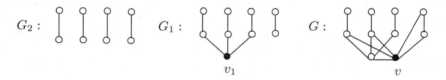

Figure 2.16: The graphs G_2, G_1 and G in Example 2.11

In Example 2.11, we stopped when we arrived at the sequence s_2, as it was clear that s_2 is a degree sequence of some graph, namely s_2 is the degree sequence of the graph G_2 of Figure 2.16. Recall that the sequence $s_2 = s_2'$ was obtained from s_1 by deleting 3 from s_1 and subtracting 1 from the next three terms of s_1. Consequently, if we add a new vertex v_1 to G_2 and join v_1 to three vertices of degree 1 in G_2, then we obtain a graph G_1 (see Figure 2.16) with degree sequence s_1.

In Example 2.11, the sequence s_1 was obtained from s_1' by rearranging its terms to produce a non-increasing sequence. The sequence s_1' was, in turn, obtained from s by deleting the first term 5 from s and subtracting 1 from the next five terms of s. Therefore, if we add a new vertex v to G_1 and join v to the vertex of degree 3, to two vertices of degree 2 and to two vertices of degree 1, then we obtain a graph G with degree sequence s. So we have just constructed a graph with degree sequence s. Informally, we have just described an efficient algorithm for deciding whether a given sequence of nonnegative integers is graphical. We present one additional example to illustrate how to proceed when the sequence is not graphical.

Example 2.12 *Decide whether the sequence* $s : 7, 7, 4, 3, 3, 3, 2, 1$ *is graphical.*

Solution. Deleting the first term 7 from s and subtracting 1 from the next seven terms of s, we obtain

$$s_1 = s_1' : 6, 3, 2, 2, 2, 1, 0.$$

Perhaps we can see that s_1 is not graphical but nevertheless let's continue. Deleting the first term 6 from s_1 and subtracting 1 from the next six terms of s_1, we obtain

$$s_2 = s_2' : 2, 1, 1, 1, 0, -1.$$

Since s_2 contains the negative number -1 and, of course, no vertex can have a negative degree, it is now very clear that s_2 is not graphical. By Theorem 2.10, the sequence s_1 is not graphical and so neither is s. ◇

Another way to show that the sequence in Example 2.12 is not graphical is to observe that if it were graphical, then there would be a graph G of order 8 with degree sequence s, implying that G has two vertices of degree 7 and so all other vertices of G have degree at least 2. Hence G can have no vertex of degree 1.

Exercises for Section 2.3

2.31 Prove that a sequence $d_1, d_2, \ldots, d_n$ is graphical if and only if $n - d_1 - 1, n - d_2 - 1, \ldots, n - d_n - 1$ is graphical.

2.32 Use Theorem 2.10 to determine which of the following sequences are graphical. For each of those that are graphical, construct a graph, as in Example 2.11, for which the given sequence is a degree sequence of the graph.

   (a) $s_1 : 5, 3, 3, 3, 3, 2, 2, 2, 1$ (b) $s_2 : 6, 3, 3, 3, 3, 2, 2, 2, 2, 1, 1$

   (c) $s_3 : 6, 5, 5, 4, 3, 2, 1$ (d) $s_4 : 7, 5, 4, 4, 4, 3, 2, 1$ (e) $s_5 : 7, 6, 5, 4, 4, 3, 2, 1.$

2.33 Prove that for every integer x with $0 \le x \le 5$, the sequence $x, 1, 2, 3, 5, 5$ is not graphical.

2.34 For which integers x $(0 \le x \le 7)$, if any, is the sequence $7, 6, 5, 4, 3, 2, 1, x$ graphical?

2.35 If the sequence $x, 7, 7, 5, 5, 4, 3, 2$ is graphical, then what are the possible values of x $(0 \le x \le 7)$?

2.36 Let $S = \{2, 6, 7\}$. Prove that there exists a positive integer k such that the sequence obtained by listing each element of S a total k times is a degree sequence of some graph. What is the minimum value of k?

2.4 Excursion: Graphs and Matrices

As we know, a graph G can be defined by two sets, namely its vertex set $V(G)$ and edge set $E(G)$ or by a diagram. A graph can also be described by a matrix and for some purposes this is especially useful.

Let G be a graph of order n and size m, where $V(G) = \{v_1, v_2, \ldots, v_n\}$ and $E(G) = \{e_1, e_2, \ldots, e_m\}$. The **adjacency matrix** of G is the $n \times n$ matrix $A = [a_{ij}]$, where

$$a_{ij} = \begin{cases} 1 & \text{if } v_iv_j \in E(G) \\ 0 & \text{otherwise;} \end{cases}$$

while the **incidence matrix** of G is the $n \times m$ matrix $B = [b_{ij}]$, where

$$b_{ij} = \begin{cases} 1 & \text{if } v_i \text{ is incident with } e_j \\ 0 & \text{otherwise.} \end{cases}$$

These matrices are shown for the graph G of Figure 2.17.

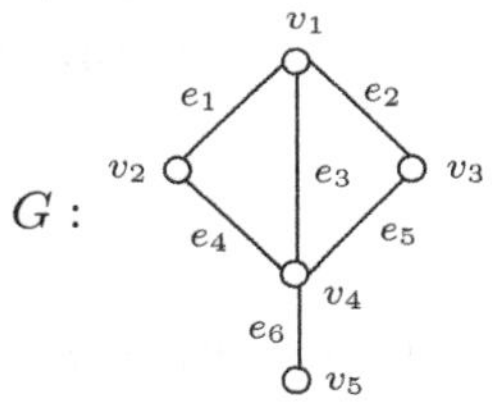

$$A = \begin{bmatrix} 0 & 1 & 1 & 1 & 0 \\ 1 & 0 & 0 & 1 & 0 \\ 1 & 0 & 0 & 1 & 0 \\ 1 & 1 & 1 & 0 & 1 \\ 0 & 0 & 0 & 1 & 0 \end{bmatrix} \qquad B = \begin{bmatrix} 1 & 1 & 1 & 0 & 0 & 0 \\ 1 & 0 & 0 & 1 & 0 & 0 \\ 0 & 1 & 0 & 0 & 1 & 0 \\ 0 & 0 & 1 & 1 & 1 & 1 \\ 0 & 0 & 0 & 0 & 0 & 1 \end{bmatrix}$$

Figure 2.17: The adjacency matrix and incidence matrix of a graph

Here are a few useful observations about the adjacency matrix and incidence matrix. First, these matrices are dependent on how the vertices and edges of G are labeled. In any case, the adjacency matrix is a symmetric $n \times n$ matrix where every entry on the main diagonal is 0. The number of 1s in row i (or column i) is the degree of the vertex v_i. While the number of 1s in row i of the incidence matrix is also the degree of v_i, the number of 1s in each of its columns is 2 since there are exactly two vertices incident with every edge.

Two $u - v$ walks are considered **equal** if, as sequences, they are identical, term by term. Let's now return to the graph G and its adjacency matrix A shown in Figure 2.17. The square A^2 and the cube A^3 of A are given in Figure 2.18. If we look at the entries along the main diagonal in A^2, we see that these are the degrees of the vertices of G. This is not only true in general but each entry in each power of A represents some characteristic of the graph G.

Theorem 2.13 *Let G be a graph with vertex set $V(G) = \{v_1, v_2, \ldots, v_n\}$ and adjacency matrix $A = [a_{ij}]$. Then the entry $a_{ij}^{(k)}$ in row i and column j of A^k is the number of distinct $v_i - v_j$ walks of length k in G.*

$$A^2 = \begin{bmatrix} 3 & 1 & 1 & 2 & 1 \\ 1 & 2 & 2 & 1 & 1 \\ 1 & 2 & 2 & 1 & 1 \\ 2 & 1 & 1 & 4 & 0 \\ 1 & 1 & 1 & 0 & 1 \end{bmatrix} \qquad A^3 = \begin{bmatrix} 4 & 5 & 5 & 6 & 2 \\ 5 & 2 & 2 & 6 & 1 \\ 5 & 2 & 2 & 6 & 1 \\ 6 & 6 & 6 & 4 & 4 \\ 2 & 1 & 1 & 4 & 0 \end{bmatrix}$$

Figure 2.18: Powers of an adjacency matrix

Proof. We proceed by induction. We begin with $A = A^1$. Each entry $a_{ij} = a_{ij}^{(1)}$ of A is either 1 or 0, according to whether $v_i v_j$ is or is not an edge of G. Hence $a_{ij}^{(1)}$ gives the number of $v_i - v_j$ walks of length 1 in G. Assume, for a positive integer k, that the number of $v_i - v_j$ walks of length k in G is given by $a_{ij}^{(k)}$. From the definition of matrix multiplication, the (i,j)-entry $a_{ij}^{(k+1)}$ in A^{k+1} is the dot product of row i of A^k and column j of A, that is,

$$a_{ij}^{(k+1)} = \sum_{t=1}^{n} a_{it}^{(k)} a_{tj} = a_{i1}^{(k)} a_{1j} + a_{i2}^{(k)} a_{2j} + \ldots + a_{in}^{(k)} a_{nj}. \qquad (2.3)$$

Every $v_i - v_j$ walk W of length $k+1$ is produced by beginning with a $v_i - v_t$ walk W' of length k for some vertex v_t adjacent to v_j and then following W' by v_j. By the induction hypothesis, the number of $v_i - v_t$ walks of length k is $a_{it}^{(k)}$, while v_t is adjacent to v_j if and only if $a_{tj} = 1$. Hence by (2.3), $a_{ij}^{(k+1)}$ does indeed provide the number of $v_i - v_j$ walks of length $k+1$ in G. ∎

In view of Theorem 2.13, $a_{ii}^{(2)} = \deg v_i$ and $a_{ii}^{(3)}$ is twice the number of triangles in G that contain v_i. Indeed, knowing what each entry of A^k represents allows us to compute powers of adjacency matrices, at least small powers of adjacency matrices of graphs of small orders without actually multiplying matrices. For example, for the graph G of Figure 2.17, the entry $a_{41}^{(3)}$ of A^3 is the number of distinct $v_4 - v_1$ walks in G of length 3. Since there are six $v_4 - v_1$ walks of length 3 in G, namely:

(1) (v_4, v_5, v_4, v_1) (2) (v_4, v_2, v_4, v_1) (3) (v_4, v_3, v_4, v_1)
(4) (v_4, v_1, v_2, v_1) (5) (v_4, v_1, v_3, v_1) (6) (v_4, v_1, v_4, v_1),

it follows that $a_{41}^{(3)} = 6$, as we have seen.

Exercises for Section 2.4

2.37 For the adjacency matrix A of the graph G_1 of Figure 2.19, determine A^2 and A^3 without computing A or performing matrix multiplication.

2.38 For the adjacency matrix A of the graph G_2 of Figure 2.19, determine A^2 and A^3 without computing A or performing matrix multiplication.

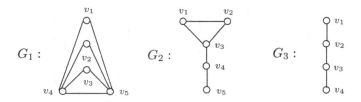

Figure 2.19: Graphs for Exercises 2.37-2.39

2.39 For the adjacency matrix A of the graph G_3 of Figure 2.19, determine A^4 without computing A or performing matrix multiplication.

2.40 For $G = K_{r,r}$ with partite sets $U = \{v_1, v_2, \ldots, v_r\}$ and $W = \{v_{r+1}, v_{r+2}, \ldots, v_{2r}\}$, determine the adjacency matrix A of G and its powers A^2, A^3 and A^4 without performing matrix multiplication.

2.41 (a) For the incidence matrix B in Figure 2.17, compute BB^t, where B^t is the transpose of B.

    (b) For a graph G with $V(G) = \{v_1, v_2, \ldots, v_n\}$, what does the (i, j)-entry of BB^t represent in G?

2.5 Exploration: Irregular Graphs

Recall that a graph G is regular if every two vertices of G have the same degree. We've already discussed the existence of regular graphs. We now consider graphs that are opposite to regular graphs and the existence of such graphs. A graph G of order at least 2 is **irregular** if every two vertices of G have distinct degrees. Before proceeding further with this line of discussion, however, let us pause a bit.

In the Fall 1988 issue (Volume 10, No. 4) of the magazine *Mathematical Intelligencer*, the British mathematics educator and writer David Wells asked the readers to evaluate 24 theorems for their beauty. Two years later (Volume 12, No. 3), he reported the results of the responses he received. The theorem that finished first was

$$e^{i\pi} = -1.$$

There was a tie for second place:

Euler's Polyhedral Formula: $V - E + F = 2$.

There are infinitely many primes.

We will visit Euler's Polyhedral Formula later (expressed in a different form), as well as two other theorems that made David Wells' list, including a theorem that was one of six theorems that tied for 15th place:

> **The Party Theorem** *At any party, there is a pair of people who have the same number of friends present.*

The so-called Party Theorem can be described in terms of graphs. Let G be a graph whose vertices are the people present at the party. Join two vertices by an edge if the two vertices (people) are friends. The number of friends that a person has at the party is then the degree of that vertex (person) in G. According to the Party Theorem, there are always two vertices with the same degree. This theorem can be restated using the terminology we introduced a short while ago.

Theorem 2.14 *No nontrivial graph is irregular.*

Proof. Assume, to the contrary, that there exists a graph G of order $n \geq 2$ whose vertices have distinct degrees. These degrees must be among the n integers $0, 1, 2, \cdots, n-1$. So G must have one vertex of each such degree. Let u and v be the vertices of G such that $\deg u = 0$ and $\deg v = n-1$. Since $\deg u = 0$, it follows that u is adjacent to no vertex of G, including v. On the other hand, since $\deg v = n-1$, it follows that v is adjacent to all other vertices of G, including u. This is impossible. ∎

In view of Theorem 2.14, there appears to be little reason to discuss irregular graphs. There are other options, however. Recall that we defined a nontrivial graph G to be **irregular** if every two vertices of G have distinct degrees. But what if we were to redefine what we mean by degree?

Let F be a nontrivial graph. For a graph G and a vertex v of G, define the F-**degree** $F \deg v$ of v in G as the number of copies (unlabeled subgraphs, induced or not, having the same structure) of F in G that contain v. For example, for $F = K_3$ and the graph G of Figure 2.20, the vertices of G are labeled with their F-degrees.

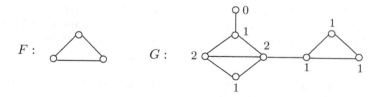

Figure 2.20: Illustrating F-degrees in a graph G

Observe that if $F = K_2$, then $F \deg v = \deg v$ for every vertex v of a graph G. So the F-degree of a vertex is a generalization of the ordinary degree of a vertex. Speaking of generalizations, we see that the following theorem generalizes the First Theorem of Graph Theory (Theorem 2.1).

Theorem 2.15 *Let F be a graph of order $k \geq 2$ and let G be a graph. If G contains m copies of the graph F, then*

$$\sum_{v \in V(G)} F \deg v = km. \tag{2.4}$$

Proof. Equality (2.4) follows since every copy of F is counted k times, once for each of the k vertices contained in this copy.   ∎

For example, the graph $F = K_3$ of Figure 2.20 has order 3 and the sum of the F-degrees of the vertices of G is 9, which implies that G contains three triangles.

Corollary 2.16 *Let F be a graph of even order and let G be a graph. Then G has an even number of vertices with odd F-degree.*

Let F be a nontrivial graph. A graph G is F-**regular** if every two vertices of G have the same F-degree, while G is F-**irregular** if every two vertices of G have distinct F-degrees. If $F = K_2$, then F-regularity and regularity are the same, as are F-irregularity and irregularity. So, for $F = K_2$, there are no nontrivial F-irregular graphs. For $F = K_3$, the F-degrees of the vertices of two graphs G_1 and G_2 are shown in Figure 2.21. The graph G_1 is F-regular but not regular, while the graph G_2 is regular but not F-regular.

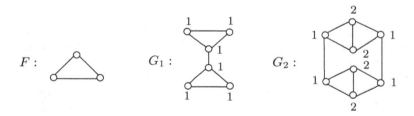

Figure 2.21: F-degrees in the graphs G_1 and G_2

For $F = P_3$, the vertices of the graph G of Figure 2.22 are also labeled with their F-degrees. The six copies of P_3 containing v are also shown in Figure 2.22. Observe that this nontrivial graph G is F-irregular. The graph G of Figure 2.22 shows that even though there exists no K_2-irregular graph, P_3-irregular graphs do exist. In fact, there is a conjecture on this topic.

Conjecture 2.17 *Let F be a nontrivial connected graph. There exists an F-irregular graph if and only if $F \neq K_2$.*

Once again, recall that we defined a graph G to be irregular if every two vertices of G have distinct degrees and showed that no nontrivial graph is irregular. We saw that if we redefined degree in a new way, then irregular graphs do exist. But what if we define degree in the standard manner and re-interpret what we mean by a graph, say we consider multigraphs instead? Since the multigraph

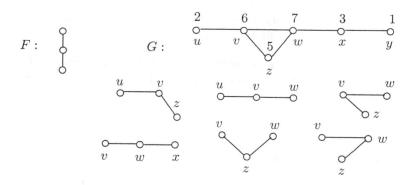

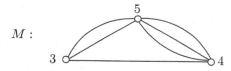

Figure 2.22: An F-irregular graph

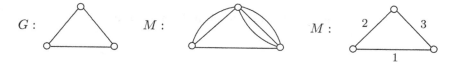

Figure 2.23: An irregular multigraph

M of Figure 2.23 is irregular (its vertices are labeled with their degrees), we see that irregular multigraphs exist.

Now that we know irregular multigraphs exist, what problems might be interesting to study? If M is a multigraph and all parallel edges joining pairs of vertices of M are replaced by a single edge, then the resulting graph G is called the **underlying graph** of M. Consider the graph $G = K_3$ of Figure 2.24, which is the underlying graph of the multigraph M of Figure 2.23. Of course, G is not irregular. In view of the multigraph M of Figure 2.23, we see that it's possible to replace one or more edges of some graph by parallel edges to produce an irregular multigraph. Which graphs have this property? That is, which graphs are the underlying graphs of irregular multigraphs? Before attempting to answer this question, we note that we can represent the multigraph M of Figure 2.23 in the simpler manner shown in Figure 2.24, where each edge of G is assigned a positive integer that represents the number of parallel edges joining a particular pair of vertices in the multigraph. This is referred to as a **weighted graph**. Irregular multigraphs can therefore be referred to as irregular weighted graphs.

Figure 2.24: A graph, a multigraph and a weighted graph

Among the nontrivial connected graphs, K_2 is the only graph that is not the underlying graph of an irregular weighted graph.

Theorem 2.18 *Let G be a connected graph of order 2 or more. Then G is the underlying graph of an irregular multigraph (weighted graph) if and only if $G \neq K_2$.*

Exercises for Section 2.5

2.42 For $F = K_4$, give an example of two graphs H_1 and H_2 such that H_1 is F-regular but not regular, while H_2 is regular but not F-regular.

2.43 Give an example of a connected graph F and a connected graph G such that G is regular and G contains vertices u and v such that $F \deg u - F \deg v \geq 2$.

2.44 For $F = P_3$, give an example of an F-irregular graph of order 7 or more.

2.45 Investigate F-degrees for a disconnected graph F of your choosing.

2.46 Find an irregular multigraph whose underlying graph is

   (a) P_3, (b) P_4, (c) C_4, (d) C_5, (e) K_4.

2.47 (a) Find an irregular multigraph (weighted graph) whose underlying graph is C_4 such that the sum s of the weights of its edges is minimum.

   (b) For the integer s in (a), investigate the following question: For which integers $s' > s$ is there an irregular weighted graph whose underlying graph is C_4 and such that the sum of the weights of its edges is s'?

2.48 Prove Theorem 2.18.

2.49 For a given graph G, color each edge of G either red or blue. A vertex v of the colored graph G has degree (a, b) if v is incident with a red edges and b blue edges. Define a graph G to be 2-color irregular if there exists a red-blue coloring of the edges of G such that no two vertices of G have the same degree. Is the graph H of Figure 2.25 2-color irregular?

H :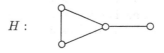

Figure 2.25: The graph in Exercise 2.49

Chapter 3

Isomorphic Graphs

3.1 The Definition of Isomorphism

Recall that two graphs G and H are equal if $V(G) = V(H)$ and $E(G) = E(H)$. We have called two graphs G and H "isomorphic" if they have the same structure and have written $G \cong H$ to indicate this. That is, $G \cong H$ if the vertices of G and H can be labeled (or relabeled) to produce two equal graphs. We now make all of this more precise.

Suppose that you are asked to give an example of three graphs having order 5 and size 5. Would the three graphs H_1, H_2 and H_3 given in Figure 3.1 be an acceptable answer to this question?

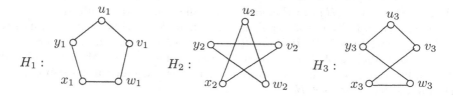

Figure 3.1: Graph(s) of order 5 and size 5

By repositioning the vertices of H_2, we have redrawn H_2 as in Figure 3.2. Similarly, we can redraw H_3 as in Figure 3.2.

It should now be clear that the graphs of Figure 3.1 differ only in the way they are labeled and in the way they are drawn, that is, they have the same structure. In a certain sense then, we have only given an example of one graph of order 5 and size 5. That is, the graphs of Figure 3.1 are simply disguised forms of the same graph, namely, the 5-cycle C_5. For example, the redrawing of H_2 shown in Figure 3.2 suggests that (1) u_2 in H_2 is playing the role of u_1 in H_1, (2) w_2 is playing the role of v_1, (3) y_2 is playing the role of w_1, (4) v_2

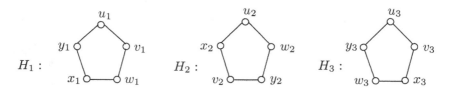

Figure 3.2: A graph of order 5 and size 5

is playing the role of x_1 and (5) x_2 is playing the role of y_1. The manner in which the vertices of H_2 correspond to the vertices of H_1 is not unique, however. Two other drawings of H_2 shown in Figure 3.3 suggest that there are other correspondences between the vertices of H_1 and the vertices of H_2. Indeed, there are several such correspondences.

Figure 3.3: Other drawings of the graph H_2

As we have said, when two graphs differ only in the way they're drawn and/or labeled, then they are said to be isomorphic. Formally, two (labeled) graphs G and H are **isomorphic** (have the same structure) if there exists a one-to-one correspondence ϕ from $V(G)$ to $V(H)$ such that $uv \in E(G)$ if and only if $\phi(u)\phi(v) \in E(H)$. In this case, ϕ is called an **isomorphism** from G to H. Thus, if G and H are isomorphic graphs, then we say that G **is isomorphic to** H and we write $G \cong H$. If G and H are unlabeled, then they are isomorphic if, under any labeling of their vertices, they are isomorphic as labeled graphs. If two graphs G and H are not isomorphic, then they are called **non-isomorphic graphs** and we write $G \not\cong H$.

As is implied by the redrawing of the graph H_2 in Figure 3.2, the function $\phi : V(H_1) \rightarrow V(H_2)$ defined by

$$\phi(u_1) = u_2, \ \phi(v_1) = w_2, \ \phi(w_1) = y_2, \ \phi(x_1) = v_2, \ \phi(y_1) = x_2$$

is an isomorphism and so $H_1 \cong H_2$. Intuitively then, two graphs are isomorphic if it is possible to redraw one of them so that the two diagrams are the same. This highly informal interpretation of isomorphism, although often suitable, is not satisfactory in all cases and so it may be necessary to rely on the formal definition. In this context then, we consider two graphs to be the "same" graph if they are isomorphic and to be "different" if they are not isomorphic. From this point of view, there is only one graph of order 1, two graphs of order 2 and four graphs of order 3. There are eleven (non-isomorphic) graphs of order 4 and these are shown in Figure 3.4.

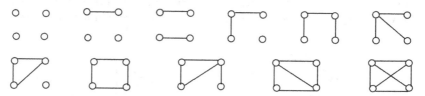

Figure 3.4: The eleven graphs of order 4

Let's look at the definition of isomorphism more closely. First, in order for two graphs G_1 and G_2 to be isomorphic, there must be a one-to-one correspondence from the vertex set of G_1 to the vertex set of G_2. This means that it must be possible to pair off the vertices of G_1 with the vertices of G_2. Therefore, $|V(G_1)| = |V(G_2)|$ and so G_1 and G_2 have the same order. It is certainly not surprising that we would want two graphs to be of the same order if we want to consider them to be the same graph.

Continuing to analyze the definition of isomorphic graphs, we see that not only must there be a one-to-one correspondence from $V(G_1)$ to $V(G_2)$ but two vertices u_1 and v_1 of G_1 are adjacent in G_1 if and only if the corresponding vertices $\phi(u_1)$ and $\phi(u_2)$ are adjacent in G_2. So adjacent vertices in G_1 are mapped to adjacent vertices in G_2, while nonadjacent vertices in G_1 are mapped to nonadjacent vertices in G_2. This implies that for G_1 and G_2 to be isomorphic, they must have the same size – again not a particularly surprising piece of information.

Hence if two graphs are isomorphic, then they must have the same order and the same size. The contrapositive of this statement says that if two graphs have different orders or different sizes, then they are not isomorphic. For example, even though the graphs F' and F'' of Figure 3.5 have the same size 6, they are not isomorphic because their orders are different. Also, the graphs H' and H'' of Figure 3.5 have the same order 6 but cannot be isomorphic since their sizes are different.

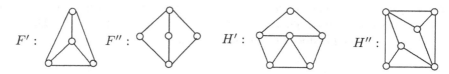

Figure 3.5: Non-isomorphic graphs

On the other hand, if two graphs have the same order and the same size, then there is no guarantee that the graphs are isomorphic. For example, the graphs G_1 and G_2 of Figure 3.6 have order 6 and size 6, yet they are not isomorphic. In order to see this, assume, to the contrary, that they are isomorphic. Then there exists an isomorphism $\phi : V(G_1) \to V(G_2)$. Hence there are three vertices of G_1 that map into u_2, v_2 and z_2 of G_2. Since u_2, v_2 and z_2 are pairwise adjacent and form a triangle, so too are the vertices of G_1 that map into these three vertices of G_2. However, G_1 doesn't contain a triangle and so a contradiction is produced.

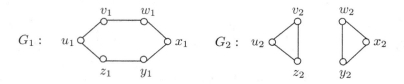

Figure 3.6: Two non-isomorphic graphs

Let's revisit the definition of isomorphism yet again. Two graphs G and H are isomorphic if there exists a one-to-one correspondence ϕ from $V(G)$ to $V(H)$ such that every two adjacent vertices of G are mapped to adjacent vertices of H and every two nonadjacent vertices of G are mapped to nonadjacent vertices of H. Recall that a function ϕ with these properties is an isomorphism. However, since $V(\overline{G}) = V(G)$ and $V(\overline{H}) = V(H)$, the same function $\phi : V(\overline{G}) \to V(\overline{H})$ also maps adjacent vertices of $\overline{G}$ to adjacent vertices of $\overline{H}$ and nonadjacent vertices of $\overline{G}$ to nonadjacent vertices of $\overline{H}$. This observation provides us with the following theorem.

Theorem 3.1 *Two graphs G and H are isomorphic if and only if their complements $\overline{G}$ and $\overline{H}$ are isomorphic.*

Let's consider the two graphs H_1 and H_2 shown in Figure 3.7. Both graphs have order 6 and size 9; so H_1 and H_2 might be isomorphic but we don't know this for sure. Since $\overline{H}_1 = G_1$ and $\overline{H}_2 = G_2$ (where G_1 and G_2 are the graphs shown in Figure 3.6) and G_1 and G_2 are not isomorphic, it follows by Theorem 3.1 that H_1 and H_2 are also not isomorphic. It is possible to see that H_1 and H_2 are not isomorphic without the aid of Theorem 3.1, however. Assume, to the contrary, that H_1 and H_2 are isomorphic. Then there exists an isomorphism $\phi : V(H_1) \to V(H_2)$. The vertices v_1, x_1 and z_1 are mutually adjacent in H_1 and form a triangle and so $\phi(v_1), \phi(x_1)$ and $\phi(z_1)$ form a triangle in H_2. However, H_2 contains no triangle and a contradiction is produced.

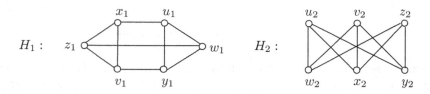

Figure 3.7: Two graphs H_1 and H_2

A graph and its complement may, in fact, be isomorphic. A graph G is **self-complementary** if $G \cong \overline{G}$. Of course, this can only occur if G and $\overline{G}$ have the same size, namely $\frac{1}{2}\binom{n}{2} = \frac{n(n-1)}{4}$. In order for $\frac{n(n-1)}{4}$ to be an integer, either $4 \mid n$ or $4 \mid (n-1)$, that is, either $n \equiv 0 \pmod 4$ or $n \equiv 1 \pmod 4$. Figure 3.8 shows four self-complementary graphs.

Figure 3.8: Self-complementary graphs

Not only are the orders the same and the sizes the same of two isomorphic graphs, so too are the degrees of their vertices.

Theorem 3.2 *If G and H are isomorphic graphs, then the degrees of the vertices of G are the same as the degrees of the vertices of H.*

Proof. Since G and H are isomorphic, there is an isomorphism $\phi : V(G) \to V(H)$. Let u be a vertex of G and suppose that $\phi(u) = v$, where v therefore belongs to H. We show that $\deg_G u = \deg_H v$. Suppose that u is adjacent to $x_1, x_2, \ldots, x_k$ in G and not adjacent to $w_1, w_2, \ldots, w_\ell$. Thus $|V(G)| = k + \ell + 1$. Then $\phi(u) = v$ is adjacent to $\phi(x_1), \phi(x_2), \ldots, \phi(x_k)$ in H and not adjacent to $\phi(w_1), \phi(w_2), \ldots, \phi(w_\ell)$. Therefore, $\deg_H v = k = \deg_G u$. ∎

Theorem 3.2 not only tells us that two isomorphic graphs G and H have the same degree sequence but the proof of this theorem also says that if ϕ is an isomorphism from $V(G)$ to $V(H)$ and u is a vertex of G, then $\deg_G u = \deg_H \phi(u)$, that is, under an isomorphism a vertex can only map into a vertex having the same degree.

So now we know that if G and H are isomorphic graphs, then their orders are the same, their sizes are the same and the degrees of their vertices are the same. On the other hand, if the degrees of their vertices are the same, then their orders must be the same and their sizes must also be the same. As with the order and size, two graphs having the same degree sequences is only a necessary condition, not a sufficient condition, for two graphs to be isomorphic. For example, the degree sequences of the non-isomorphic graphs G_1 and G_2 of Figure 3.6 are both $2, 2, 2, 2, 2, 2$; while the degree sequences of the non-isomorphic graphs of Figure 3.7 are both $3, 3, 3, 3, 3, 3$.

Therefore, the challenge for determining whether two graphs are isomorphic is when the two graphs have the same degree sequence. Let's consider some examples of this.

Example 3.3 *Determine whether the graphs F_1 and F_2 of Figure 3.9 are isomorphic.*

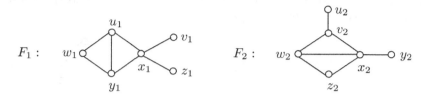

Figure 3.9: The two graphs in Example 3.3

Solution. Since both F_1 and F_2 have the degree sequence $4, 3, 3, 2, 1, 1$, they may be isomorphic. But they are not. Assume, to the contrary, that $F_1 \cong F_2$. Then there exists an isomorphism $\phi : V(F_1) \rightarrow V(F_2)$. The vertex x_1 is the only vertex of F_1 having degree 4. Thus $\phi(x_1)$ has degree 4 in F_2. Since x_2 is the only vertex of F_2 having degree 4, it follows that $\phi(x_1) = x_2$. Since both v_1 and z_1 are adjacent to x_1 in F_1, both $\phi(v_1)$ and $\phi(z_1)$ are adjacent to $\phi(x_1) = x_2$ in F_2. Because $\deg_{F_1} v_1 = \deg_{F_1} z_1 = 1$, it is also the case that $\deg_{F_2} \phi(v_1) = \deg_{F_2} \phi(z_1) = 1$. But this says that x_2 is adjacent to two end-vertices in F_2, which is not the case. This is a contradiction. $\diamond$

The argument just given to show that the graphs F_1 and F_2 of Figure 3.9 are not isomorphic can be simplified. The vertex of degree 4 in F_1 is adjacent to two end-vertices; while the vertex of degree 4 in F_2 is not. Therefore, F_1 and F_2 are not isomorphic.

Indeed, let G_1 and G_2 be two graphs. We might as well assume that G_1 and G_2 have the same degree sequence; otherwise, we know immediately that $G_1 \ncong G_2$. If G_1 has some property that doesn't depend on how G_1 is drawn or how G_1 is labeled and G_2 does not have this property, then $G_1 \ncong G_2$. For example, if G_1 contains two vertices of degree 3 that are mutually adjacent to a vertex of degree 2 and G_2 does not, then $G_1 \ncong G_2$. If G_1 contains two triangles that have a common vertex and G_2 doesn't, then $G_1 \ncong G_2$. If G_1 contains eight triangles and G_2 contains only seven triangles, then $G_1 \ncong G_2$. This last statement brings up an important point, however. If the explanation that was given as to why two graphs are not isomorphic is that one has eight triangles and the other has seven triangles, then this is probably not convincing since it may not be clear that all triangles in both graphs have been accounted for. It would be preferable to locate a property that is easier to justify (assuming that the two graphs are in fact not isomorphic).

Example 3.4 *Determine whether the graphs H_1 and H_2 of Figure* 3.10 *are isomorphic.*

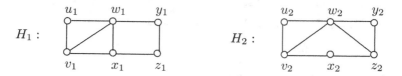

Figure 3.10: The two graphs in Example 3.4

Solution. First, observe that H_1 and H_2 have the degree sequence $4, 3, 3, 2, 2, 2$. Hence further consideration is needed. Because these two graphs do not "appear" to be isomorphic, we seek some structural difference. Observe that H_1 contains two adjacent vertices of degree 2 (namely y_1 and z_1), while H_2 does not. Thus $H_1 \not\cong H_2$.   ◇

What we are observing is that if G and H are isomorphic graphs, $\phi : V(G) \to V(H)$ is an isomorphism and the vertex u of G is mapped by ϕ to the vertex v of H, then any property that u has in G must be a property that v has in H provided this property doesn't depend on how the graphs are drawn or labeled. More generally, any structural property of G must also be possessed by H. For example,

(1) if G contains a k-cycle for some integer $k \geq 3$, then so does H and

(2) if G contains a $u - v$ path of length k, then H contains a $\phi(u) - \phi(v)$ path of length k.

These remarks give the following theorem.

Theorem 3.5 *Let G and H be isomorphic graphs. Then*

(a) *G is bipartite if and only if H is bipartite and*

(b) *G is connected if and only if H is connected.*

As expected, two digraphs D_1 and D_2 are **isomorphic** if there exists a one-to-one correspondence $\phi : V(D_1) \to V(D_2)$ such that $(u_1, v_1) \in E(D_1)$ if and only if $(\phi(u_1), \phi(v_1)) \in E(D_2)$. Digraphs will be studied in detail in Chapter 7.

Exercises for Section 3.1

3.1 Give an example of three different (non-isomorphic) graphs of order 5 and size 5.

3.2 Give an example of three graphs of the same order, same size and same degree sequence such that no two of these graphs are isomorphic.

3.3 For each of the pairs G_1, G_2 of graphs in Figures 3.11(a) and 3.11(b), determine (with careful explanation) whether G_1 and G_2 are isomorphic.

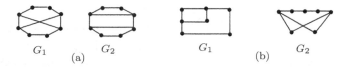

Figure 3.11: The graphs in Exercise 3.3

3.4 Which pairs of graphs in Figure 3.12 are isomorphic? Explain your answer.

Figure 3.12: The graphs in Exercise 3.4

3.5 Let G_1 and G_2 be two graphs with $V(G_1) = \{u_1, v_1, w_1, x_1, y_1, z_1\}$ and $V(G_2) = \{u_2, v_2, w_2, x_2, y_2, z_2\}$. If v_1 has degree 3 and is adjacent to a vertex of degree 2, while v_2 has degree 3 and is not adjacent to a vertex of degree 2, can we conclude that $G_1 \not\cong G_2$? Explain your answer.

3.6 Let G_1 and G_2 be two graphs having the same degree sequence. If G_1 contains a vertex of degree 2 that is adjacent to a vertex of degree 3 and a vertex of degree 4, while G_2 contains a vertex of degree 2 that is adjacent to two vertices of degree 3, can we conclude that $G_1 \not\cong G_2$? Explain your answer.

3.7 Is the solution of the following problem correct?

Problem: Determine whether the graphs G_1 and G_2 of Figure 3.13 are isomorphic.

Figure 3.13: The two graphs in Exercise 3.7

Solution. The graph G_1 has a 5-cycle C. Two vertices of C are connected by a path of length 2, lying inside C. The graph G_2 does not contain such a 5-cycle, however. Therefore, $G_1 \not\cong G_2$.   ◇

3.8 Which pairs of graphs in Figure 3.14 are isomorphic? Explain your answer.

3.9 Determine whether the graphs G_1 and G_2 of Figure 3.15 are isomorphic. Explain your answer.

Figure 3.14: Graphs in Exercise 3.8

Figure 3.15: Two graphs in Exercise 3.9

3.10 Does there exist a disconnected self-complementary graph?

3.11 Let G be a self-complementary graph of order $n = 4k$, where $k \geq 1$. Let $U = \{v : \deg v \leq n/2\}$ and $W = \{v : \deg v \geq n/2\}$. Prove that if $|U| = |W|$, then G contains no vertex v such that $\deg v = n/2$.

3.12 Let G and H be two self-complementary graphs with disjoint vertex sets, where H has even order n. Let F be the graph obtained from $G \cup H$ by joining each vertex of G to every vertex of degree less than $n/2$ in H. Show that F is self-complementary.

3.13 Suppose that there exist two connected graphs G and H and a one-to-one function ϕ from $V(G)$ onto $V(H)$ such that $d_G(u, v) = d_H(\phi(u), \phi(v))$ for every two vertices u and v of G. Prove or disprove: G and H are isomorphic.

3.14 Prove or disprove: Let G and H be two connected graphs of order n, where $V(G) = \{v_1, v_2, \ldots, v_n\}$. If there exists a one-to-one correspondence $\phi : V(G) \to V(H)$ such that $d_G(v_i, v_{i+1}) = d_H(\phi(v_i), \phi(v_{i+1}))$ for all i ($1 \leq i \leq n - 1$), then $G \cong H$.

3.15 Prove or disprove: Let G and H be two connected graphs. If there exists a one-to-one correspondence $\phi : V(G) \to V(H)$ and two distinct vertices $u, v \in V(G)$ such that $d_G(u, v) \neq d_H(\phi(u), \phi(v))$, then $G \ncong H$.

3.2 Isomorphism as a Relation

Let us state once again that a graph G_1 is isomorphic to a graph G_2 if there exists an isomorphism $\phi : V(G_1) \to V(G_2)$. Isomorphism therefore produces a relation on any set of graphs, namely, a graph G_1 is related to a graph G_2 if G_1 is isomorphic to G_2. This relation is an equivalence relation. This says that isomorphism is reflexive (every graph is isomorphic to itself), isomorphism is symmetric (if G_1 is isomorphic to G_2, then G_2 is isomorphic to G_1) and isomorphism is transitive (if G_1 is isomorphic to G_2 and G_2 is isomorphic to G_3, then G_1 is isomorphic to G_3).

The proof that isomorphism is an equivalence relation relies on three fundamental properties of bijective functions (functions that are one-to-one and onto): (1) every identity function is bijective, (2) the inverse of every bijective function is also bijective, (3) the composition of two bijective functions is bijective. (See Appendix 2 for a review of these terms and facts.)

Theorem 3.6 *Isomorphism is an equivalence relation on the set of all graphs.*

Proof. First, we show that isomorphism is reflexive, that is, every graph is isomorphic to itself. Let G be a graph and consider the identity function $\epsilon : V(G) \to V(G)$ defined by $\epsilon(v) = v$ for each vertex v of G. Thus ϵ is bijective. Certainly, two vertices u and v of G are adjacent if and only if $\epsilon(u) = u$ and $\epsilon(v) = v$ are adjacent. Therefore, ϵ is an isomorphism and so G is isomorphic to G.

Next, we show that isomorphism is symmetric. Let G_1 and G_2 be graphs and assume that G_1 is isomorphic to G_2. Therefore, there exists an isomorphism $\phi : V(G_1) \to V(G_2)$. Since ϕ is a bijective function, its inverse $\phi^{-1} : V(G_2) \to V(G_1)$ exists and is a bijective function. Let u_2 and v_2 be any two vertices in G_2 and suppose that $\phi^{-1}(u_2) = u_1$ and $\phi^{-1}(v_2) = v_1$. Hence $\phi(u_1) = u_2$ and $\phi(v_1) = v_2$. If u_2 and v_2 are adjacent vertices in G_2, then u_1 and v_1 are adjacent vertices in G_1 since ϕ is an isomorphism. On the other hand, if u_2 and v_2 are not adjacent, then u_1 and v_1 are not adjacent. Therefore, u_2 and v_2 are adjacent in G_2 if and only if $\phi^{-1}(u_2)$ and $\phi^{-1}(v_2)$ are adjacent in G_1. Hence ϕ^{-1} is an isomorphism and so G_2 is isomorphic to G_1.

Finally, we show that isomorphism is transitive. For graphs G_1, G_2 and G_3, assume that G_1 is isomorphic to G_2 and G_2 is isomorphic to G_3. Hence there exist isomorphisms $\alpha : V(G_1) \to V(G_2)$ and $\beta : V(G_2) \to V(G_3)$. Consider the composition $\beta \circ \alpha : V(G_1) \to V(G_3)$. Since α and β are bijective, so is $\beta \circ \alpha$. Since α is an isomorphism, vertices u_1 and v_1 of G_1 are adjacent if and only if $\alpha(u_1)$ and $\alpha(v_1)$ are adjacent in G_2. Because β is an isomorphism, $\alpha(u_1)$ and $\alpha(v_1)$ are adjacent in G_2 if and only if $\beta(\alpha(u_1))$ and $\beta(\alpha(v_1))$ are adjacent in G_3. Therefore, u_1 and v_1 are adjacent in G_1 if and only if $(\beta \circ \alpha)(u_1)$ and $(\beta \circ \alpha)(v_1)$ are adjacent in G_3 and so $\beta \circ \alpha$ is an isomorphism. Hence G_1 is isomorphic to G_3. ∎

One of the major consequences of knowing that isomorphism is an equivalence relation on a set of graphs is that this produces a partition of this set into equivalence classes (subsets) which are **isomorphism classes** here. Every two graphs in the same isomorphism class are isomorphic and every two graphs in different isomorphism classes are not isomorphic.

Suppose that we are asked for all graphs with degree sequence s: 2, 2, 2, 2, 2, 2, 2, 2, 2. What we are clearly seeking here are non-isomorphic graphs. The answer to this question is given in Figure 3.16. (There are four such graphs!)

Figure 3.16: All graphs with degree sequence $s : 2, 2, 2, 2, 2, 2, 2, 2, 2$

Consider the graphs H and G of Figure 3.17. We defined a graph H to be a subgraph of a graph G, written $H \subseteq G$, if $V(H) \subseteq V(G)$ and $E(H) \subseteq E(G)$. This, in fact, *is* the definition if the graphs H and G are labeled (that is, if the vertex sets of H and G have been specified). The graphs H and G of Figure 3.17 are not labeled, however. For unlabeled graphs H and G, we say that H **is isomorphic to a subgraph** of G if for any labeling of the vertices of H and G, the labeled graph H is isomorphic to a subgraph of the labeled graph G. Consequently, for the graphs H and G of Figure 3.17, H *is* isomorphic to a subgraph of G.

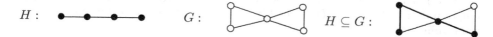

Figure 3.17: A subgraph H of a graph G

Exercises for Section 3.2

3.16 How many (non-isomorphic) graphs have the degree sequence s: 6, 6, 6, 6, 6, 6, 6, 6, 6?

3.17 Consider the (unlabeled) graphs H_1, H_2, H_3 and G of Figure 3.18.

    (a) Is H_1 isomorphic to a subgraph of G?

    (b) Is H_2 isomorphic to a subgraph of G?

    (c) Is H_3 isomorphic to a subgraph of G?

Figure 3.18: Graphs in Exercise 3.17

3.18 Does there exist a graph with exactly three components, exactly two of which are not isomorphic?

3.19 We are given a collection of n graphs $G_1, G_2, \ldots, G_n$, some pairs of which
are isomorphic and some pairs of which are not. Show that there is an
even number of these graphs that are isomorphic to an odd number of
graphs. [Hint: Construct a graph G with $V(G) = \{v_1, v_2, \ldots, v_n\}$, where
$v_i v_j \in E(G)$ if and only if G_i is isomorphic to G_j.]

3.3 Excursion: Graphs and Groups

While it may be quite difficult to determine that two isomorphic graphs G_1
and G_2 are in fact isomorphic if G_1 and G_2 are drawn differently or labeled
differently, there is no difficulty in showing that G_1 and G_2 are isomorphic if
they are drawn and labeled identically. In this case, we then have a single graph,
say G and surely the identity function $\epsilon : V(G) \to V(G)$, where $\epsilon(v) = v$ for all
$v \in V(G)$, is an isomorphism. Consequently, for the graph H of Figure 3.19, the
function $\alpha_1 : V(H) \to V(H)$ defined by $\alpha_1(v) = v$ for every vertex v of H is an
isomorphism.

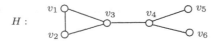

Figure 3.19: A graph H

There are other isomorphisms from the graph H of Figure 3.19 to itself,
however. For example, the function $\alpha_2 : V(H) \to V(H)$ defined by

$$\alpha_2(v_1) = v_2, \ \alpha_2(v_2) = v_1 \text{ and } \alpha_2(v) = v \text{ if } v \neq v_1, v_2$$

is an isomorphism as well. There are two other isomorphisms from the graph H
to itself, namely α_3 and α_4, defined by

$$\alpha_3(v_5) = v_6, \ \alpha_3(v_6) = v_5 \text{ and } \alpha_3(v) = v \text{ if } v \neq v_5, v_6$$

and

$$\alpha_4(v_1) = v_2, \ \alpha_4(v_2) = v_1, \ \alpha_4(v_5) = v_6,$$
$$\alpha_4(v_6) = v_5 \text{ and } \alpha_4(v) = v \text{ if } v = v_3, v_4.$$

An isomorphism from a graph G to itself is called an **automorphism** of
G. Since composition is associative, the identity function is an automorphism,
the inverse of an automorphism is an automorphism and the composition of two
automorphisms is an automorphism, it follows that the set of all automorphisms
of a graph G forms a group under the operation of composition. This group is

denoted by $\mathrm{Aut}(G)$ and is called the **automorphism group** of G. For example, for the graph H of Figure 3.19, $\mathrm{Aut}(H) = \{\alpha_1, \alpha_2, \alpha_3, \alpha_4\}$.

Since every automorphism of a graph is a permutation on $V(G)$, automorphisms can be expressed, more simply, in terms of permutation cycles. (See Appendix 2 for a review of permutations.) For the graph H of Figure 3.19, the elements of $\mathrm{Aut}(H)$ can be expressed as

$$\alpha_1 = \epsilon \text{ (the identity)}, \ \alpha_2 = (v_1 \ v_2), \ \alpha_3 = (v_5 \ v_6), \ \alpha_4 = (v_1 \ v_2)(v_5 \ v_6).$$

For example, expressing α_4 as the "product" of the permutation cycles $(v_1 \ v_2)$ and $(v_5 \ v_6)$ means that (1) α_4 maps v_1 into v_2 and v_2 into v_1, (2) α_4 maps v_5 into v_6 and v_6 into v_5 and (3) α_4 fixes all other vertices of H (that is, α_4 maps any other vertex of H into itself).

The group table for the automorphism group of the graph H of Figure 3.19 is shown in Figure 3.20. The reason for the entry α_2 in row 4, column 3 of the group table is because the "product" of α_4 and α_3 is α_2. That is, since

$$
\begin{aligned}
(\alpha_4\alpha_3)(v_1) &= \alpha_4(\alpha_3(v_1)) = \alpha_4(v_1) = v_2 \\
(\alpha_4\alpha_3)(v_2) &= \alpha_4(\alpha_3(v_2)) = \alpha_4(v_2) = v_1 \\
(\alpha_4\alpha_3)(v_3) &= \alpha_4(\alpha_3(v_3)) = \alpha_4(v_3) = v_3 \\
(\alpha_4\alpha_3)(v_4) &= \alpha_4(\alpha_3(v_4)) = \alpha_4(v_4) = v_4 \\
(\alpha_4\alpha_3)(v_5) &= \alpha_4(\alpha_3(v_5)) = \alpha_4(v_6) = v_5 \\
(\alpha_4\alpha_3)(v_6) &= \alpha_4(\alpha_3(v_6)) = \alpha_4(v_5) = v_6,
\end{aligned}
$$

it follows that $\alpha_4\alpha_3 = \alpha_2$. However, this can be seen more easily when the automorphisms are expressed in terms of permutation cycles. Since

$$\alpha_4\alpha_3 = [(v_1 \ v_2)(v_5 \ v_6)](v_5 \ v_6),$$

we can see, reading from right to left, that (1) v_1 is mapped into v_2 by $\alpha_4\alpha_3$ and v_2 is mapped into v_1 by $\alpha_4\alpha_3$, (2) both v_3 and v_4 are fixed by $\alpha_4\alpha_3$ and (3) v_5 is mapped into v_6 by α_3 and v_6 is mapped into v_5 by α_4, resulting in v_5 being fixed by $\alpha_4\alpha_3$. Similarly, (4) v_6 is fixed by $\alpha_4\alpha_3$. That is,

$$\alpha_4\alpha_3 = (v_1 \ v_2)(v_5 \ v_6)(v_5 \ v_6) = (v_1 \ v_2) = \alpha_2.$$

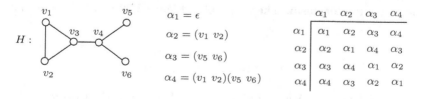

	α_1	α_2	α_3	α_4
α_1	α_1	α_2	α_3	α_4
α_2	α_2	α_1	α_4	α_3
α_3	α_3	α_4	α_1	α_2
α_4	α_4	α_3	α_2	α_1

$\alpha_1 = \epsilon$

$\alpha_2 = (v_1 \ v_2)$

$\alpha_3 = (v_5 \ v_6)$

$\alpha_4 = (v_1 \ v_2)(v_5 \ v_6)$

Figure 3.20: The group table for $\mathrm{Aut}(H)$

As another illustration of an automorphism group, consider the graph F of Figure 3.21. The elements of $\mathrm{Aut}(F)$ and the group table are given in that figure as well. For example, the automorphism β_5 maps u to v, maps v to w and maps w to u, leaving all other vertices fixed. If we write β for β_5, then $\beta^2 = \beta\beta = \beta_6$. Furthermore, if we write α for β_2, then $\beta_3 = \beta_5\beta_2 = \beta\alpha$ and $\beta_4 = \alpha\beta$. Consequently, each of the elements of $\text{Aut}(F)$ can be expressed in terms of α and β, namely,

$$\beta_1 = \alpha^2 = \beta^3, \qquad \beta_2 = \alpha, \qquad \beta_3 = \beta\alpha = \alpha\beta^2,$$
$$\beta_4 = \alpha\beta = \beta^3\alpha, \qquad \beta_5 = \beta, \qquad \beta_6 = \beta^2.$$

Because of this property, α and β are **generators** for the group $\text{Aut}(F)$ and $\{\alpha, \beta\}$ is a **generating set** for this group.

	β_1	β_2	β_3	β_4	β_5	β_6
β_1	β_1	β_2	β_3	β_4	β_5	β_6
β_2	β_2	β_1	β_6	β_5	β_4	β_3
β_3	β_3	β_5	β_1	β_6	β_2	β_4
β_4	β_4	β_6	β_5	β_1	β_3	β_2
β_5	β_5	β_3	β_4	β_2	β_6	β_1
β_6	β_6	β_4	β_2	β_3	β_1	β_5

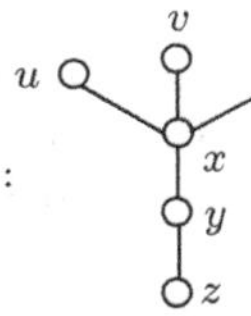

F :

$$\beta_1 = \epsilon$$
$$\alpha = \beta_2 = (u\ v) \qquad \beta = \beta_5 = (u\ v\ w)$$
$$\beta_3 = (u\ w) \qquad \beta_6 = (u\ w\ v)$$
$$\beta_4 = (v\ w)$$

Figure 3.21: The graph F and the group table for $\text{Aut}(F)$

For a vertex v of a graph G, the set of all vertices into which v can be mapped by some automorphism of G is an **orbit** of G. In fact, if a relation R is defined on $V(G)$ by $x\,R\,y$ if $\alpha(x) = y$ for some $\alpha \in \text{Aut}(G)$, then R is an equivalence relation on $V(G)$. The distinct equivalence classes resulting from this relation are the orbits of G. Two vertices u and v are **similar** if they belong to the same orbit. For the graph H of Figure 3.20, there are four orbits, namely, $\{v_1, v_2\}$, $\{v_3\}$, $\{v_4\}$ and $\{v_5, v_6\}$; while for the graph F of Figure 3.21, there are also four orbits: $\{u, v, w\}$, $\{x\}$, $\{y\}$, $\{z\}$. There can be great advantages to knowing the orbits of a graph G. If it is useful to have some structural information about each vertex of G, then it may not be necessary to consider all vertices of G because of the similarity of certain vertices. In such a case, we need only consider one vertex from each orbit as a representative of the orbit. If a graph G of order n has n distinct orbits, then $\text{Aut}(G)$ consists of a single automorphism, namely, the identity automorphism. The graph G of Figure 3.22 is one such graph.

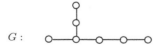

G :

Figure 3.22: A graph of order 7 with seven distinct orbits

On the other hand, if a graph G contains a single orbit, then every two vertices of G are similar and G is called **vertex-transitive**. The graph G_1 of Figure 3.23 is vertex-transitive. Since an automorphism can only map a vertex into a vertex of the same degree, every vertex-transitive graph is regular. The converse is not true, however. For example, the 3-regular graph G_2 of Figure 3.23 is not vertex-transitive since, for example, u belongs to a triangle of G_2 (in fact, two triangles), while w belongs to no triangle of G_2. Therefore, no automorphism of G_2 maps u into w. Among the well-known vertex-transitive graphs are the complete graphs, the cycles, the complete bipartite graphs $K_{s,s}$ and the Petersen graph.

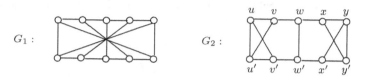

Figure 3.23: A vertex-transitive graph and
a regular graph that is not vertex-transitive

A few fundamental ideas from group theory are useful to review, beginning with the definition of a group itself. A **group** is a nonempty set A (finite or infinite) together with an associative binary operation $\circ$ on A containing an identity element e (necessarily unique) such that $e \circ a = a \circ e = a$ for every element $a \in A$ and having the added property that for every element $a \in A$, there exists an inverse element b (necessarily unique) in A with $b \circ a = a \circ b = e$. Such a group is often denoted by $(A, \circ)$. Because the operation $\circ$ is associative, $(x \circ y) \circ z = x \circ (y \circ z)$ for all $x, y, z \in A$. If $a \circ b = b \circ a$ for all $a, b \in A$, then $(A, \circ)$ is an **abelian group**. The group $\mathrm{Aut}(H)$ for the graph H of Figure 3.20 is abelian, while $\mathrm{Aut}(F)$ is a nonabelian group for the graph F of Figure 3.21. (If the main topic of this text had been group theory, then G would have been used to denote a group. However, graphs have the priority here!)

If $(A, \circ)$ is a group and B is a subset of A such that $(B, \circ)$ is a group, then $(B, \circ)$ is called is a **subgroup** of A. Two groups $(A, \circ)$ and $(B, *)$ are **isomorphic** if there exists a bijective function $\phi : A \to B$ such that $\phi(x \circ y) = \phi(x) * \phi(y)$ for all $x, y \in A$. It is often common to refer to the operation $\circ$ on a group $(A, \circ)$ as multiplication and write $a \circ b$ as ab instead. In this case, we commonly denote the group by A, with the operation understood.

A common type of group to consider is a group of permutations, where the operation is composition. In fact, a well-known theorem of Arthur Cayley states that every group is isomorphic to a group of permutations. As we have mentioned, the automorphism group of a graph is a permutation group. The group of *all* permutations on a set of cardinality n is called the **symmetric group** S_n and its order is $n!$. The automorphism group of a graph G of order n is a group of permutations on the vertex set $V(G)$, that is, the automorphism group of a graph G of order n is a subgroup of S_n. By a theorem of Joseph-Louis Lagrange, the order of $\mathrm{Aut}(G)$ divides $n!$ (the order of S_n).

Since the automorphism group of every graph has finite order, we will only be interested in finite groups. If a group of order n has a single generator, then the group is **cyclic** of order n. Neither $\mathrm{Aut}(H)$ nor $\mathrm{Aut}(F)$ for the graphs H and F of Figures 3.20 and 3.21, respectively, are cyclic. In particular, $\mathrm{Aut}(H)$ is the so-called **Klein four group** (named for Felix Klein), while $\mathrm{Aut}(F)$ is the symmetric group S_3 of order 6. There is another interesting class of groups that appears often in group theory. First, we consider a member of this class.

Example 3.7 *Determine* $\mathrm{Aut}(C_5)$.

Solution. Let $G = C_5$, where the vertices of G are labeled as in Figure 3.24.

$$G :$$

Figure 3.24: The graph G of Example 3.7

One of the automorphisms of G is $\alpha = (v_1\ v_2\ v_3\ v_4\ v_5)$, which can be thought of as a "rotation" of G. Another automorphism of G is $\beta_1 = (v_2\ v_5)(v_3\ v_4)$, which can be thought of as a "reflection" of G. The automorphism group of G consists of the identity ϵ, the four rotations

$$\alpha = (v_1\ v_2\ v_3\ v_4\ v_5), \qquad \alpha^2 = (v_1\ v_3\ v_5\ v_2\ v_4),$$
$$\alpha^3 = (v_1\ v_4\ v_2\ v_5\ v_3), \qquad \alpha^4 = (v_1\ v_5\ v_4\ v_3\ v_2)$$

and the five reflections

$$\beta_1 = (v_2\ v_5)(v_3\ v_4), \qquad \beta_2 = (v_1\ v_3)(v_4\ v_5), \qquad \beta_3 = (v_1\ v_5)(v_2\ v_4),$$
$$\beta_4 = (v_1\ v_2)(v_3\ v_5), \qquad \beta_5 = (v_1\ v_4)(v_2\ v_3).$$

Letting $\beta = \beta_1$, we see in the group table of $\mathrm{Aut}(G)$ shown in Figure 3.25 that α and β are generators since

$$\epsilon = \alpha^5 = \beta^2, \qquad \beta_1 = \beta, \qquad \beta_2 = \alpha^2\beta = \beta\alpha^3,$$
$$\beta_3 = \alpha^4\beta = \beta\alpha, \qquad \beta_4 = \alpha\beta = \beta\alpha^4, \qquad \beta_5 = \alpha^3\beta = \beta\alpha^2.$$

In general, the automorphism group of the cycle C_n has order $2n$. This group is called a **dihedral group** and is commonly denoted by D_n. Thus the dihedral group D_5 has order 10 and its group table is shown in Figure 3.25. Furthermore, $D_3 = S_3$. $\Diamond$

A finite group A can have, indeed may require, a large generating set, although it is customary not to include the identity element of A as a generator. There is a digraph that one commonly associates with a group $A = \{a_1, a_2, \ldots, a_n\}$ and a generating set Δ for A. The **Cayley color digraph** $D_\Delta(A)$, named for the famous mathematician Arthur Cayley, has A as its vertex set where (a_i, a_j), $i \neq j$, is an arc of $D_\Delta(A)$ if $a_j = a_i b$ for some generator $b \in \Delta$. Furthermore, we label (or color) this arc by b. Consequently, every vertex a_i of $D_\Delta(A)$ is adjacent to the vertex $a_i b$ for each $b \in \Delta$. In addition, every

$$G:$$

$$\alpha = (v_1\ v_2\ v_3\ v_4\ v_5)$$
$$\beta_1 = (v_2\ v_5)(v_3\ v_4)$$
$$\beta_2 = (v_1\ v_3)(v_4\ v_5)$$
$$\beta_3 = (v_1\ v_5)(v_2\ v_4)$$
$$\beta_4 = (v_1\ v_2)(v_3\ v_5)$$
$$\beta_5 = (v_1\ v_4)(v_2\ v_3)$$

	ϵ	α	α^2	α^3	α^4	β_1	β_2	β_3	β_4	β_5
ϵ	ϵ	α	α^2	α^3	α^4	β_1	β_2	β_3	β_4	β_5
α	α	α^2	α^3	α^4	ϵ	β_4	β_5	β_1	β_2	β_3
α^2	α^2	α^3	α^4	ϵ	α	β_2	β_3	β_4	β_5	β_1
α^3	α^3	α^4	ϵ	α	α^2	β_5	β_1	β_2	β_3	β_4
α^4	α^4	ϵ	α	α^2	α^3	β_3	β_4	β_5	β_1	β_2
β_1	β_1	β_3	β_5	β_2	β_4	ϵ	α^3	α	α^4	α^2
β_2	β_2	β_4	β_1	β_3	β_5	α^2	ϵ	α^3	α	α^4
β_3	β_3	β_5	β_2	β_4	β_1	α^4	α^2	ϵ	α^3	α
β_4	β_4	β_1	β_3	β_5	β_2	α	α^4	α^2	ϵ	α^3
β_5	β_5	β_2	β_4	β_1	β_3	α^3	α	α^4	α^2	ϵ

Figure 3.25: The group table of Aut(G) in Example 3.7

vertex a_j of $D_\Delta(A)$ is adjacent from the vertex $a_j b^{-1}$ for each $b \in \Delta$. That is, the outdegree *and* the indegree of every vertex of $D_\Delta(A)$ is $|\Delta|$. We now consider some examples of Cayley color digraphs.

Example 3.8 *For the Klein four group* $A = \{a_1, a_2, a_3, a_4\}$, *whose group table is given in Figure 3.26, the Cayley color digraph is shown in the same figure for the generating set* $\Delta = \{a_2, a_4\}$.

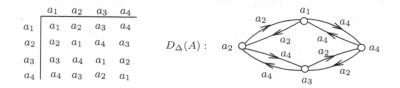

	a_1	a_2	a_3	a_4
a_1	a_1	a_2	a_3	a_4
a_2	a_2	a_1	a_4	a_3
a_3	a_3	a_4	a_1	a_2
a_4	a_4	a_3	a_2	a_1

Figure 3.26: The group table and the Cayley color digraph in Example 3.8

Example 3.9 *For the cyclic group* $A = \{e, a, a^2, a^3\}$ *whose group table is given in Figure 3.27, the Cayley color digraphs are shown in the same figure for the two generating sets* $\Delta_1 = \{a\}$ *and* $\Delta_2 = \{a, a^2\}$.

Just as every graph has an automorphism group, so too does every digraph. In the case of a Cayley color digraph D, those automorphisms ϕ of D that preserve colors (that is, such that the arcs (u, v) and $(\phi u, \phi v)$ have the same color (generator) for every arc (u, v) of D) form a subgroup of the automorphism group of the digraph D (whose arcs are not colored).

The topic of automorphism groups of graphs appeared in the first book on graph theory, in fact, early in the book. On page 5 of the first section (*The Basic Concepts*) of the first chapter (*Foundations*) of his 1936 book, Dénes König posed the following question (translated from German):

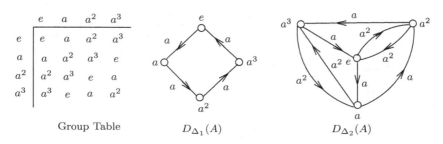

Figure 3.27: The group table and Cayley color digraphs in Example 3.9

When can a given abstract group be interpreted as the group of a graph and if this is the case, how can the corresponding graph be constructed?

In the early 1900s, Germany was known for its mathematicians who excelled in group theory. One of the best known was Ferdinand Georg Frobenius (1849-1917), who made numerous important contributions to several areas of mathematics, but especially to group theory. One of Frobenius' doctoral students was Issai Schur (1875-1941), who died on his 66th birthday (January 10).

Schur, a gifted mathematician, became well known for his work in groups, particularly on representation theory, although Schur did research in many areas of mathematics. Indeed, through Schur's efforts, Berlin became known for the study of groups. Schur was very popular with students; often his lectures at the University of Berlin were given to overflow audiences. However, beginning in 1933, events in Germany made life very difficult for Schur, who was Jewish. In 1935 Schur was dismissed from his position at the university and the remainder of his life was often unbearable.

While at the University of Berlin, Schur supervised several doctoral students, including Richard Rado, whom we will meet in Chapter 11, Richard Brauer (1901-1977), well known for his work in algebra and number theory and Helmut Wielandt (1910-2001), who made major contributions to the study of permutation groups and linear algebra. Another of Schur's students was Roberto Frucht (1906-1997).

Roberto Frucht entered the University of Berlin in 1924 at the age of 18. Although Frucht's favorite mathematical area was tensor calculus, he could not find a doctoral advisor in that area. Being an admirer of Schur, Frucht asked him if he would agree to be his advisor and Schur agreed – provided Frucht would write his thesis in an area of interest to Schur. Consequently, Frucht switched to group theory and received his Ph.D. in 1930.

At that time, Frucht's father lost his job and earning a living became a top priority for Frucht. Finding a job as a mathematician in Germany was difficult during that period, except as a high school teacher. However, German citizenship was required for that and Frucht was a Czechoslovakian citizen. Consequently, Frucht moved to Trieste, Italy to work at an Italian insurance company. Frucht stayed in Italy until 1938. During that period, his life was relatively inactive, mathematically speaking. However, one day in 1936 he received a catalog from Akademische Verlagsgesellschaft advertising a book in graph theory (by Dénes König). Frucht immediately ordered the book and he became an enthusiastic graph theorist the very day that the book arrived.

König's question on automorphism groups attracted the attention of Frucht immediately. After being unsuccessful for several months trying to solve the problem, he found a solution that seemed rather easy (after he had found it). In 1939 Frucht escaped from Italy to South America shortly before the outbreak of World War II. After working as an actuary in Argentina for a while, he was successful in acquiring a position at the Universidad Santa Maria in Valparaiso, Chile. He continued his interest in graph theory and remained in Chile the rest of his life.

Theorem 3.10 (**Frucht's Theorem**) *For every finite group A, there exists a graph G such that $\text{Aut}(G)$ is isomorphic to A.*

To prove Theorem 3.10, Frucht used his result that for every finite abstract group A, the group of color-preserving automorphisms of the Cayley color digraph $D_\Delta(A)$ is isomorphic to A. The next step for Frucht was to convert $D_\Delta(A)$ into a graph G so that the automorphisms of G correspond to the color-preserving automorphisms of $D_\Delta(A)$ in an appropriate manner.

Although this can be accomplished in a number of ways, one way is to replace each arc (u, v) labeled $b_1 \in \Delta$ in $D_\Delta(A)$ by a path (u, u_1, v_1, v) where u_1 and v_1 are new vertices and place a pendant edge at u_1 and attach a path of length 2 at v_1. If there is another generator, say b_2, then replace any arc (u, v) labeled b_2 in $D_\Delta(A)$ by a path (u, u_2, v_2, v) and attach a path of length 3 at u_2 and a path of length 4 at v_2. This procedure is continued if there are additional generators. See Figure 3.28.

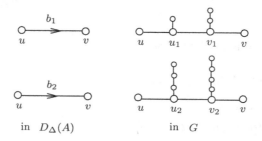

Figure 3.28: Constructing a graph G from $D_\Delta(A)$

For example, consider the Cayley color digraph $D_\Delta(A)$ in Example 3.8, which is redrawn in Figure 3.29. The associated graph G is also shown in that figure, where $b_1 = a_2$ and $b_2 = a_4$. By Frucht's theorem, $\text{Aut}(G)$ is isomorphic to A (the Klein four group).

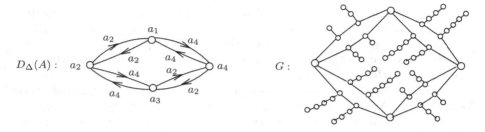

Figure 3.29: Constructing a graph G from $D_\Delta(A)$

Exercises for Section 3.3

3.20 For the graph H of Figure 3.19, give an example of a permutation on $V(H)$ that preserves degrees but which is not an automorphism of H.

3.21 Determine the automorphism group of K_3.

3.22 Determine the automorphism group of $K_{1,3}$.

3.23 Determine the automorphism group of P_n for $n \geq 2$.

3.24 Determine the automorphism group of C_4.

3.25 For each of the graphs H_1 and H_2 in Figure 3.30, determine

   (a) the orbits of the graph.

   (b) the automorphism group of the graph.

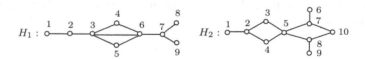

Figure 3.30: The graphs in Exercise 3.25

3.26 For the graph G_2 in Figure 3.23, determine

   (a) the orbits of G_2.

   (b) the automorphism group of G_2.

3.27 Prove that $\text{Aut}(G)$ and $\text{Aut}(\overline{G})$ are isomorphic for every graph G.

3.28 For the group $A = \{e, a, b\}$ whose group table is given in Figure 3.31,

   (a) find a generating set Δ and the corresponding Cayley color digraph.

   (b) use the Cayley color digraph in (a) to construct a graph G such that $\text{Aut}(G) \cong A$.

	e	a	b
e	e	a	b
a	a	b	e
b	b	e	a

Figure 3.31: The group table in Exercise 3.28

(c) find a graph H of order 12 such that $\text{Aut}(H) \cong A$.

3.29 Consider the group $A = \{e, a, b, c\}$ whose group table is given in Figure 3.32, where $a^2 = b^2 = c^2 = e$.

(a) For $\Delta = \{a, b\}$, find the corresponding Cayley color digraph.

(b) Use the Cayley color digraph in (a) to construct a graph G such that $\text{Aut}(G) \cong A$.

	e	a	b	c
e	e	a	b	c
a	a	e	c	b
b	b	c	e	a
c	c	b	a	e

Figure 3.32: The group table in Exercise 3.29

3.30 Consider the group $A = \{e, a, b, c, d\}$ whose group table is given in Figure 3.33.

(a) For $\Delta = \{a\}$, find the corresponding Cayley color digraph.

(b) Use the Cayley color digraph in (a) to construct a graph G such that $\text{Aut}(G) \cong A$.

	e	a	b	c	d
e	e	a	b	c	d
a	a	b	c	d	e
b	b	c	d	e	a
c	c	d	e	a	b
d	d	e	a	b	c

Figure 3.33: The group table in Exercise 3.30

3.31 Let $A = S_3$ be the symmetric group on the set $\{1, 2, 3\}$. For $\Delta = \{(123), (12)\}$,

(a) find the corresponding Cayley color digraph,

(b) use the Cayley color digraph in (a) to construct a graph G such that $\text{Aut}(G) \cong A$.

3.32 Use each of the Cayley color digraphs in Example 3.9 to construct a graph
      G such that $\mathrm{Aut}(G) \cong A$.

3.4 Excursion: Reconstruction and Solvability

Figure 3.34 shows a deck of five cards, each with a drawing of a graph.

Figure 3.34: A deck of cards

Since we can't see the graph drawn on each card, we separate the cards,
which are shown in Figure 3.35. We've also numbered the cards now. What
do you notice about the graphs on these five cards? Of course, the graphs on
cards 1 and 2 are isomorphic, as are the graphs on cards 3 and 4. However, the
observation we are looking for is that all graphs have order 4.

Figure 3.35: A deck of five cards

It turns out that there exists a certain graph G of some order n such that
for each $v \in V(G)$, the unlabeled subgraph $G - v$ is drawn on one of the cards
in the deck given in Figure 3.35. We can see that $n = 5$ in two ways. First, n
must equal 5 since there are five cards. Also, each subgraph $G - v$, $v \in V(G)$,
has order $n - 1 = 4$ and so $n = 5$.

If the value of a parameter for a graph G or whether a graph G has a certain
property can be determined from the (unlabeled) graphs $G - v$, $v \in V(G)$, then
this parameter or property is said to be **recognizable** for G. The observations
we have made above provide a proof in general for the following.

Theorem 3.11 *The order of every graph is recognizable.*

Before continuing, let's look at the (very small) deck of cards in Figure 3.36. Since there are only two cards in the deck and each card is the (trivial) graph of order 1, the graph G in question has order 2. Actually, there are two (non-isomorphic) graphs of order 2, namely K_2 and $\overline{K}_2$. Of course, the size of K_2 is 1 and the size of $\overline{K}_2$ is 0. But, whether $G = K_2$ or $G = \overline{K}_2$, the two graphs $G - v$ in each case are K_1. That is, if $G = K_2$ or $G = \overline{K}_2$, then there is no way to determine the size of G from the subgraphs $G - v$, $v \in V(G)$. Therefore, the sizes of K_2 and $\overline{K}_2$ are not recognizable. This is the exception, however, rather than the rule.

Figure 3.36: A deck of two cards

Theorem 3.12 *The size of every graph of order at least 3 is recognizable.*

Proof. Suppose that G is a graph of order $n \geq 3$ and size m. Let $V(G) = \{v_1, v_2, \ldots, v_n\}$. Of course, in the deck consisting of the subgraphs $G - v_i$, $1 \leq i \leq n$, the vertices will not be labeled. Let e be an edge of G, say $e = v_1 v_2$. The edge e then appears in each of the subgraphs $G - v_i$ for $3 \leq i \leq n$ but appears in neither $G - v_1$ nor $G - v_2$. Suppose that the size of $G - v_i$ is m_i ($1 \leq i \leq n$). In the sum $\sum_{i=1}^{n} m_i$, every edge is counted $n - 2$ times. That is,

$$\sum_{i=1}^{n} m_i = m(n - 2) \quad \text{and so} \quad m = \frac{\sum_{i=1}^{n} m_i}{n - 2}. \qquad \blacksquare$$

Let card i ($1 \leq i \leq 5$) in Figure 3.35 display the subgraph $G - v_i$, which has size m_i. Thus

$$\sum_{i=1}^{5} m_i = 4 + 4 + 3 + 3 + 1 = 15.$$

By Theorem 3.12, $m = 15/(5 - 2) = 5$. Hence the size of the graph G in question described by the deck in Figure 3.35 is 5.

Now that we know that the size of every graph G of order at least 3 is recognizable, we can show that one additional feature of G is recognizable.

Theorem 3.13 *A degree sequence of every graph of order at least 3 is recognizable.*

Proof. Let G be a graph of order $n \geq 3$ and size m, where $V(G) = \{v_1, v_2, \ldots, v_n\}$. From Theorem 3.12, m can be determined from the subgraphs $G - v_i$ ($1 \leq i \leq n$). Suppose that the size of $G - v_i$ is m_i for $1 \leq i \leq n$. That is, the size of G is m but when the vertex v_i is removed from G, the size of the resulting subgraph $G - v_i$ is m_i. Consequently, $\deg v_i = m - m_i$ and so $m - m_1, m - m_2, \ldots, m - m_n$ is a degree sequence for G.   ∎

Returning to the subgraphs $G - v_i$ $(1 \leq i \leq 5)$ shown on the deck of the cards in Figure 3.35, we see that $m_1 = m_2 = 4$, $m_3 = m_4 = 3$ and $m_5 = 1$. Since we have already seen that the size m of G is 5, it follows that $1, 1, 2, 2, 4$ is a degree sequence of G. In particular, $\deg v_5 = 4$ so that v_5 is adjacent to each of the four vertices of the graph $G - v_5$ given in card 5. Hence we now know precisely (up to isomorphism) what the mystery graph G is for the deck of cards in Figure 3.35. The graph G is shown in Figure 3.37.

G :

Figure 3.37: The graph G for the deck of cards in Figure 3.35

Consequently, from the deck of cards in Figure 3.35 that gives the subgraphs $G - v_i$ $(1 \leq i \leq 5)$, we have not only been able to determine the order of G, the size of G and a degree sequence for G, we have been able to determine G itself.

A graph G of order $n \geq 2$ is **reconstructible** if G can be uniquely determined (up to isomorphism) from its subgraphs $G - v_i$ $(1 \leq i \leq n)$. Thus, the graph G of Figure 3.37 is reconstructible. From our earlier remarks, neither graph of order 2 is reconstructible. It is believed by many but has not been verified by any that every graph of order 3 or more is reconstructible.

The Reconstruction Conjecture *Every graph of order 3 or more is reconstructible.*

This conjecture is believed to have been made in 1941 and is often attributed jointly to Paul J. Kelly (1915-1995) and Stanislaw M. Ulam. Kelly, who spent many years as a faculty member at the University of California at Santa Barbara, obtained a number of results on this topic. Ulam was born on April 3, 1909 in Lemberg, Poland (now Lvov, Ukraine) and became interested in astronomy, physics and mathematics while a teenager and learned calculus on his own. He entered the Polytechnic Institute in Lvov in 1927. One of his professors there was Kazimierz Kuratowski, whom we will visit again in Chapter 9. Ulam studied under Stefan Banach and received his Ph.D. in 1933.

In 1940, Ulam acquired a position as an assistant professor at the University of Wisconsin (where Kelly was studying for his Ph.D.). Three years later, John von Neumann asked to meet Ulam at a railroad station in Chicago. This led to Ulam going to the Los Alamos National Laboratory in New Mexico to work on the hydrogen bomb with the physicist Edward Teller. While at Los Alamos, Ulam developed the well-known Monte Carlo method for solving mathematical problems using a statistical sampling method with random numbers. Throughout his life, he made important contributions in many areas of mathematics. Ulam died on May 13, 1984.

The Reconstruction Problem is to determine whether the Reconstruction Conjecture is true. Solving the Reconstruction Problem requires verifying that there do not exist two non-isomorphic graphs G and H such that the set of subgraphs $G - v$, $v \in V(G)$, and the set of subgraphs $H - v$, $v \in V(H)$, are the same. If there are two non-isomorphic graphs G and H with this property, then these graphs must have the same order, the same size and the same degree sequence, as all of these parameters and properties are recognizable. Furthermore, any two such graphs G and H must both be connected or must both be disconnected. This is a consequence of Theorem 1.10, which states that a graph G of order 3 or more is connected if and only if G contains two distinct vertices u and v such that $G - u$ and $G - v$ are connected.

Theorem 3.14 *For all graphs of order at least 3, both connectedness and disconnectedness are recognizable properties.*

Proof. By Theorem 1.10, at least two of the subgraphs $G - v$, $v \in V(G)$, are connected if and only if G is connected. ∎

The Reconstruction Problem concerns providing some information about a certain graph (or certain graphs) and asks us to show that only one graph G satisfies the given information, as well as to identify what this graph G is. In this case, the given information is a collection of subgraphs of G, namely all subgraphs of the type $G - v$, where $v \in V(G)$. However, we could be given a wide variety of information concerning a graph. Proceeding in a manner we discussed earlier, let's assume that we are given pieces of information concerning a graph G, where each item is written on a card. The set of all such cards is our **deck**. Any graph G that satisfies all information in the deck is called a **solution of the deck**. The question then becomes to determine all solutions of the deck. A deck may then have a unique solution, two or more solutions or no solution at all. If, for a given graph G of order $n \geq 3$, a deck consists of all n subgraphs of the type $G - v$, where $v \in V(G)$ and if the the Reconstruction Conjecture is true, then the deck has a unique solution, namely G.

Example 3.15 *Find all graphs G for which the deck of cards shown in Figure 3.38 gives the subgraphs $G - v$, where $v \in V(G)$.*

Figure 3.38: The deck of cards for Example 3.15

Solution. First, observe that the order of any solution G of this deck is 6. The sum of the sizes of the graphs on the deck is $6 + 5 + 5 + 4 + 4 + 6 = 30$. Then the size of a solution G is $30/(6-2) = 7.5$. This is impossible and so the graphs on the deck in Figure 3.38 are not the subgraphs $G - v$, $v \in V(G)$, of any graph G. Thus this deck has no solution. $\diamond$

Example 3.16 *Determine the solutions of the deck of cards shown in Figure 3.39.*

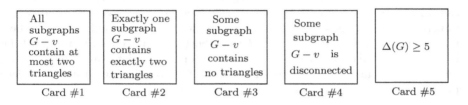

All subgraphs $G - v$ contain at most two triangles	Exactly one subgraph $G - v$ contains exactly two triangles	Some subgraph $G - v$ contains no triangles	Some subgraph $G - v$ is disconnected	$\Delta(G) \geq 5$
Card #1	Card #2	Card #3	Card #4	Card #5

Figure 3.39: The deck of cards for Example 3.16

Solution. Suppose that G is a solution. By Card #1, all subgraphs $G - v$ of a solution G contain at most two triangles and by Card #2 exactly one subgraph $G - v$, say $G - v_1$, contains exactly two triangles. The two triangles in $G - v_1$ are either disjoint, have one vertex in common or have an edge in common, as in Figure 3.40.

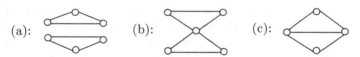

Figure 3.40: Possible subgraphs of a solution

We can eliminate the subgraph in Figure 3.40(a) because of Card #3. If the subgraph in Figure 3.40(c) occurs, then by Card #5, G must contain at least two additional vertices x and y. In that case, both $G - x$ and $G - y$ contain at least two triangles, contradicting Card #2. Necessarily then G contains the subgraph in Figure 3.40(b) with one additional vertex z. Furthermore, this subgraph must be an induced subgraph since $G - v_1$ has exactly two triangles. Certainly, z must be adjacent to the vertex of degree 4 because of Card #5. If z is adjacent to any other vertices in the subgraph in Figure 3.40(b), then Card #2 is contradicted. Hence we arrive at only one graph G, namely, the graph G of Figure 3.41. Card #4 is not needed. (It was a Joker!) $\diamond$

Exercises for Section 3.4

3.33 Give an example of two non-isomorphic graphs G and H of order 3 or more containing vertices u and v, respectively, such that $G - u$ and $H - v$ are isomorphic or explain why no such example exists.

Figure 3.41: The unique solution to the deck in Figure 3.39

3.34 For the deck D of cards given in Figure 3.42, where card i contains the subgraph $G_i = G - v_i$, $v_i \in V(G)$, for some graph G, answer the following with explanation.

(a) What is the order n of G?

(b) What is the size m of G?

(c) What are the degrees of the vertices of G?

(d) Is G connected?

(e) What are the solutions of D?

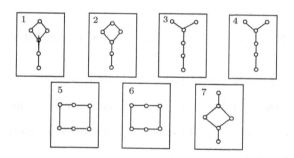

Figure 3.42: The deck of cards for Exercise 3.34

3.35 For a graph G of order n and size m, the subgraphs $G - v$ for $v \in V(G)$ are given on the deck of cards in Figure 3.43. Answer the following with explanation.

(a) What is n?

(b) What is m?

(c) Is G connected?

(d) What is a degree sequence of G?

(e) Find all solutions of the deck.

3.36 Determine the solutions G of the deck of cards shown in Figure 3.44.

3.37 Determine the solutions of the deck of cards shown in Figure 3.45.

3.38 Determine the solutions of the deck of two cards shown in Figure 3.46.

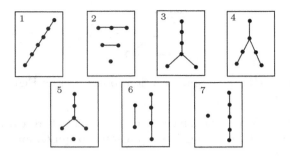

Figure 3.43: The deck of cards for Exercise 3.35

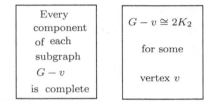

Figure 3.44: The deck of cards for Exercise 3.36

3.39 Determine the solutions of the deck of two cards shown in Figure 3.47.

3.40 Determine the solutions of the deck of one card shown in Figure 3.48.

3.41 Determine the solutions of the deck of two cards shown in Figure 3.49.

3.42 Let $G = 3K_2 + K_1$.

    (a) Give an example of a deck of cards for which G is the unique solution. Show your work.

    (b) Find a deck with a small number of cards for which G is the unique solution. Show your work.

    (c) Give an example of a deck D of cards for which G and one other graph are the only two solutions of D. Show your work.

3.43 Give an example of a deck of cards having exactly two solutions of order 3 or more.

3.44 Give an example of a deck of three cards having no solution, where any subdeck consisting of two of the three cards has at least one solution.

G is connected	$\Delta(G - v) \leq 2$ for every vertex v of G	There is no subgraph $G - v$ with $\Delta(G - v) = 1$	$\Delta(G - v) = 2$ for some vertex v of G

Figure 3.45: The deck of cards for Exercise 3.37

For every two distinct vertices u and v of G, the sizes of $G - u$ and $G - v$ are different	Some subgraph $G - v$ has order 5

Figure 3.46: The deck of cards for Exercise 3.38

only one subgraph $G - v$ contains exactly two vertices that are not adjacent	Two different subgraphs $G - v$ contain only odd vertices

Figure 3.47: The deck of cards for Exercise 3.39

Every subgraph $G - v$, where $v \in V(G)$, is regular

Figure 3.48: The deck of cards for Exercise 3.40

Figure 3.49: The deck of cards for Exercise 3.41

Chapter 4

Trees

4.1 Bridges

Suppose that there are some villages in a sparsely populated region where country roads allow us to travel directly between certain pairs of these villages. Since the traffic along these roads is ordinarily light, it is not surprising that very few roads have been built in this region. In fact, suppose that we have the situation illustrated in Figure 4.1, where there are seven villages denoted by $v_1, v_2, \ldots, v_7$ and six roads. Not only can this map be modeled by the graph G of Figure 4.1, the map essentially *is* a graph.

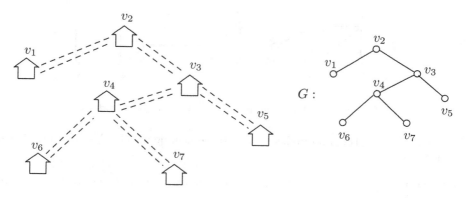

Figure 4.1: A graph model of villages and roads

There are two interesting features of the map (and the graph) of Figure 4.1. First, you may have heard of the traveler who stops someplace during his trip and asks a local resident for directions to some location: "How do you get there?" only to get the response "You can't get there from here." Well, fortunately, we don't have *that* situation with the villages in Figure 4.1. Indeed, it is possible to

85

travel along country roads from each of the seven villages to all other villages. In other words, the graph G of Figure 4.1 is connected. Although this is a very positive feature (an essential feature, one might say), the map and graph also have a negative feature. Namely, if it ever became necessary to close any of the roads due to road construction, flooding or a major snowstorm, then it would no longer be possible to travel between every two villages. In terms of the graph G of Figure 4.1, this says that if we were to remove any edge of G, then the resulting graph would no longer be connected. An edge with this property plays an important role in graph theory.

Recall that if e is an edge of a graph G, then $G - e$ is the subgraph of G having the same vertex set as G and whose edge set consists of all edges of G except e. Also, if X is a set of edges of G, then $G - X$ is the subgraph possessing the same vertex set as G and all edges of G except those in X. If G has order n, then $G - E(G)$ is the empty graph $\overline{K}_n$.

An edge $e = uv$ of a connected graph G is called a **bridge** of G if $G - e$ is disconnected. In this case, $G - e$ necessarily contains exactly two components, one containing u and the other containing v. If the vertex v, say, has degree 1, then the component of $G - e$ containing v is a single vertex, in which case $G - v$ has only one component; that is, if v is an end-vertex of a connected graph G, then $G - v$ is connected. An edge e is a **bridge** of a disconnected graph if e is a bridge of some component of G. Recall that $k(G)$ is the number of components of a graph G. Thus an edge e is a bridge of a graph G if and only if $k(G - e) = k(G) + 1$. In the disconnected graph G of Figure 4.2, the edges u_2u_5, v_3v_4, v_4v_5 and w_1w_2 are bridges (which are indicated in bold). No other edges of G are bridges.

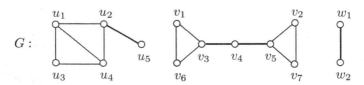

Figure 4.2: A disconnected graph with four bridges

The following theorem makes it easy to determine which edges in a graph are bridges.

Theorem 4.1 *An edge e of a graph G is a bridge if and only if e lies on no cycle of G.*

Proof. First, suppose that $e = uv$ is an edge of G that is not a bridge and that e lies in the component G_1 of G. (Of course $G_1 = G$ if G is connected.) Then $G_1 - e$ is connected. Hence there exists a $u - v$ path P in $G_1 - e$. However, P together with e form a cycle containing e in G_1 and therefore in G as well.

We now verify the converse. Suppose that $e = uv$ lies on a cycle C of G and that e (and C) belong to the component G_1 of G. Then there is a $u - v$ path P'

in G_1 not containing e. We show that $G_1 - e$ is connected. Let x and y be any two vertices of $G_1 - e$. We show that x and y are connected in $G_1 - e$. Since G_1 is connected, G_1 contains an $x - y$ path Q. If e is not on Q, then Q is an $x - y$ path in $G_1 - e$ as well. On the other hand, if e lies on Q, then replacing e in Q by the $u - v$ path P' produces an $x - y$ walk. By Theorem 1.6, $G_1 - e$ contains an $x - y$ path. ∎

The graph G of Figure 4.1 is connected and has no cycles. Therefore, every edge of G is a bridge. Graphs with these two properties are especially important and will be the main subject of this chapter.

Exercises for Section 4.1

4.1 Give an example of a nontrivial connected graph G with the properties that (1) every bridge of G is adjacent to an edge that is not a bridge, (2) every edge of G that is not a bridge is adjacent to a bridge, (3) G contains two nonadjacent bridges and (4) every two edges of G that are not bridges are adjacent.

4.2 Prove that every connected graph all of whose vertices have even degrees contains no bridges.

4.3 Prove that if uv is a bridge in a graph G, then there is a unique $u - v$ path in G.

4.4 Let G be a connected graph and let e_1 and e_2 be two edges of G. Prove that $G - e_1 - e_2$ has three components if and only if both e_1 and e_2 are bridges in G.

4.5 (a) Let G be a connected graph of order n, where every edge of G is a bridge. What is the size of G?

(b) Let G be a disconnected graph of order n having k components, where every edge of G is a bridge. What is the size of G?

4.6 Let G be a connected graph of order $n \geq 3$ without bridges. Suppose that for every edge e of G, each edge of $G - e$ is a bridge. What is G? Justify your answer.

4.2 Trees

A graph G is called **acyclic** if it has no cycles. A **tree** is an acyclic connected graph. Therefore, the graph G of Figure 4.1 is a tree. When dealing with trees, we often use T rather than G to denote a tree. By Theorem 4.1, every edge in a tree is a bridge. Indeed, we could define a tree as a connected graph, every edge of which is a bridge. Figure 4.3 shows all six trees of order 6. The tree $T_1 = K_{1,5}$ is a star and $T_6 = P_6$ is a path. The number of end-vertices in the trees of Figure 4.3 ranges from 2 to 5. We'll have more to say about this shortly. A tree containing exactly two vertices that are not end-vertices (which are necessarily adjacent) is called a **double star**. The trees T_2 and T_3 in Figure 4.3 are double stars.

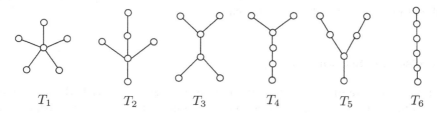

T_1 T_2 T_3 T_4 T_5 T_6

Figure 4.3: The trees of order 6

Another common class of trees consists of the "caterpillars." A **caterpillar** is a tree of order 3 or more, the removal of whose end-vertices produces a path called the **spine** of the caterpillar. Thus every path and star (of order at least 3) and every double star is a caterpillar, as is every tree shown in Figure 4.3. The trees T' and T'' of Figure 4.4 are also caterpillars but T''' is not.

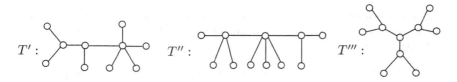

T' : T'' : T''' :

Figure 4.4: Two caterpillars and a tree that is not a caterpillar

There are occasions when it is convenient to select a vertex of a tree T under discussion and designate this vertex as the **root** of T. The tree T then becomes a **rooted tree**. Often the rooted tree T is drawn with the root r at the top and the other vertices of T drawn below, in levels, according to their distances from r. An example is given in Figure 4.5.

Acyclic graphs are also referred to as **forests**. Therefore, each component of a forest is (not surprisingly) a tree. Of course, the one fact that distinguishes trees from forests is that a tree is required to be connected, while a forest is not required to be connected. Since a tree is connected, every two vertices in a tree are connected by a path. In fact, we can say more.

Theorem 4.2 *A graph G is a tree if and only if every two vertices of G are connected by a unique path.*

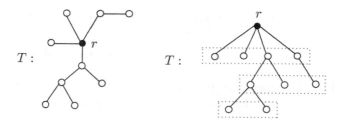

Figure 4.5: A rooted tree

Proof. First, let G be a tree. Then G is connected by definition. Thus every two vertices of G are connected by a path. Assume, to the contrary, that there are two vertices of G that are connected by two distinct paths. Then a cycle is produced from some or all of the edges of these two paths. This is a contradiction.

For the converse, suppose that every two distinct vertices of G are connected by a unique path. Certainly then, G is connected. Assume, to the contrary, that G has a cycle C. Let u and v be two distinct vertices of C. Then C determines two distinct $u - v$ paths, producing a contradiction. Thus G is acyclic and so G is a tree.   ∎

As we have already observed, each tree in Figures 4.3 and 4.4 has two or more end-vertices. All nontrivial trees have this property.

Theorem 4.3 *Every nontrivial tree has at least two end-vertices.*

Proof. Let T be a nontrivial tree and among all paths in T, let P be a path of greatest length. Suppose that P is a $u-v$ path, say $P = (u = u_0, u_1, \ldots, u_k = v)$, where $k \geq 1$. We show that u and v are end-vertices of G. Necessarily, neither u nor v is adjacent to any vertex not on P, for otherwise, a path of greater length would be produced. Certainly, u is adjacent to u_1 on P and v is adjacent to u_{k-1} on P. Moreover, since T contains no cycles, neither u nor v is adjacent to any other vertices in P. Therefore, $\deg u = \deg v = 1$.   ∎

One major consequence of this result is that if T is a tree of order $k + 1 \geq 2$, then for each end-vertex v of T, the subgraph $T - v$ is a tree of order k. This fact is useful for induction proofs of results concerning trees. We illustrate this idea now by showing that the size of every tree is one less than its order, another useful property of trees.

Theorem 4.4 *Every tree of order n has size $n - 1$.*

Proof. We proceed by induction on n. There is only one tree of order 1, namely K_1, which has size 0. Thus the result is true for $n = 1$. Assume for a positive integer k that the size of every tree of order k is $k - 1$. Let T be a tree of order $k + 1$. By Theorem 4.3, T contains at least two end-vertices. Let v be one of them. Then $T' = T - v$ is a tree of order k. By the induction hypothesis, the size of T' is $m = k - 1$. Since T has exactly one more edge than T', the size of T is $m + 1 = (k - 1) + 1 = (k + 1) - 1$, as desired. ∎

Let's illustrate some of the ideas that we've just discussed.

Example 4.5 *The degrees of the vertices of a certain tree T of order 13 are 1, 2 and 5. If T has exactly three vertices of degree 2, how many end-vertices does it have?*

Solution. Since T has three vertices of degree 2, it has ten vertices of degree 1 or 5. Let x denote the number of end-vertices of T. So T has $10 - x$ vertices of degree 5. Since T has 13 vertices, T has 12 edges by Theorem 4.4. Summing the degrees of the vertices of T and applying the First Theorem of Graph Theory, we obtain

$$\begin{aligned}
1 \cdot x + 2 \cdot 3 + 5 \cdot (10 - x) &= 2 \cdot 12 \\
x + 6 + 50 - 5x &= 24 \\
x &= 8. \quad \Diamond
\end{aligned}$$

Note that drawing a tree of order 13 with three vertices of degree 2, two vertices of degree 5 and eight end-vertices does not answer the question. It only says that the tree we drew has eight end-vertices, not necessarily that the tree T in Example 4.5 has eight end-vertices. Of course, *our* solution tells us that *every* tree with the property described in Example 4.5 has eight end-vertices.

We can now determine the size of a forest in terms of its order and the number of components it has.

Corollary 4.6 *Every forest of order n with k components has size $n - k$.*

Proof. Suppose that the size of a forest F is m. Let $G_1, G_2, \ldots, G_k$ be the components of F, where $k \geq 1$. Furthermore, suppose that G_i has order n_i and size m_i for $1 \leq i \leq k$. Then $n = \sum_{i=1}^{k} n_i$ and $m = \sum_{i=1}^{k} m_i$. Since each component G_i $(1 \leq i \leq k)$ is a tree, it follows by Theorem 4.4 that $m_i = n_i - 1$. Therefore,

$$m = \sum_{i=1}^{k} m_i = \sum_{i=1}^{k} (n_i - 1) = n - k. \qquad \blacksquare$$

Again by Theorem 4.4, a tree of order n is a connected graph containing $n-1$ edges. Indeed, every connected graph of order n contains at least $n - 1$ edges. Although this can be verified in several ways, we establish this fact using a proof by minimum counterexample in order to illustrate this useful method of proof. (Proof by minimum counterexample is reviewed in Appendix 3.)

Theorem 4.7 *The size of every connected graph of order n is at least $n - 1$.*

Proof. It is not difficult to see that the theorem is true for connected graphs of order 1, 2 or 3. Assume that the theorem is false however. Then there exists a connected graph G of smallest order n whose size m is at most $n-2$. Necessarily, $n \geq 4$. Since G is a nontrivial connected graph, G contains no isolated vertices.

We claim that G contains an end-vertex; for assume, to the contrary, that the degree of every vertex of G is at least 2. Then the sum of the degrees of the vertices of G is $2m \geq 2n$; so $m \geq n \geq m + 2$, which is impossible. So, as we claimed, G contains an end-vertex.

Let v be an end-vertex of G. Since G is connected, has order n and size $m \leq n - 2$, it follows that $G - v$ is connected and has order $n - 1$ and size $m - 1 \leq n - 3$, contradicting the assumption that G is a connected graph of smallest order whose size is at least 2 less than its order. ∎

Let G be a tree of order n and size m. By the definition of a tree and Theorem 4.4, G has the following three properties: (1) G is connected, (2) G is acyclic, (3) $m = n-1$. In fact, if G is a graph of order n and size m that satisfies *any* two of these three properties, then G is a tree.

Theorem 4.8 *Let G be a graph of order n and size m. If G satisfies any two of the properties:*

(1) *G is connected,* (2) *G is acyclic,* (3) *$m = n - 1$,*

then G is a tree.

Proof. First, if G satisfies (1) and (2), then G is a tree by definition. Thus, we may assume that G satisfies (1) and (3) or G satisfies (2) and (3). We consider these two cases.

Case 1. G satisfies (1) and (3). Since G is connected, it suffices to show that G is acyclic. Assume, to the contrary, that G contains a cycle C. Let e be an edge of C. Then e is not a bridge of G by Theorem 4.1. So $G - e$ is a connected graph of order n and size $n - 2$, which contradicts Theorem 4.7. Therefore, G is acyclic and so is a tree.

Case 2. G satisfies (2) and (3). Since G is acyclic, it suffices to show that G is connected. Since G satisfies (2) and (3), it follows that G is a forest of order n and size $m = n - 1$. By Corollary 4.6, the size of G is $n - k$, where k is the number of components of G. Hence $n - 1 = n - k$ and so $k = 1$. Therefore, G is connected. ∎

If T is a tree of order k, then it should be clear that T is isomorphic to a subgraph of K_k. Of course, $\delta(K_k) = k - 1$. Not only is T isomorphic to a subgraph of K_k, the tree T is isomorphic to a subgraph of *every* graph having minimum degree at least $k - 1$.

Theorem 4.9 *Let T be a tree of order k. If G is a graph with $\delta(G) \geq k - 1$, then T is isomorphic to some subgraph of G.*

Proof. We proceed by induction on k. The result is certainly true for $k = 1$ since every graph contains a vertex. It is also true for $k = 2$ since every graph without isolated vertices contains edges.

Assume for every tree T' of order $k - 1$, where $k \geq 3$, and for every graph H with minimum degree at least $k - 2$ that T' is isomorphic to a subgraph of H. Now, let T be a tree of order k and let G be a graph with $\delta(G) \geq k - 1$. We show that T is isomorphic to a subgraph of G.

Let v be an end-vertex of T and u the vertex of T adjacent to v. Then $T - v$ is a tree of order $k - 1$. Since $\delta(G) \geq k - 1 > k - 2$, it follows by the induction hypothesis that $T - v$ is isomorphic to a subgraph F of G. Let u' denote the vertex of F corresponding to u in T. Since $\deg_G u' \geq k - 1$ and the order of F is $k - 1$, the vertex u' is adjacent to a vertex w of G that does not belong to F (see Figure 4.6). Therefore, T is isomorphic to the subgraph of G obtained by adding the vertex w and the edge $u'w$ to F.   ■

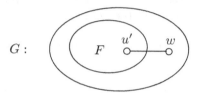

Figure 4.6: The subgraph $T - v$ of G in the proof of Theorem 4.9

Exercises for Section 4.2

4.7 (a) Draw all trees of order 5.   (b) Draw all forests of order 6.

4.8 Prove that if every vertex of a graph G has degree at least 2, then G contains a cycle.

4.9 Show that a graph of order n and size $n - 1$ need not be a tree.

4.10 Give an example of each of the following or explain why no such example exists.

  (a) A graph that is not a tree in which every edge is a bridge.

  (b) A tree of order 4 whose complement is not a tree.

  (c) A tree T containing exactly three vertices that are not end-vertices and such that T is not a caterpiller.

4.11 For $k = 2, 3, 4$, give an example of a tree T_k with $\Delta(T_k) = k$ such that no two vertices of the same degree are adjacent to the same vertex.

4.12 (a) Give an example of a tree T and an edge e of T such that the two components of $T - e$ are isomorphic.

(b) Show that there exists no tree T containing two distinct edges e_1 and e_2 such that the two components of $T - e_1$ are isomorphic *and* the two components of $T - e_2$ are isomorphic.

(c) Show that there exists a tree T containing two distinct edges e_1 and e_2 such that (1) $T - e_1 \cong T - e_2$ and (2) for the two components T_1 and T_1' of $T - e_1$ and the two components T_2 and T_2' of $T - e_2$, we have $T_1 \cong T_2$ and $T_1' \cong T_2'$.

4.13 A certain tree T of order 21 has only vertices of degree 1, 3, 5 and 6. If T has exactly 15 end-vertices and one vertex of degree 6, how many vertices of T have degree 5?

4.14 A certain tree T of order 35 is known to have 25 vertices of degree 1, two vertices of degree 2, three vertices of degree 4, one vertex of degree 5 and two vertices of degree 6. It also contains two vertices of the same (unknown) degree x. What is x?

4.15 A tree T with 50 end-vertices has an equal number of vertices of degree 2, 3, 4 and 5 and no vertices of degree greater than 5. What is the order of T?

4.16 (a) Give an example of a tree of order 6 containing four vertices of degree 1 and two vertices of degree 3. (Only one tree has these properties.)

(b) Find all trees T where two-thirds of the vertices of T have degree 1 and the remaining one-third of the vertices have degree 3.

4.17 (a) Give an example of a tree of order 8 containing six vertices of degree 1 and two vertices of degree 4. (Only one tree has these properties.)

(b) Find all trees T where 75% of the vertices of T have degree 1 and the remaining 25% of the vertices have degree 4.

(c) Find all trees T where 75% of the vertices of T have degree 1 and the remaining 25% of the vertices have another degree (a fixed degree).

(d) Find all trees T where 25% of the vertices of T have degree 1 and the remaining 75% of the vertices of T have another degree (a fixed degree).

4.18 A certain tree T of order n contains only vertices of degree 1 and 3. Show that T contains $(n - 2)/2$ vertices of degree 3.

4.19 Let T be a tree of order n and size m having n_i vertices of degree i for $i \geq 1$. Then $n = \sum_i n_i$ and $2(n - 1) = 2m = \sum_i in_i$.

(a) Prove that $n_1 = 2 + n_3 + 2n_4 + 3n_5 + 4n_6 + \ldots$.

(b) A tree T has three vertices of degree 2, five vertices of degree 3, two vertices of degree 4 and no vertices of degree 5 or more. According to the formula in (a), how many end-vertices does T have?

4.20 Prove or disprove:

    (a) If G is a graph of order n and size m with three cycles, then $m \geq n+2$.

    (b) There exist exactly two regular trees.

4.21 Let T be a tree of order n. Prove that T is isomorphic to a subgraph of $\overline{C}_{n+2}$.

4.22 Let T be a tree of order n. Show that the size of the complement $\overline{T}$ of T is the same as the size of K_{n-1}.

4.23 Find all trees T such that $\overline{T}$ is also a tree.

4.24   (a) Find all those graphs G of order $n \geq 4$ such that the subgraph induced by every three vertices of G is a tree or show that no such graph exists.

    (b) State and solve a generalization of the problem in (a).

4.3   The Minimum Spanning Tree Problem

If the seven villages illustrated in Figure 4.1 really did exist, quite possibly the villages developed one by one; and as a new village developed, a new road was constructed that connected this village to the previously developed villages. For example, suppose that the three villages v_1, v_2 and v_3 already existed with a road between v_1 and v_2 and a road between v_2 and v_3. Furthermore, suppose that the settlement v_4 developed into a village. Then it would be logical to construct a paved road between v_4 and one of v_1, v_2 and v_3. Of course, it might be preferable to construct a road between v_4 and an intermediate location along an existing road, which in turn may lead to a new development at this junction. But let's assume that this doesn't occur. However, just as with many decisions in life, the decision as to which road should be built would most likely be a financial one.

On the other hand, suppose that initially no roads existed between any pair of the villages $v_1, v_2, \ldots, v_7$ (as might be the case if these are the seven Olympic dormitories that are to be constructed to house the athletes at a forthcoming Olympic Games). Then we need to construct roads between pairs of dormitories. Which roads will be constructed is quite possibly a financial decision here as well. Before proceeding further, let us consider a new concept.

If a connected graph G of order n has no cycles, then, of course, G is a tree. On the other hand, suppose that G contains cycles. Let e_1 be an edge lying on a cycle of G. By Theorem 4.1, e_1 is not a bridge and $G - e_1$ is connected. If

$G - e_1$ contains cycles, then let e_2 be an edge lying on a cycle of $G - e_1$. Then $G - e_1 - e_2$ is connected. Eventually, we arrive at a set $X = \{e_1, e_2, \ldots, e_k\}$ of edges of G such that $G - X$ is a tree. The tree $G - X$ just constructed is a subgraph of G that has the same vertex set as G.

We can look at the observation we just made in another way. Let G be a connected graph. Consider the empty graph H with vertex set $V(G)$. Add an edge f_1 of G to H. Then add another edge f_2 of G to H. Next, add another edge f_3 of G to H, where $f_3 \notin \{f_1, f_2\}$ and such that no cycle is produced. We continue this until we have added edges $f_1, f_2, \ldots, f_{n-1}$ of G to H, producing a graph F of order n, size $n - 1$ and no cycles. By Theorem 4.8, F is a tree with $V(F) = V(G)$. We know it's possible to construct a tree in this manner as we can always choose the edges of $G - X$ mentioned above.

We have just described two ways of producing trees T that are subgraphs of a given connected graph G such that $V(T) = V(G)$. Recall that a subgraph H of a graph G is a spanning subgraph of G if H contains every vertex of G. A spanning subgraph H of a connected graph G such that H is a tree is called a **spanning tree** of G. For the connected graph G of Figure 4.7, two different spanning trees T_1 and T_2 of G are also shown in Figure 4.7. We have now observed the following.

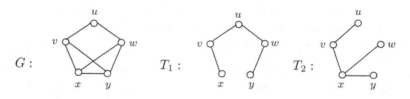

Figure 4.7: Two spanning trees in a graph

Theorem 4.10 *Every connected graph contains a spanning tree.*

Once again, let's return to the example we considered in Figure 4.1, where there are seven villages and six roads. The graph describing this situation is a tree. Hence it is possible to travel between every two villages. Indeed, there is a unique path between every two villages. To travel between villages v_1 and v_6, we are forced to pass through v_2, v_3 and v_4, even if we didn't want to. Therefore, the trip between v_1 and v_6 may be inconvenient. Of course, to make the trip between pairs of the villages more convenient, we could always build a new road (between v_1 and v_6, say). However, this would cost more money (possibly a great deal of money). But how was it decided initially that the six roads in Figure 4.1 were the ones to be constructed? Certainly, whichever roads were chosen should produce a connected graph. If the resulting graph contains a cycle, then there are edges in the graph that are not bridges. That is, if producing a connected graph was our primary goal, then we could have accomplished this for less money by constructing roads so that the resulting graph is a tree. But how did we choose those *particular* six roads to build?

Suppose that we have a number of villages (such as the villages $v_1, v_2, \ldots, v_7$) and we would like to build roads as cheaply as possible so that, at the conclusion, the resulting graph is connected. How do we do this? Assume that we have an accurate estimate of the cost of building a road between each pair of villages. If the cost of building a road between some pair of villages is exorbitant (because any such road would have to pass through quicksand, private property or through or over a mountain, for example), then we do not even consider building such a road. This problem can be stated in terms of graphs.

Let G be a connected graph each of whose edges is assigned a number (called the **cost** or **weight** of the edge). We denote the weight of an edge e of G by $w(e)$. Recall that such a graph is called a weighted graph. For each subgraph H of G, the **weight** $w(H)$ of H is defined as the sum of the weights of its edges, that is,

$$w(H) = \sum_{e \in E(H)} w(e).$$

We seek a spanning tree of G whose weight is minimum among all spanning trees of G. Such a spanning tree is called a **minimum spanning tree**. The problem of finding a minimum spanning tree in a connected weighted graph is called the **Minimum Spanning Tree Problem**.

The importance of the Minimum Spanning Tree Problem is due to its applications in the design of computer, communications and transportation networks. The history of this problem was researched by Ronald L. Graham and Pavol Hell in 1985. (We will encounter Graham again in Chapter 11.) They concluded that the Minimum Spanning Tree Problem was initially formulated by Otakar Borůvka in 1926 because of his interest in the most economical layout of a power-line network. He also gave the first solution of the problem. Prior to Borůvka, however, the anthropologist Jan Czekanowski's work on classification schemes led him to consider ideas closely related to the Minimum Spanning Tree Problem.

Over the years, this problem has been solved in a variety of ways using a number of algorithms. One of the best known was discovered by Joseph Bernard Kruskal (1928-2010). Kruskal was from a family of five children, three boys and two girls. All boys became mathematicians. Kruskal received his Ph.D. from Princeton in 1954. His advisors were Paul Erdős and Roger Lyndon. Throughout his life he was an active researcher, with much of his work in mathematics and linguistics. He spent much of his life working at Bell Laboratories. However, it was only two years after completing his doctoral degree that the paper was published containing the algorithm that bears his name.

Kruskal's Algorithm: For a connected weighted graph G, a spanning tree T of G is constructed as follows: For the first edge e_1 of T, we select any edge of G of minimum weight and for the second edge e_2 of T, we select any remaining edge of G of minimum weight. For the third edge e_3 of T, we choose any remaining edge of G of minimum weight that does not produce a cycle with the previously selected edges. We continue in this manner until a spanning tree is produced.

Figure 4.8 shows how a spanning tree of a connected weighted graph is constructed using Kruskal's Algorithm. We now show that Kruskal's Algorithm produces a minimum spanning tree in every connected weighted graph.

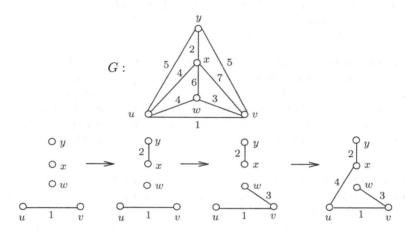

Figure 4.8: Constructing a spanning tree by Kruskal's Algorithm

Theorem 4.11 *Kruskal's Algorithm produces a minimum spanning tree in a connected weighted graph.*

Proof. Let G be a connected weighted graph of order n and let T be a spanning tree obtained by Kruskal's Algorithm, where the edges of T are selected in the order $e_1, e_2, \ldots, e_{n-1}$. Necessarily then, $w(e_1) \leq w(e_2) \leq \ldots \leq w(e_{n-1})$ and the weight of T is

$$w(T) = \sum_{i=1}^{n-1} w(e_i).$$

We show that T is a minimum spanning tree of G. Assume, to the contrary, that T is not a minimum spanning tree. Among all minimum spanning trees of G, let H be one that has a maximum number of edges in common with T. Since H and T are not identical, there is at least one edge of T that is not in H. Let e_i be the first edge of T that is not in H. Therefore, if $i > 1$, then the edges e_1, $e_2, \ldots, e_{i-1}$ belong to both H and T. Now define $G_0 = H + e_i$. Then G_0 has a cycle C. Since T has no cycle, there is an edge e_0 on C that is not in T. The graph $T_0 = G_0 - e_0$ is therefore a spanning tree of G and

$$w(T_0) = w(H) + w(e_i) - w(e_0).$$

Since H is a minimum spanning tree of G, it follows that $w(H) \leq w(T_0)$. Consequently, $w(H) \leq w(H) + w(e_i) - w(e_0)$ and so $w(e_0) \leq w(e_i)$. By Kruskal's Algorithm, certainly $w(e_0) = w(e_i)$ if $i = 1$. Suppose then that $i > 1$. By

Kruskal's Algorithm, e_i is an edge of minimum weight that can be added to the edges $e_1, e_2, \ldots, e_{i-1}$ without producing a cycle. However, e_0 can also be added to $e_1, e_2, \ldots, e_{i-1}$ without producing a cycle. Thus $w(e_i) \leq w(e_0)$, which implies that $w(e_i) = w(e_0)$ when $i > 1$ as well. Therefore, $w(T_0) = w(H)$ and so T_0 is also a minimum spanning tree of G. However, T_0 has more edges in common with T than H does, which is a contradiction.   ∎

Another well-known algorithm for finding a minimum spanning tree in a connected weighted graph was developed by Robert Clay Prim (born in 1921). Like Kruskal, he received his Ph.D. from Princeton (in 1949). He was vice president of research at the Sandia Corporation. The paper containing the algorithm that bears his name was published in 1957. This algorithm was originally discovered by the Czech mathematician Vojtěch Jarnik in 1930 but the algorithm is named for Prim.

Prim's Algorithm: For a connected weighted graph G, a spanning tree T of G is constructed as follows: For an arbitrary vertex u for G, an edge of minimum weight incident with u is selected as the first edge e_1 of T. For subsequent edges $e_2, e_3, \cdots, e_{n-1}$, we select an edge of minimum weight among those edges having exactly one of its vertices incident with an edge already selected.

Figure 4.9 illustrates how to construct a spanning tree of a connected weighted graph by Prim's Algorithm. Again, a tree obtained by Prim's Algorithm is also a minimum spanning tree, as we show next.

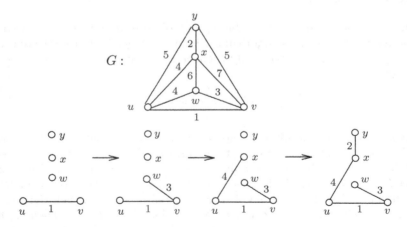

Figure 4.9: Constructing a spanning tree by Prim's Algorithm

Theorem 4.12 *Prim's Algorithm produces a minimum spanning tree in a connected weighted graph.*

Proof. Let G be a nontrivial connected weighted graph of order n and let T be a spanning tree obtained by Prim's Algorithm, where the edges of T are selected in the order e_1, e_2, ..., e_{n-1} and where e_1 is incident with a given vertex u. Thus the weight of T is

$$w(T) = \sum_{i=1}^{n-1} w(e_i).$$

Assume, to the contrary, that T is not a minimum spanning tree. Let $\mathcal{H}$ be the set of all minimum spanning trees of G having a maximum number of edges in common with T. If no tree in $\mathcal{H}$ contains e_1, then let $k = 0$ and let H be any tree in $\mathcal{H}$; otherwise, let k be the maximum positive integer for which there is a tree $H \in \mathcal{H}$ such that H contains e_1, e_2, ..., e_k. Hence no tree in $\mathcal{H}$ contains all of the edges e_1, e_2, ..., e_{k+1}, where $0 \leq k < n-1$. If H does not contain e_1 and so $k = 0$, then let $U = \{u\}$. If $k \geq 1$, let U be the vertex set of the tree with edge set $\{e_1, e_2, \ldots, e_k\}$. So U consists of the $k + 1$ vertices that are incident with one or more of the edges $e_1, e_2, \ldots, e_k$. By Prim's Algorithm, e_{k+1} joins a vertex of U and a vertex of $V(T) - U$.

The subgraph $H + e_{k+1}$ therefore contains a cycle C and e_{k+1} is on C. Necessarily, C contains an edge e_0 distinct from e_{k+1} such that e_0 also joins a vertex of U and a vertex of $V(T) - U$. By the construction of T from Prim's Algorithm, $w(e_{k+1}) \leq w(e_0)$. Now $T' = H + e_{k+1} - e_0$ is a spanning tree of G whose weight is $w(T') = w(H) + w(e_{k+1}) - w(e_0)$. Since H is a minimum spanning tree, $w(H) \leq w(T')$ and so $w(H) \leq w(H) + w(e_{k+1}) - w(e_0)$, which implies that $w(e_0) \leq w(e_{k+1})$. Consequently, $w(e_0) = w(e_{k+1})$ and $w(H) = w(T')$. Therefore, T' is also a minimum spanning tree of G. If e_0 does not belong to T, then T' is a minimum spanning tree having more edges in common with T than H does, which is impossible since $H \in \mathcal{H}$. Hence e_0 belongs to T, which implies that T' has the same number of edges in common with T that H does and so $T' \in \mathcal{H}$. Necessarily, $e_0 = e_j$ for some $j > k + 1$. Since T' contains all of the edges e_1, e_2, ..., e_{k+1}, this contradicts the defining property of H. ■

Exercises for Section 4.3

4.25 Determine all spanning trees for the graphs G and H in Figure 4.10. Which of these spanning trees are isomorphic?

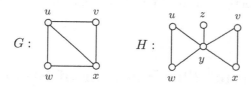

Figure 4.10: The graphs in Exercise 4.25

4.26 Prove that an edge e of a connected graph is a bridge if and only if e belongs to every spanning tree of G.

4.27 Apply both Kruskal's and Prim's Algorithms to find a minimum spanning
tree in the weighted graph in Figure 4.11. In each case, show how this tree
is constructed, as in Figures 4.8 and 4.9.

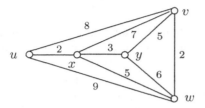

Figure 4.11: The weighted graph in Exercise 4.27

4.28 Apply both Kruskal's and Prim's Algorithms to find a minimum spanning
tree in the weighted graph in Figure 4.12. In each case, show how this tree
is constructed, as in Figures 4.8 and 4.9.

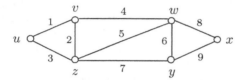

Figure 4.12: The weighted graph in Exercise 4.28

4.29 Let G be a connected weighted graph whose edges have distinct weights.
Prove that G has a unique minimum spanning tree.

4.30 Let G be a connected weighted graph and T a minimum spanning tree of
G. Show that T is a unique minimum spanning tree of G if and only if
the weight of each edge e of G that is not in T exceeds the weight of every
other edge on the cycle in $T + e$.

4.31 Show, for each integer $k \geq 2$, that there exists a connected weighted graph
containing exactly k unequal minimum spanning trees.

4.4   Excursion: The Number of Spanning Trees

We have already mentioned that every connected graph G contains a spanning
tree. It is essential, of course, that G is connected, for otherwise G has no
spanning trees. Also, if G is itself a tree, then G contains exactly one spanning

tree, namely G itself. On the other hand, if G is a connected labeled graph that is not a tree, then G contains more than one spanning tree. But how many? In this section, we are concerned with counting the number of (unequal) spanning trees of a labeled connected graph. To simplify the consideration of a number of graphs, we intend that vertex labels are present even if none are shown. Spanning trees with different edge sets are therefore different. We now consider two examples.

Example 4.13 *Determine the number of spanning trees of the graph G of Figure 4.13.*

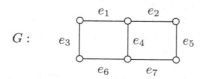

Figure 4.13: The graph in Example 4.13

Solution. Observe that at least one edge of each cycle of G must be absent from every spanning tree of G. We consider the number of spanning trees of G that (1) do not contain e_4 and (2) contain e_4. First, any spanning tree that does not contain e_4 must contain exactly five of the six edges e_1, e_2, e_3, e_5, e_6, e_7. Hence there are six spanning trees of G that do not contain e_4. Second, any spanning tree that contains e_4 must not contain exactly one of e_1, e_3, e_6 and must not contain exactly one of e_2, e_5, e_7. Therefore, there are $3 \cdot 3 = 9$ spanning trees that contain e_4. So there are $6 + 9 = 15$ spanning trees of G. $\diamond$

Example 4.14 *Determine the number of spanning trees of the graph G of Figure 4.14.*

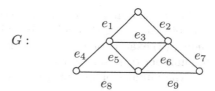

Figure 4.14: The graph in Example 4.14

Solution. At least one of the edges e_3, e_5, e_6 is absent in each spanning tree of G. This divides the spanning trees of G into three mutually disjoint categories.

Category 1 consists of those spanning trees containing none of the edges e_3, e_5, e_6. Then a spanning tree is obtained by deleting any of the remaining six edges e_1, e_2, e_4, e_7, e_8, e_9. Hence there are 6 such spanning trees.

Category 2 *consists of those spanning trees containing exactly one of the edges* e_3, e_5, e_6. Suppose first that e_3 belongs to a spanning tree but e_5 and e_6 do not. Then exactly one of e_1 or e_2 does not belong to a spanning tree and exactly one of e_4, e_7, e_8, e_9 does not belong to a spanning tree. Therefore, the number of spanning trees of G containing e_3 but neither e_5 nor e_6 is $2 \cdot 4 = 8$. By the symmetry of G, the number of spanning trees of G containing exactly one of the edges e_3, e_5, e_6 is $3 \cdot 8 = 24$.

Category 3 *consists of those spanning trees containing exactly two of the edges* e_3, e_5, e_6. Suppose first that e_3 and e_5 belong to a spanning tree but e_6 does not. Then such a spanning tree is obtained by deleting exactly one edge in each of the following three pairs of edges: (1) e_1, e_2; (2) e_4, e_8; (3) e_7, e_9. Therefore, the number of spanning trees of G containing e_3 and e_5 but not e_6 is $2^3 = 8$. Again, by symmetry, the number of spanning trees of G containing exactly two of the three edges e_3, e_5, e_6 is $3 \cdot 8 = 24$.

Therefore, the total number of spanning trees of G is $6 + 24 + 24 = 54$. $\diamondsuit$

We now turn to the problem of determining the number of spanning trees in a complete graph K_n. Since K_n is a tree for $n = 1$ or $n = 2$, we need only consider $n \geq 3$. Since K_3 is a cycle of length 3, each spanning tree is obtained by deleting one of the three edges, that is, the number of spanning trees of K_3 is 3 (see Figure 4.15).

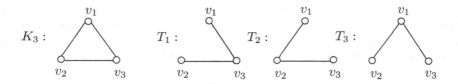

Figure 4.15: The spanning trees of K_3

Determining the number of spanning trees of K_4 (see Figure 4.16) is more troublesome, but one way of computing this is by observing that any spanning tree of K_4 contains (1) none, (2) exactly one or (3) exactly two of the edges $v_1 v_2$, $v_1 v_3$ and $v_2 v_3$. Since the number of spanning trees in each of these three cases is 1, 6 and 9, respectively, the number of spanning trees of K_4 is $1 + 6 + 9 = 16$.

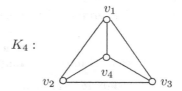

Figure 4.16: The complete graph K_4

Actually, computing the total number of of spanning trees of the graph $G = K_n$, where $V(G) = \{v_1, v_2, \ldots, v_n\}$, is the same as computing the number of distinct trees with vertex set $\{v_1, v_2, \ldots, v_n\}$. The following formula was established in 1889 by Arthur Cayley and is often referred to as the **Cayley Tree Formula**. As a consequence of this formula, it becomes clear why there are 16 spanning trees of K_4.

Theorem 4.15 *The number of distinct trees of order n with a specified vertex set is n^{n-2}.*

We have encountered Cayley several times already. Arthur Cayley was born on August 16, 1821 in Richmond, Surrey, England. He showed great skill with numerical calculations as a youngster. In 1838 he entered Trinity College, Cambridge and had three papers published while still an undergraduate. He graduated in 1842 and then spent four years teaching as a Cambridge fellow, at which time he continued to publish at a high rate. When his fellowship expired, Cayley found himself without a position. He then studied law and became a lawyer in 1849, an occupation he continued for the next 14 years. Although a skilled lawyer, Cayley considered this as a way to make money so that he could do what he really enjoyed: mathematics. During his 14 years as a lawyer, Cayley authored more than 200 mathematical papers.

In 1849 Cayley wrote a paper on permutations connecting his ideas with those of Augustin Louis Cauchy. In 1854 he wrote two remarkable papers on abstract groups. At that time the only known groups were permutation groups. Cayley defined an abstract group and gave a table to display the multiplication in the group. He also recognized that matrices formed groups.

In 1863 Cayley was appointed a professor of Pure Mathematics at Cambridge. Although Cayley was earning only a fraction of what he earned as a lawyer, he was happier as he was able to work on mathematics full-time. He was a prolific researcher his entire life and when he died on January 26, 1895 in Cambridge, he had published over 900 papers, a number exceeded only by Leonhard Euler, Paul Erdős and Cauchy.

Although we will not be presenting a proof of Theorem 4.15, there are several quite different proofs of this formula. In fact, John W. Moon wrote a paper in 1967 in which he describes ten different proofs of this famous theorem (and even more proofs have been found since). Moon, a professor at the University of Alberta in Canada throughout much of his academic career, did a great deal of research on trees (and on a class of digraphs that we will visit in Chapter 7). With an interest in music, as has been the case with a number of mathematicians, Moon has a philosophy about the creative aspect of mathematics that is no doubt not unlike that of many mathematicians:

> *The sense of pleasure and satisfaction that comes when one has discovered something new is hard to describe to non-mathematicians; and sometimes there is almost a sense of awe, that one has been privileged to have a peek at what lies behind the mystery of things.*

And when you can share the experience with a co-worker, so much the better.

There is a method of determining the number of spanning trees of any graph. The next theorem is implicit in the work of Gustav Kirchhoff who was born in Königberg, Prussia in 1824. (We will visit that city again in Chapter 6.) Kirchhoff is well known for his research on electrical currents, which he announced in 1845. This led to Kirchhoff's laws, the first of which states that the sum of the currents into a vertex equals the sum of the currents out of the vertex. Two years later, in 1847, he graduated from the University of Königberg. It was during that year that he published the paper that led to his theorem on counting spanning trees. Kirchhoff spent much of his life working on experimental physics. When his health began to fail, he accepted a position as chair of mathematical physics in 1875 (12 years before he died) in Berlin as this did not present the problems his poor health was causing him in carrying out experiments. Since the proof of the theorem on counting spanning trees is complex, we will not include it.

By a **cofactor** of an $n \times n$ matrix $M = [m_{ij}]$, we mean $(-1)^{i+j} \det(M_{ij})$, where $\det(M_{ij})$ indicates the determinant of the $(n-1) \times (n-1)$ submatrix M_{ij} of M, obtained by deleting row i and column j of M. The following result is often called the **Matrix Tree Theorem**.

Theorem 4.16 *Let G be a graph with $V(G) = \{v_1, v_2, \ldots, v_n\}$, let $A = [a_{ij}]$ be the adjacency matrix of G and let $C = [c_{ij}]$ be the $n \times n$ matrix, where*

$$c_{ij} = \begin{cases} \deg v_i & \text{if } i = j \\ -a_{ij} & \text{if } i \neq j. \end{cases}$$

Then the number of spanning trees of G is the value of any cofactor of C.

We illustrate the Matrix Tree Theorem with a simple example. The graph G of Figure 4.17 clearly contains three spanning trees. The adjacency matrix A and the matrix C are also shown in Figure 4.17.

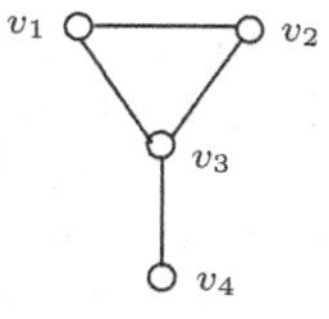

$$A = \begin{bmatrix} 0 & 1 & 1 & 0 \\ 1 & 0 & 1 & 0 \\ 1 & 1 & 0 & 1 \\ 0 & 0 & 1 & 0 \end{bmatrix} \qquad C = \begin{bmatrix} 2 & -1 & -1 & 0 \\ -1 & 2 & -1 & 0 \\ -1 & -1 & 3 & -1 \\ 0 & 0 & -1 & 1 \end{bmatrix}$$

Figure 4.17: Illustrating the Matrix Tree Theorem

The $(3,3)$-cofactor of C is expanded along the 3rd row, obtaining

$$\begin{vmatrix} 2 & -1 & 0 \\ -1 & 2 & 0 \\ 0 & 0 & 1 \end{vmatrix} = 0 \begin{vmatrix} -1 & 0 \\ 2 & 0 \end{vmatrix} - 0 \begin{vmatrix} 2 & 0 \\ -1 & 0 \end{vmatrix} + 1 \begin{vmatrix} 2 & -1 \\ -1 & 2 \end{vmatrix}$$

$$= 0 \cdot 0 - 0 \cdot 0 + 1 \cdot 3 = 3.$$

Exercises for Section 4.4

4.32 Show that there is only one positive integer k such that no graph contains exactly k spanning trees.

4.33 Let F be a subgraph of a connected graph G. Prove that F is a subgraph of some spanning tree of G if and only if F contains no cycles.

4.34 (a) Find the number of spanning trees in the graph G of Figure 4.18.

　　(b) Find the number of spanning trees in the graph G_k for $k \geq 5$ of Figure 4.18. [Note that (a) is the case where $k = 4$.]

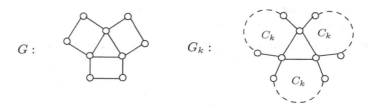

Figure 4.18: The graphs in Exercise 4.34

4.35 (a) Find the number of spanning trees in the graph G of Figure 4.19.

　　(b) Find the number of spanning trees in the graph G_k for $k \geq 6$ of Figure 4.19. [Note that (a) is the case where $k = 5$.]

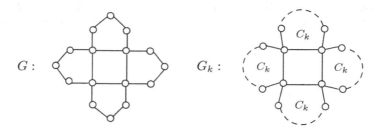

Figure 4.19: The graphs in Exercise 4.35

4.36 Find the number of spanning trees in the graph G of Figure 4.20.

4.37 (a) According to the Cayley Tree Formula, how many distinct trees are there with vertex set $S = \{u, v, w, x, y\}$?

　　(b) Divide the trees in (a) into classes so that two trees are in the same class if and only if they are isomorphic. Determine the number of such classes and the number of trees in each class by considering the number of ways each can be labeled with the elements of S.

$G:$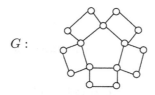

Figure 4.20: The graph in Exercise 4.36

4.38 Use the Matrix Tree Theorem to confirm the number of distinct trees with vertex set $\{v_1, v_2, v_3, v_4\}$.

4.39 (a) Use the Matrix Tree Theorem to determine the number of spanning trees of the graph of Figure 4.21.

   (b) Draw all spanning trees in the graph G of Figure 4.21.

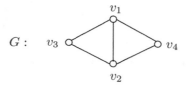

Figure 4.21: The graph in Exercise 4.39

4.40 Let T and T' be two spanning trees of a connected graph G of order n. Show that there exists a sequence $T = T_0, T_1, \ldots, T_k = T'$ of spanning trees of G such that T_i and T_{i+1} have $n-2$ edges in common for each i with $1 \le i \le k-1$.

Chapter 5

Connectivity

5.1 Cut-Vertices

It is probably clear by now that one of the most important properties that a graph can possess is that of being connected. Figure 5.1 shows seven graphs of order 7. The graph G_1 is a tree, $G_4 = C_7$ and $G_7 = K_7$. Obviously, all of these graphs are connected. However, some appear to be "more connected" than others. Indeed, the main goal of this chapter is the introduction of measures of how connected a graph is.

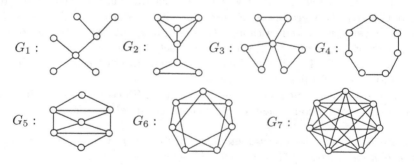

Figure 5.1: Connected graphs

Some graphs are so slightly connected that the removal of a single edge results in a disconnected graph. We have already seen this and an edge with this property is a bridge. The graph G_2 has a bridge. So does G_1. In fact, *every* edge of G_1 is a bridge since G_1 is a tree. We now turn from connected graphs containing an edge whose removal results in a disconnected graph to connected graphs containing a *vertex* whose removal results in a disconnected graph.

Recall that if v is a vertex of a nontrivial graph G, then by $G - v$ we mean the (induced) subgraph of G whose vertex set consists of all vertices of G except

v and whose edge set consists of all edges of G except those incident with v. This concept is illustrated in Figure 5.2. In fact, if U is a proper subset of the vertex set of G, then $G - U$ is the (induced) subgraph of G whose vertex set is $V(G) - U$ and whose edge set consists of all edges of G joining two vertices in $V(G) - U$. A vertex v in a connected graph G is a **cut-vertex** of G if $G - v$ is disconnected. More generally, a vertex v is a cut-vertex in a graph G if v is a cut-vertex of a component of G. In the graph G of Figure 5.2, v and x are the only cut-vertices. In the graph $G - v$, the vertex x is not a cut-vertex; however, s is a cut-vertex of $G - v$. Consequently, for $U = \{s, v\}$, the graph $G - U$ is disconnected. The graphs G_1, G_2 and G_3 of Figure 5.1 also contain cut-vertices but no other graphs in Figure 5.1 contain cut-vertices.

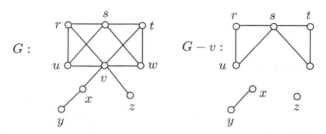

Figure 5.2: The graphs G and $G - v$

Notice that the graph G of Figure 5.2 not only contains the cut-vertex v, it contains three bridges, two of which are incident with v. We have already noticed that the two graphs of Figure 5.1 with bridges, namely G_1 and G_2, contain cut-vertices as well (although G_3 contains a cut-vertex but no bridges). In fact, except for the graph K_2, every connected graph with bridges contains cut-vertices as well. Recall by Theorem 4.1 that an edge e in a graph G is a bridge if and only if e lies on no cycle of G.

We now establish some facts about cut-vertices. Since a cut-vertex in a graph G is a cut-vertex of a component of G, we restrict ourselves to connected graphs.

Theorem 5.1 *Let v be a vertex incident with a bridge in a connected graph G. Then v is a cut-vertex of G if and only if $\deg v \geq 2$.*

Proof. Suppose that uv is a bridge of G. Then $\deg v \geq 1$. Assume that $\deg v = 1$. Since v is an end-vertex of G, the graph $G - v$ is connected and so v is not a cut-vertex of G.

For the converse, assume that $\deg v \geq 2$. Then there is a vertex w different from u that is adjacent to v. Assume, to the contrary, that v is not a cut-vertex. Thus $G - v$ is connected and so there is a $u - w$ path P in $G - v$. However then, P together with v and the two edges uv and vw form a cycle containing the bridge uv. This contradicts Theorem 4.1.   ∎

One immediate consequence of Theorem 5.1 is that if a vertex v of a nontrivial tree T is not an end-vertex of T, then v is a cut-vertex of T. Another immediate consequence of Theorem 5.1 is stated next.

Corollary 5.2 *Let G be a connected graph of order 3 or more. If G contains a bridge, then G contains a cut-vertex.*

If v is a cut-vertex in a connected graph G, then, of course, $G-v$ contains two or more components. If u and w are vertices in distinct components of $G - v$, then u and w are not connected in $G - v$. On the other hand, u and w are necessarily connected in G. These observations lead us to the following theorem.

Theorem 5.3 *Let v be a cut-vertex in a connected graph G and let u and w be vertices in distinct components of $G - v$. Then v lies on every $u - w$ path in G.*

We can now present a characterization of vertices in a connected graph G that are cut-vertices of G.

Corollary 5.4 *A vertex v of a connected graph G is a cut-vertex of G if and only if there exist vertices u and w distinct from v such that v lies on every $u-w$ path of G.*

Proof. Suppose that v is a cut-vertex of G. Then $G - v$ is disconnected. Let u and w be vertices in different components of $G-v$. It then follows by Theorem 5.3 that every $u - w$ path in G contains v.

On the other hand, if G contains two vertices u and w such that every $u - w$ path in G contains v, then there is no $u - w$ path in $G - v$. Thus u and w are not connected in $G - v$ and so $G - v$ is disconnected. Therefore, v is a cut-vertex of G. ∎

By Corollary 5.4, if v is a vertex in a connected graph G that is not a cut-vertex of G, then for *every* two vertices u and w of G that are distinct from v, there is a $u - w$ path that does not contain v.

While the concepts of bridges and cut-vertices are parallel concepts and have a number of similarities, they have some major differences as well. We have seen that it's possible for every edge of a connected graph G to be a bridge. Of course, G must then be a tree. On the other hand, every nontrivial connected graph must contain vertices that are not cut-vertices.

Theorem 5.5 *Let G be a nontrivial connected graph and let $u \in V(G)$. If v is a vertex that is farthest from u in G, then v is not a cut-vertex of G.*

Proof. Assume, to the contrary, that v is a cut-vertex of G. Let w be a vertex belonging to a component of $G - v$ that does not contain u. Since every $u - w$ path contains v, it follows that $d(u,w) > d(u,v)$, which is a contradiction. ∎

It is now immediate that every nontrivial connected graph contains at least *two* vertices that are not cut-vertices.

Corollary 5.6 *Every nontrivial connected graph contains at least two vertices that are not cut-vertices.*

Proof. Let u and v be vertices of a nontrivial connected graph G such that $d(u, v) = \text{diam}(G)$. Since each of u and v is farthest from the other, it follows by Theorem 5.5 that both u and v are not cut-vertices of G.   ∎

Exercises for Section 5.1

5.1 Give an example of a graph that

   (a) contains more bridges than cut-vertices.

   (b) contains more cut-vertices than bridges.

5.2 (a) For each integer $k \geq 2$, give an example of a connected graph G containing a vertex v such that $G - v$ has k components.

   (b) Give an example of a connected graph G of order 3 or more containing vertices u and v such that $G - u - v$ has fewer components than $G - u$.

5.3 Prove or disprove:

   (a) If a vertex v of a graph G lies on a cycle of G, then v is not a cut-vertex.

   (b) If a vertex v of a graph G does not lie on any cycle of G, then v is a cut-vertex.

   (c) A tree of order 3 or more has more cut-vertices than end-vertices.

   (d) A tree of order 3 or more has more cut-vertices than bridges.

5.4 Prove that if v is a cut-vertex of a graph G, then v is not a cut-vertex of the complement $\overline{G}$ of G.

5.5 Find a counterexample to each of the following statements.

   (a) If G is a connected graph of order 13 and v is a cut-vertex of G, then there exists a component of $G - v$ containing at least 7 vertices.

   (b) If G is a connected graph containing only even vertices, then G contains no cut-vertices.

   (c) If G is a connected graph with a cut-vertex, then G contains a bridge.

   (d) If G is a connected graph with a bridge, then G contains a cut-vertex.

5.6 Prove that a 3-regular graph G has a cut-vertex if and only if G has a bridge.

5.7 Prove that if T is a tree of order at least 3, then T contains a cut-vertex v such that every vertex adjacent to v, with at most one exception, is an end-vertex.

5.8 (a) Let G be a nontrivial connected graph. Prove that if v is an end-vertex of a spanning tree of G, then v is not a cut-vertex of G.

   (b) Use (a) to give an alternative proof of the fact that every nontrivial connected graph contains at least two vertices that are not cut-vertices.

   (c) Let v be a vertex in a nontrivial connected graph G. Show that there exists a spanning tree of G that contains all edges of G that are incident with v.

   (d) Prove that if a connected graph G has exactly two vertices that are not cut-vertices, then G is a path. [Recall that if a tree contains a vertex of degree exceeding 2, then T has more than two end-vertices.]

5.2 Blocks

We now turn our attention from connected graphs containing cut-vertices to connected graphs that contain no cut-vertices. A nontrivial connected graph with no cut-vertices is called a **nonseparable graph**. Hence all of the graphs G_4, G_5, G_6 and G_7 of Figure 5.1 are nonseparable. In addition, K_2 is a nonseparable graph; indeed, K_2 is the only nonseparable graph of order 2. Since nonseparable graphs of order 3 or more contain no cut-vertices, they contain no bridges; that is, every edge lies on a cycle. In fact, more can be said.

Theorem 5.7 *A graph of order at least 3 is nonseparable if and only if every two vertices lie on a common cycle.*

Proof. First, suppose that G is a graph of order at least 3 such that every two vertices of G lie on a common cycle. Assume, to the contrary, that G is not nonseparable. Since every two vertices lie on a common cycle, every two vertices are connected and so G is connected. Because G is not nonseparable, G must contain a cut-vertex, say v. Let u and w be two vertices that belong to different components of $G - v$. By assumption, u and w lie on a common cycle C on G. However then, C determines two distinct $u - w$ paths of G, at least one of which does not contain v, contradicting Theorem 5.3. Therefore, G contains no cut-vertices and so G is nonseparable.

We now verify the converse. Let G be a nonseparable graph of order at least 3. Since G contains no cut-vertices, it follows by Corollary 5.2 that G contains no bridges. Assume, to the contrary, that there are pairs of vertices of G that do not lie on a common cycle. Among all such pairs, let u, v be a pair for which $d(u, v)$ is minimum. If $d(u, v) = 1$, then $uv \in E(G)$ and uv must lie on a cycle, as we observed earlier. Therefore, $d(u, v) = k \geq 2$.

Let $P = (u = v_0, v_1, \ldots, v_{k-1}, v_k = v)$ be a $u - v$ path of length k in G. Since $d(u, v_{k-1}) = k - 1 < k$, there is a cycle C containing u and v_{k-1}. By assumption, v is not on C. Since v_{k-1} is not a cut-vertex of G and u and v are distinct from v_{k-1}, it follows that there is a $v - u$ path Q that does not contain v_{k-1}. Since u is on C, there is a first vertex x of Q that is on C. Let Q' be the $v - x$ subpath of Q (see Figure 5.3) and let P' be a $v_{k-1} - x$ path on C that contains u. (If $x \neq u$, then the path P' is unique.) However, the cycle C' produced by proceeding from v to its neighbor v_{k-1}, along P' to x and then along Q' to v contains both u and v, a contradiction.   ∎

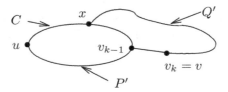

Figure 5.3: The cycle C and paths P', Q' in the proof of Theorem 5.7

If a nontrivial connected graph is not nonseparable, then it must contain cut-vertices as well as certain nonseparable subgraphs that are of special interest. A maximal nonseparable subgraph of a graph G is called a **block** of G. That is, a block of G is a nonseparable subgraph of a graph G that is not a proper subgraph of any other nonseparable subgraph in G. Each block of G is an induced subgraph of G. Therefore, if G itself is nonseparable, then G has only one block, namely the graph G. On the other hand, if G is connected and has cut-vertices, then G has two or more blocks. Figure 5.4 shows a connected graph G with two cut-vertices u and v and four blocks B_1, B_2, B_3 and B_4. So, just as a disconnected graph has special connected subgraphs called components, a connected graph with cut-vertices contains special nonseparable subgraphs called blocks.

There is another way to look at blocks. First, we describe an equivalence relation that is defined on the edge set of a nontrivial connected graph. (Again, recall that equivalence relations are reviewed in Appendix 2.)

Theorem 5.8 *Let R be the relation defined on the edge set of a nontrivial connected graph G by $e\ R\ f$, where $e, f \in E(G)$, if $e = f$ or e and f lie on a common cycle of G. Then R is an equivalence relation.*

Proof. It is immediate that R is reflexive and symmetric, so we need only show that R is transitive. Let $e, f, g \in E(G)$ such that $e\ R\ f$ and $f\ R\ g$. If $e = f$ or $f = g$, then it follows that $e\ R\ g$. Hence we may assume that e and f lie on a cycle C and that f and g lie on a cycle C'. If e lies on C' or g lies on C, then $e\ R\ g$. Thus we may assume that this does not occur.

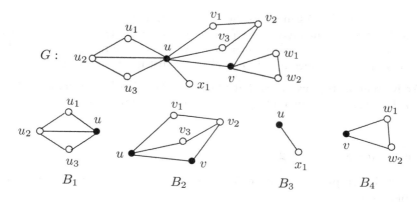

Figure 5.4: A connected graph and its blocks

Let $e = uv$ and suppose that P is the $u - v$ path on C not containing e. Let x be the first vertex of P belonging to C' and y the last vertex of P belonging to C'. Furthermore, let P' be the $x - y$ path on C' that contains g and let P'' be the $x - y$ path on C that contains e. Then P' and P'' produce a cycle C'' containing both e and g. Therefore, $e \; R \; g$.   ∎

The equivalence relation R described in Theorem 5.8 produces a partition of the edge set of every nontrivial connected graph G into equivalence classes. Each subgraph of G induced by the edges in an equivalence class is in fact a block of G. The following corollary of Theorem 5.8 provides properties of blocks in a nontrivial connected graph.

Corollary 5.9 *Every two distinct blocks B_1 and B_2 in a nontrivial connected graph G have the following properties:*

(a) *The blocks B_1 and B_2 are edge-disjoint.*

(b) *The blocks B_1 and B_2 have at most one vertex in common.*

(c) *If B_1 and B_2 have a vertex v in common, then v is a cut-vertex of G.*

Proof. That every two distinct blocks are edge-disjoint is an immediate consequence of Theorem 5.8. We now verify (b). Assume, to the contrary, that B_1 and B_2 have two distinct vertices u and v in common. Since B_1 and B_2 are connected subgraphs of G, there is a $u - v$ path P' in B_1 and a $u - v$ path P'' in B_2. Furthermore, since B_1 and B_2 are edge-disjoint, so too are P' and P''. Let w be the first vertex that P' and P'' have in common after u (possibly $w = v$). The $u - w$ subpath Q' of P' and $u - w$ subpath Q'' of P'' form a cycle in G containing an edge e_1 of B_1 and an edge e_2 of B_2. Hence e_1 and e_2 belong to the same block of G, which is impossible. This verifies (b).

It remains to verify (c). Suppose that two blocks B_1 and B_2 of G have a vertex v in common. Then v is incident with an edge $e_1 = vv_1$ in B_1 and an edge $e_2 = vv_2$ in B_2. Assume, to the contrary, that v is not a cut-vertex of G. By Corollary 5.4, G has a $v_1 - v_2$ path P not containing v. Then P together with v and the edges e_1 and e_2 produce a cycle containing e_1 and e_2. This is impossible, however, since e_1 and e_2 belong to distinct blocks of G.   ∎

We have indicated two ways that blocks can be described. Exercise 5.15 asks you to show that these two interpretations of blocks are equivalent.

Exercises for Section 5.2

5.9 For the graph G of Figure 5.5, determine the cut-vertices, bridges and blocks of G.

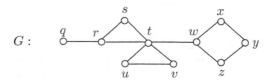

Figure 5.5: The graph G in Exercise 5.9

5.10 Prove that a connected graph G of size at least 2 is nonseparable if and only if any two adjacent edges of G lie on a common cycle of G.

5.11 Prove that if G is a graph of order $n \geq 3$ such that $\deg v \geq n/2$ for every vertex v of G, then G is nonseparable.

5.12 If a connected graph G contains three blocks and k cut-vertices, what are the possible values for k? Explain your answer.

5.13 Prove or disprove: If G is a connected graph with cut-vertices and u and v are vertices of G such that $d(u, v) = \text{diam}(G)$, then no block of G contains both u and v.

5.14 Let G be a connected graph containing a cut-vertex v and let G_1 be a component of $G - v$.

   (a) Show that the induced subgraph $G[V(G_1) \cup \{v\}]$ of G is connected.

   (b) Show that the induced subgraph $G[V(G_1) \cup \{v\}]$ of G need not be a block of G.

5.15 For a nontrivial connected graph G, a block of G has been defined as (1) a nonseparable subgraph of G that is not a proper subgraph of any other nonseparable subgraph of G and has been described as (2) a subgraph of G induced by the edges in an equivalence class resulting from the equivalence relation defined in Theorem 5.8. Show that these two interpretations of blocks are equivalent.

5.16 Give an example of a graph G with the following properties:

(a) every two adjacent vertices lie on a common cycle,

(b) there exist two adjacent edges that do not lie on a common cycle.

5.3 Connectivity

A connected graph G with cut-vertices has the property that G can be disconnected by the removal of a single vertex. On the other hand, nonseparable graphs contain no cut-vertices. In a sense then, nonseparable graphs are more highly connected than connected graphs with cut-vertices. This suggests a way of measuring the connectedness of graphs. By a **vertex-cut** in a graph G, we mean a set U of vertices of G such that $G - U$ is disconnected. A vertex-cut of minimum cardinality in G is called a **minimum vertex-cut**. If G is not complete, then G contains two nonadjacent vertices. The removal of all vertices of G except these two nonadjacent vertices produces a disconnected graph. In other words, every graph that is not complete has a vertex-cut. On the other hand, the removal of any proper subset of vertices of a complete graph results in another complete graph. Therefore, a connected graph contains a vertex-cut if and only if G is not complete.

If U is a minimum vertex-cut in a noncomplete connected graph G, then $G - U$ is disconnected and contains components $G_1, G_2, \ldots, G_k$, where $k \geq 2$. Every vertex $u \in U$ is adjacent to at least one vertex in G_i for each i $(1 \leq i \leq k)$; for otherwise $U - \{u\}$ is also a vertex-cut, which is impossible. The structure of such a graph G is indicated in Figure 5.6, where there are no edges between any two distinct components of $G - U$.

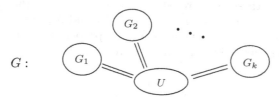

G :

Figure 5.6: A minimum vertex-cut in a graph

For a graph G that is not complete, the **vertex-connectivity** (or simply the **connectivity**) $\kappa(G)$ of G is defined as the cardinality of a minimum vertex-cut of G; if $G = K_n$ for some positive integer n, then $\kappa(G)$ is defined to be $n - 1$. (The symbol κ is the Greek letter kappa.) In general then, the **connectivity** $\kappa(G)$ of a graph G is the minimum value of $|U|$ among all subsets U of $V(G)$ such that $G - U$ is either disconnected or trivial. Therefore, for every graph G of order n,

$$0 \leq \kappa(G) \leq n - 1.$$

Thus a nontrivial graph G has connectivity 0 if and only if G is disconnected; a graph G has connectivity 1 if and only if $G = K_2$ or G is a connected graph with cut-vertices; and a graph G has connectivity 2 or more if and only if G is a nonseparable graph of order 3 or more. For the graphs of Figure 5.1, $\kappa(G_1) = \kappa(G_2) = \kappa(G_3) = 1$, $\kappa(G_4) = \kappa(G_5) = 2$, $\kappa(G_6) = 4$ and $\kappa(G_7) = 6$.

As it turns out, we will often be more interested in graphs that cannot be disconnected by removing some prescribed number of vertices. For a nonnegative integer k, a graph G is said to be k-**connected** if $\kappa(G) \geq k$. Therefore, a k-connected graph is also ℓ-connected for every integer ℓ with $0 \leq \ell \leq k$. In particular, the graphs G_4, G_5, G_6 and G_7 of Figure 5.1 are all 2-connected. The graphs G_6 and G_7 are also 3-connected, although neither G_4 nor G_5 is 3-connected. Thus G is 1-connected if and only if G is nontrivial and connected, while G is 2-connected if and only if G is nonseparable and has order at least 3. In general, a graph G is k-connected if and only if the removal of fewer than k vertices does not result in a disconnected or trivial graph.

The connectivity of a graph G provides a measure of "how connected" G is. There are other measures, including a common one involving the edges of G. An **edge-cut** in a nontrivial graph G is a set X of edges of G such that $G - X$ is disconnected. An edge-cut X of a connected graph G is **minimal** if no proper subset of X is an edge-cut of G. If X is a minimal edge-cut of a connected graph G, then $G - X$ contains exactly two components G_1 and G_2. Necessarily then, X consists of all those edges of G joining G_1 and G_2.

If X is an edge-cut of a connected graph G that is not minimal, then there is a proper subset Y of X that is a minimal edge-cut. An edge-cut of minimum cardinality is called a **minimum edge-cut**. While every minimum edge-cut is a minimal edge-cut, the converse is not true. For the graph H of Figure 5.7, consider the sets $X_1 = \{e_3, e_4, e_5\}$, $X_2 = \{e_1, e_2, e_6\}$ and $X_3 = \{e_1, e_6\}$ of edges of H. All three of these sets are edge-cuts because all of the graphs $H - X_1$, $H - X_2$ and $H - X_3$ are disconnected. Both X_1 and X_3 are minimal edge-cuts, while X_2 is not a minimal edge-cut as X_3 is a proper subset of X_2. The set X_3 is a minimum edge-cut, while X_1 and X_2 are not.

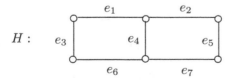

Figure 5.7: Illustrating edge-cuts in a graph

The **edge-connectivity** $\lambda(G)$ of a nontrivial graph G is the cardinality of a minimum edge-cut of G, while we define $\lambda(K_1) = 0$. (The symbol λ is the Greek letter lambda.) Thus $\lambda(G)$ is the minimum value of $|X|$ among all subsets X of

$E(G)$ such that $G - X$ is either a disconnected or trivial graph. For every graph G of order n,

$$0 \leq \lambda(G) \leq n - 1.$$

Note that $\lambda(G) = 0$ if and only if G is disconnected or G is trivial, while $\lambda(G) = 1$ if and only if G is connected and contains a bridge. For the graph H of Figure 5.7, $\lambda(G) = 2$ since $X_3 = \{e_1, e_6\}$ is a minimum edge-cut in H.

For a nonnegative integer k, a graph G is k-**edge-connected** if $\lambda(G) \geq k$. Consequently, every k-edge-connected graph is ℓ-edge-connected for every integer ℓ with $0 \leq \ell \leq k$. Therefore, every 1-edge-connected graph is nontrivial and connected and every 2-edge-connected graph is a connected graph of order 3 or more that contains no bridges. For example, the graphs G_3, G_4 and G_5 of Figure 5.1 are 2-edge-connected graphs that are not 3-edge-connected.

Theorem 5.10 *For every positive integer n, $\lambda(K_n) = n - 1$.*

Proof. By definition, $\lambda(K_1) = 0$. Let $G = K_n$ for $n \geq 2$. Since every vertex of G has degree $n - 1$, if we remove the $n - 1$ edges incident with a vertex, then a disconnected graph results. Thus $\lambda(G) \leq n - 1$. Now let X be a minimum edge-cut of G. So $|X| = \lambda(G)$. Then $G - X$ has exactly two components G_1 and G_2, where G_1 has order k, say, and G_2 has order $n - k$. Since (1) X consists of all edges joining G_1 and G_2 and (2) G is complete, it follows that $|X| = k(n-k)$. Because $k \geq 1$ and $n - k \geq 1$, we have $(k-1)(n-k-1) \geq 0$ and so

$$(k - 1)(n - k - 1) = k(n - k) - n + 1 \geq 0.$$

Hence $\lambda(G) = |X| = k(n - k) \geq n - 1$. Therefore, $\lambda(K_n) = n - 1$. ∎

As we noted earlier, complete graphs do not contain vertex-cuts. But for a graph G that is not complete, the cardinality of a minimum vertex-cut in G can never exceed the cardinality of any edge-cut in G. Indeed, the following theorem provides us with inequalities concerning the connectivity, edge-connectivity and minimum degree of a graph. The proof is similar, in part, to the argument we just used to show that $\lambda(K_n) = n - 1$ in Theorem 5.10.

Theorem 5.11 *For every graph G,*

$$\kappa(G) \leq \lambda(G) \leq \delta(G).$$

Proof. If G is disconnected or trivial, then $\kappa(G) = \lambda(G) = 0$ and the inequalities hold; while if $G = K_n$ for some integer $n \geq 2$, then $\kappa(G) = \lambda(G) = \delta(G) = n-1$. Thus we may assume that G is a connected graph of order $n \geq 3$ that is not complete. Hence $\delta(G) \leq n - 2$.

First, we show that $\lambda(G) \leq \delta(G)$. Let v be a vertex of G with $\deg v = \delta(G)$. Since the set of the $\delta(G)$ edges incident with v in G is an edge-cut of G, it follows that

$$\lambda(G) \leq \delta(G) \leq n - 2.$$

It remains to show that $\kappa(G) \leq \lambda(G)$. Let X be a minimum edge-cut of G. Then $|X| = \lambda(G) \leq n - 2$. Necessarily, $G - X$ contains exactly two components G_1 and G_2. Suppose that the order of G_1 is k. Thus the order of G_2 is $n - k$, where $k \geq 1$ and $n - k \geq 1$. Consequently, every edge in X joins a vertex of G_1 and a vertex of G_2. We consider two cases.

Case 1. *Every vertex of G_1 is adjacent in G to every vertex of G_2.* Thus $|X| = k(n - k)$. Since $(k - 1)(n - k - 1) \geq 0$, it follows that

$$(k - 1)(n - k - 1) = k(n - k) - n + 1 \geq 0$$

and so $\lambda(G) = |X| = k(n - k) \geq n - 1$. However, $\lambda(G) \leq n - 2$; so this case cannot occur.

Case 2. *There exist vertices u in G_1 and v in G_2 such that u and v are not adjacent in G.* We now define a set U of vertices of G. For each $e \in X$, we select a vertex for U in the following way. If u is incident with e, then choose the other vertex in G_2 that is incident with e as an element of U; otherwise, select the vertex that is incident with e and belongs to G_1 as an element of U. Then $|U| \leq |X|$. Since $u, v \notin U$ and there is no $u - v$ path in $G - U$, it follows that $G - U$ is disconnected and so U is a vertex-cut. Hence

$$\kappa(G) \leq |U| \leq |X| = \lambda(G),$$

as desired.   ∎

Theorem 5.11 is due to Hassler Whitney. Although a mathematician who made a number of important contributions to graph theory, Whitney was primarily interested in topology. Whitney was born in New York on March 23, 1907 and received his Ph.D. from Harvard in 1932 under the direction of George David Birkhoff. Whitney's doctoral dissertation was written in graph theory. He stayed on at Harvard until 1952, when he accepted an offer from the Institute for Advanced Study at Princeton. He remained there until he retired in 1977. During his life he was the recipient of many mathematical awards: the National Medal of Science (1976), the Wolf Prize (1983) and the Steel Prize (1985).

Although research was a large part of Whitney's professional life, he contributed to mathematics in many ways. During 1944-1949 he edited the *American Journal of Mathematics* and during 1949-1954 he edited *Mathematical Reviews*. During 1953-1956, he chaired the National Science Foundation mathematical panel. On the personal side, Whitney was an avid mountain climber. In fact, the Whitney-Gilman Ridge on Cannon Cliff in Franconia, New Hampshire was named for him and his cousin, who were the first to climb it (on August 3, 1939). Whitney died on May 10, 1989.

Both inequalities in Theorem 5.11 can be strict as the graph G in Figure 5.8 shows, where $\kappa(G) = 1$, $\lambda(G) = 2$ and $\delta(G) = 3$. For cubic graphs, however, the connectivity and edge-connectivity are always equal.

Theorem 5.12 *If G is a cubic graph, then $\kappa(G) = \lambda(G)$.*

$$G:$$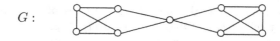

Figure 5.8: A graph G with $\kappa(G) = 1$, $\lambda(G) = 2$ and $\delta(G) = 3$

Proof. For a cubic graph G, it follows that $\kappa(G) = \lambda(G) = 0$ if and only if G is disconnected. If $\kappa(G) = 3$, then $\lambda(G) = 3$ by Theorem 5.11. So two cases remain, namely $\kappa(G) = 1$ or $\kappa(G) = 2$. Let U be a minimum vertex-cut of G. Then $|U| = 1$ or $|U| = 2$. So $G - U$ is disconnected. Let G_1 and G_2 be two components of $G - U$. Since G is cubic, for each $u \in U$, at least one of G_1 and G_2 contains exactly one neighbor of u.

Case 1. $\kappa(G) = |U| = 1$. Thus U consists of a cut-vertex u of G. Since some component of $G - U$ contains exactly one neighbor w of u, the edge uw is a bridge of G and so $\lambda(G) = \kappa(G) = 1$.

Case 2. $\kappa(G) = |U| = 2$. Let $U = \{u, v\}$. Assume that each of u and v has exactly one neighbor, say u' and v', respectively, in the same component of $G - U$. (This is the case that holds if $uv \in E(G)$.) Then $X = \{uu', vv'\}$ is an edge-cut of G and $\lambda(G) = \kappa(G) = 2$. (See Figure 5.9(a) for the situation when u and v are not adjacent.)

(a): 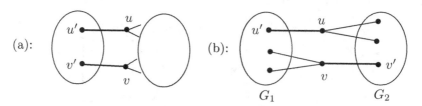 (b):

Figure 5.9: A step in the proof of Case 2

Hence we may assume that u has one neighbor u' in G_1 and two neighbors in G_2; while v has two neighbors in G_1 and one neighbor v' in G_2 (see Figure 5.9(b)). Therefore, $uv \notin E(G)$ and $X = \{uu', vv'\}$ is an edge-cut of G; so $\lambda(G) = \kappa(G) = 2$. ∎

The connectivity of a graph G of a given order n and size m can only be so large. For example, if $m < n - 1$, then G is disconnected by Theorem 4.7 and so $\kappa(G) = 0$. If $m \geq n - 1$, then there is a sharp upper bound for $\kappa(G)$, which we present next. (See Appendix 1 for a review of the floor and ceiling of a real number.)

Theorem 5.13 *If G is a graph of order n and size $m \geq n - 1$, then*

$$\kappa(G) \leq \left\lfloor \frac{2m}{n} \right\rfloor.$$

Proof. Since the sum of the degrees of the vertices of G is $2m$, the average degree of the vertices of G is $2m/n$ and so $\delta(G) \leq 2m/n$. Since $\delta(G)$ is an integer, $\delta(G) \leq \lfloor 2m/n \rfloor$. By Theorem 5.11, $\kappa(G) \leq \lfloor 2m/n \rfloor$. ∎

The bound given in Theorem 5.13 is sharp in the sense that for every two integers n and m with $1 \leq n - 1 \leq m \leq \binom{n}{2}$, there exists a graph G of order n and size m such that $\kappa(G) = \lfloor 2m/n \rfloor$. If $m = n - 1$, then every tree T of order n has the desired property as

$$\kappa(T) = \left\lfloor \frac{2m}{n} \right\rfloor = \left\lfloor \frac{2n-2}{n} \right\rfloor = 1.$$

Hence we may assume that $3 \leq n \leq m \leq \binom{n}{2}$.

It is useful to describe a class of graphs $H_{r,n}$ for integers r and n with $2 \leq r < n$ such that $H_{r,n}$ has order n, is "nearly" r-regular, has size m and $\kappa(H_{r,n}) = r = \lfloor 2m/n \rfloor$. These graphs are referred to as the **Harary graphs**, named for Frank Harary. (We will visit Harary again in Chapter 6). For the purpose of describing these graphs, we introduce a new concept.

Let G be a connected graph of diameter d. For an integer k with $1 \leq k \leq d$, the kth **power** G^k of G is the graph with $V(G^k) = V(G)$ such that uv is an edge of G^k if $1 \leq d_G(u,v) \leq k$. The graphs G^2 and G^3 are referred to as the **square** and **cube**, respectively, of G. The square C_8^2 of C_8 and the cube P_6^3 of P_6 are shown in Figure 5.10.

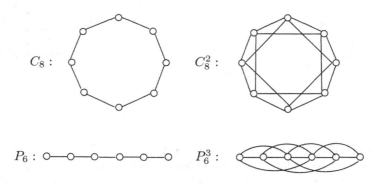

Figure 5.10: The square and cube of two graphs

If G is a connected graph, then not only is G^2 connected, it is 2-connected.

Theorem 5.14 *If G is a connected graph of order at least 3, then its square G^2 is 2-connected.*

We are now prepared to describe the Harary graphs $H_{r,n}$ for integers r and n with $2 \leq r < n$. First, let $V(H_{r,n}) = \{v_1, v_2, \ldots, v_n\}$; in fact, each graph $H_{r,n}$ contains the n-cycle $C = (v_1, v_2, \ldots, v_n, v_1)$. Suppose first that r is even, say $r = 2k$. Then $H_{r,n} = C^k$. In fact, the graph C_8^2 shown in Figure 5.10 is the Harary graph $H_{4,8}$. Hence for r is even, $H_{r,n}$ is an r-regular graph of order n.

Suppose next that r is odd, say $r = 2k + 1 \geq 3$. First assume that n is even, say $n = 2\ell$. Then $H_{r,n}$ consists of the graph $H_{r-1,n} = C^k$ together with all edges $v_i v_{i+\ell}$ for $1 \leq i \leq \ell$. Hence for r odd and n even, the Harary graph $H_{r,n}$ is also an r-regular graph of order n. The graph $H_{5,8}$ is shown in Figure 5.11. Next assume that n is odd, say $n = 2\ell + 1$, where then $\ell > k$. In this case, $H_{r,n}$ is obtained from $H_{r-1,n} = C^k$ by adding the edges $v_i v_{i+\ell+1}$ for $1 \leq i \leq \ell$ together with the edge $v_1 v_{1+\ell}$. Hence for r odd and n odd, the Harary graph $H_{r,n}$ is a graph of order n containing $n - 1$ vertices of degree r and one vertex of degree $r + 1$. The graph $H_{5,9}$ is shown in Figure 5.11. Our interest in the Harary graphs comes from the following fact.

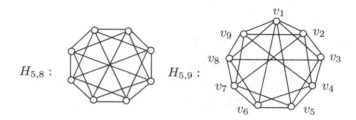

Figure 5.11: The Harary graphs $H_{5,8}$ and $H_{5,9}$

Theorem 5.15 *For every two integers r and n with $2 \leq r < n$,*

$$\kappa(H_{r,n}) = r.$$

If r is even or if r is odd and n is even, then $H_{r,n}$ is an r-regular graph of order n and so has size $m = rn/2$. Thus $\lfloor 2m/n \rfloor = r$. On the other hand, if r and n are both odd, then $H_{r,n}$ contains $n - 1$ vertices of degree r and one vertex of degree $r + 1$ and so $m = (rn + 1)/2$. In this case as well, $\lfloor 2m/n \rfloor = r$ and so by Theorem 5.13 $\kappa(H_{r,n}) \leq r$. In fact, $\kappa(H_{r,n}) \leq \delta(H_{r,n}) = r$. While it is a bit tedious to show that $\kappa(H_{r,n}) = r$ in general at this point, we will discuss this further in the next section.

Exercises for Section 5.3

5.17 Does it make sense to define the concept of a minimal vertex-cut in a graph? If so, how would this be defined and what would be a natural question to ask? If not, why is this the case?

5.18 Let PG be the Petersen graph. Give an example of

(a) a minimum vertex-cut in PG.

(b) a vertex-cut U in PG such that U is not a minimum vertex-cut of PG and no proper subset of U is a vertex-cut of PG.

5.19 Prove or disprove: Let G be a nontrivial graph. For every vertex v of G, $\kappa(G - v) = \kappa(G)$ or $\kappa(G - v) = \kappa(G) - 1$.

5.20 Let G be a connected graph of order $n \geq 4$ and let k be an integer with $2 \leq k \leq n - 2$.

    (a) Prove that if G is not k-connected, then G contains a vertex-cut U with $|U| = k - 1$.

    (b) Prove that if G is not k-edge-connected, then G contains an edge-cut X with $|X| = k - 1$.

5.21 Give an example of a graph with the following properties or explain why no such example exists.

    (a) a 2-connected graph that is not 3-connected.

    (b) a 3-connected graph that is not 2-connected.

    (c) a 2-edge-connected graph that is not 3-edge-connected.

    (d) a 3-edge-connected graph that is not 2-edge-connected.

5.22 (a) Prove that if G is a k-connected graph and e is an edge of G, then $G - e$ is $(k - 1)$-connected.

    (b) Prove that if G is a k-edge-connected graph and e is an edge of G, then $G - e$ is $(k - 1)$-edge-connected.

5.23 (a) Prove that if G is a k-connected graph, then $G + K_1$ is $(k + 1)$-connected.

    (b) Prove that if G is a k-edge-connected graph, then $G + K_1$ is $(k + 1)$-edge-connected.

5.24 Let G be a graph of order n and let k be an integer with $1 \leq k \leq n - 1$. Prove that if $\delta(G) \geq (n + k - 2)/2$, then G is k-connected.

5.25 Give an example of a graph G with the following properties or explain why no such example exists.

    (a) $\kappa(G) = 2$, $\lambda(G) = 3$ and $\delta(G) = 4$

    (b) $\kappa(G) = 3$, $\lambda(G) = 2$ and $\delta(G) = 4$

    (c) $\kappa(G) = 3$, $\lambda(G) = 3$ and $\delta(G) = 2$

    (d) $\kappa(G) = 2$, $\lambda(G) = 2$ and $\delta(G) = 3$

5.26 Prove that if G is a graph of order n such that $\delta(G) \geq (n - 1)/2$, then $\lambda(G) = \delta(G)$.

5.27 Prove or disprove:

    (a) If G is a graph with $\kappa(G) = k \geq 1$, then $G - U$ is disconnected for every set U of k vertices of G.

(b) If G is a graph with $\lambda(G) = k \geq 1$, then $G - X$ is disconnected for every set X of k edges of G.

(c) If G is a connected graph and U is a minimum vertex-cut, then $G - U$ contains exactly two components.

(d) If G is a graph of order n that is not complete and contains a vertex v of degree $n - 1$, then v belongs to every vertex-cut of G.

(e) If G is a graph of order n that contains a vertex v of degree $n - 1$, then every edge-cut of G contains an edge incident with v.

5.28 (a) Show that if G is a 0-regular graph, then $\kappa(G) = \lambda(G)$.

(b) Show that if G is a 1-regular graph, then $\kappa(G) = \lambda(G)$.

(c) Show that if G is a 2-regular graph, then $\kappa(G) = \lambda(G)$.

(d) By (a) – (c) and Theorem 5.12, if G is r-regular, where $0 \leq r \leq 3$, then $\kappa(G) = \lambda(G)$. Find the minimum positive integer r for which there exists an r-regular graph G such that $\kappa(G) \neq \lambda(G)$. Verify your answer.

(e) Find the minimum positive integer r for which there exists an r-regular graph G such that $\lambda(G) \geq \kappa(G) + 2$.

(f) The problem in (e) should suggest another question to you. Ask and answer such a question.

5.29 Give an example of

(a) a connected graph G such that every vertex of G belongs to a minimum vertex-cut but some edge of G belongs to no minimum edge-cut.

(b) a connected graph H such that every edge of H belongs to a minimum edge-cut but some vertex of H belongs to no minimum vertex-cut.

5.30 For a graph G, define $\overline{\kappa}(G) = \max\{\kappa(H)\}$ and $\overline{\lambda}(G) = \max\{\lambda(H)\}$, each maximized over all subgraphs H of G. How are $\overline{\kappa}(G)$ and $\overline{\lambda}(G)$ related to $\kappa(G)$ and $\lambda(G)$, respectively, and to each other?

5.31 In the graph G of Figure 5.12, the vertices represent street intersections and the edges represent roads.

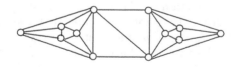

Figure 5.12: The graph in Exercise 5.31

(a) What is the maximum number k such that if road repairs are done at the same time to any k roads (making use of these roads impossible), then it is still possible to travel between every two intersections?

(b) What is the maximum number k such that if intersection repairs are done at the same time to any k intersections (making use of these intersections impossible), then it is still possible to travel between every two intersections that are not under repair?

5.32 Verify Theorem 5.14: *If G is a connected graph of order at least 3, then its square G^2 is 2-connected.*

5.4   Menger's Theorem

We have now seen that one measure of the connectedness of a graph is its vertex-connectivity, which depends on the minimum number of vertices that must be removed to result in a disconnected or trivial graph. We will see that connectivity can be looked at in another manner.

A set S of vertices of a graph G is said to **separate** two vertices u and v of G if $G - S$ is disconnected and u and v belong to different components of $G - S$. Thus, if S separates u and v, then surely u and v are nonadjacent vertices and S is a vertex-cut of G. Certainly, the cardinality of S must be at least as large as $\kappa(G)$. Such a set S is called a $u - v$ **separating set**. A $u - v$ separating set of minimum cardinality is called a **minimum $u - v$ separating set**. An **internal vertex** of a $u - v$ path P is a vertex of P different from u and v. A collection $\{P_1, P_2, \ldots, P_k\}$ of $u - v$ paths is called **internally disjoint** if every two of these paths have no vertices in common other than u and v.

There are many theorems in mathematics which state that the minimum number of elements in some set equals the maximum number of elements in some other set. The following theorem is such a "min-max" theorem. It is referred to as **Menger's theorem.**

Theorem 5.16 (**Menger's Theorem**) *Let u and v be nonadjacent vertices in a graph G. The minimum number of vertices in a $u - v$ separating set equals the maximum number of internally disjoint $u - v$ paths in G.*

Proof. We proceed by induction on the size of the graph. Certainly, the result is true vacuously for all empty graphs. Assume that the result is true for all graphs of size less than m, where m is a positive integer, and let G be a graph of size m. Let u and v be two nonadjacent vertices of G. Suppose that there are k vertices in a minimum $u - v$ separating set. Certainly, G can contain no more than k internally disjoint $u - v$ paths. We show, in fact, that G contains k internally disjoint $u - v$ paths. Since the result is true for $k = 0$ and $k = 1$, we may assume that $k \geq 2$. We consider three cases.

Case 1. There exists a minimum $u - v$ separating set U in G containing a vertex x that is adjacent to both u and v. Then the size of the subgraph $G - x$ is less than m and $U - \{x\}$ is a minimum $u - v$ separating set in $G - x$ consisting of $k - 1$ vertices. By the induction hypothesis, there are $k - 1$ internally disjoint $u - v$ paths in $G - x$. These paths together with the path (u, x, v) constitute k internally disjoint $u - v$ paths in G.

Case 2. There exists a minimum $u - v$ separating set W in G containing a vertex in W that is not adjacent to u and a vertex in W that is not adjacent to v. Let $W = \{w_1, w_2, \ldots, w_k\}$. Let G_u be the subgraph of G consisting of all $u - w_i$ paths in G, where only $w_i \in W$ for each i $(1 \le i \le k)$ and let G'_u be the graph obtained from G_u by adding a new vertex v' and joining v' to each vertex w_i for $1 \le i \le k$. Let G_v and G'_v be defined similarly, where G'_v is obtained from G_v by adding the new vertex u'. Representations of the graphs G_u, G'_u, G_v and G'_v are shown in Figure 5.13.

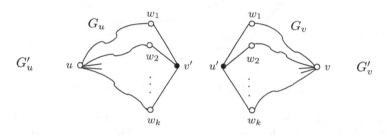

Figure 5.13: The graphs G_u, G'_u, G_v and G'_v in Case 2

Since W contains a vertex that is not adjacent to u and a vertex that is not adjacent to v, the size of each of the graphs G'_u and G'_v is less than m. Since W is a minimum $u - v$ separating set in G'_u, it follows by the induction hypothesis that G'_u contains k internally disjoint $u - v'$ paths, each consisting of a $u - w_i$ path P_i followed by the edge $w_i v'$. Similarly, there are k internally disjoint $u' - v$ paths in G'_v, each consisting of a $w_i - v$ path Q_i preceded by the edge $u' w_i$. Since W is a $u - v$ separating set in G, the two graphs G_u and G_v have only the vertices of W in common. Therefore, the k paths obtained by following P_i by Q_i for each i $(1 \le i \le k)$ are internally disjoint $u - v$ paths in G.

Case 3. For each minimum $u - v$ separating set S in G, either every vertex of S is adjacent to u and not adjacent to v or every vertex of S is adjacent to v and not adjacent to u. Let $P = (u, x, y, \ldots, v)$ be a $u - v$ geodesic in G and let $e = xy$. Consider the subgraph $G - e$ in G. Certainly, every minimum $u - v$ separating set in $G - e$ contains at least $k - 1$ vertices. We claim, in fact, that a minimum $u - v$ separating set in $G - e$ contains k vertices, for assume, to the contrary, that $G - e$ contains a minimum $u - v$ separating set $Z = \{z_1, z_2, \ldots, z_{k-1}\}$. Then $Z \cup \{x\}$ is a minimum $u - v$ separating set in G. Since x is adjacent to u, each vertex z_i $(1 \le i \le k - 1)$ is also adjacent to u and not adjacent to v. On the other hand, $Z \cup \{y\}$ is a minimum $u - v$ separating set in G and each z_i $(1 \le i \le k - 1)$ is adjacent to u. This implies that y is also adjacent to u, which contradicts the fact that P is a $u - v$ geodesic. Therefore, as claimed, k is the minimum number of vertices in a $u - v$ separating set in $G - e$. By the induction hypothesis, there are k internally disjoint $u - v$ paths in $G - e$. Hence there are k internally disjoint $u - v$ paths in G as well.   ∎

In the graph G of Figure 5.14, $U = \{w_1, w_2, w_3\}$ is a $u - v$ separating set. Since there is no $u - v$ separating set with fewer than three vertices, U is a minimum $u - v$ separating set. By Theorem 5.16, there are three internally disjoint $u - v$ paths in G (as indicated in bold in Figure 5.14).

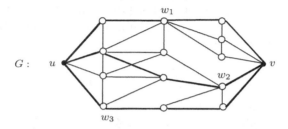

Figure 5.14: Illustrating Menger's theorem

As we mentioned above, Theorem 5.16 is referred to as Menger's theorem, named for Karl Menger who was born in Vienna, Austria on January 13, 1902. Menger developed a talent for mathematics and physics at an early age and entered the University of Vienna in 1920 to study physics. The following year he attended a lecture by Hans Hahn on *Neueres über den Kurvenbegriff* (*What's new concerning the concept of a curve*) and Menger's interests turned towards mathematics. In the lecture it was mentioned that there was no satisfactory definition of a curve (at that time), despite unsuccessful attempts to do so by a number of distinguished mathematicians, including Georg Cantor, Camille Jordan and Giuseppe Peano. Some mathematicians including Felix Hausdorff and Ludwig Bieberbach felt that it was unlikely that this problem would ever be solved. Despite being an undergraduate with limited mathematical background, Menger solved the problem and presented his solution to Hahn. This led Menger to work on curve and dimension theory.

Menger became quite ill while a student and his studies were interrupted. During this time both of his parents died. He eventually returned to the university where he completed his studies under Hahn in 1924. The next year Menger went to Amsterdam to work with Luitzen Brouwer and Menger broadened his mathematical interests. In 1927 Menger returned to the University of Vienna to accept the position of Chair of Geometry. It was during that year that he published the paper "Zur allgemeinen Kurventheorie" (Menger's theorem). Menger himself referred to this result as the "n-arc theorem" and proved it as a lemma for a theorem in curve theory.

In the spring of 1930, Karl Menger traveled to Budapest and met many Hungarian mathematicians, including Dénes König. Menger had read some of

König's papers. During his visit, Menger learned that König was working on a book that would contain what was known at that time about graph theory. Menger was pleased to hear this and mentioned his theorem to König, which had only been published three years earlier. König was not aware of Menger's work and, in fact, didn't believe that the theorem was true. Indeed, the very evening of their meeting, König set out to construct a counterexample. When the two met again the next day, König greeted Menger with "A sleepless night!". König then asked Menger to describe his proof, which he did. After that, König said that he would add a final section to his book on the theorem, which he did. This was a major reason why Menger's theorem became so widely known among those interested in graph theory.

Because of the political situation in Austria in 1938, Menger left for a position in the United States at the University of Notre Dame. While there, his work emphasized geometry, which didn't have the impact of his earlier work since geometry was not a subject of great interest to many, especially in the United States. Menger went to the Illinois Institute of Technology in 1948 and spent the rest of his life in the Chicago area. Menger was considered one of the leading mathematicians of the 20th century. He died on October 5, 1985.

With the aid of Menger's theorem, Hassler Whitney was able to present a characterization of k-connected graphs.

Theorem 5.17 *A nontrivial graph G is k-connected for some integer $k \geq 2$ if and only if for each pair u, v of distinct vertices of G there are at least k internally disjoint $u - v$ paths in G.*

Proof. Since the result holds if G is complete, we may assume that G is not complete. Assume that G is a k-connected graph, where $k \geq 2$. Let u and v be two distinct vertices of G. Suppose first that u and v are not adjacent and let U be a minimum $u - v$ separating set. Then $|U| \geq \kappa(G) \geq k$. By Theorem 5.16, G contains at least k internally disjoint $u - v$ paths. Next, suppose that u and v are adjacent, where $e = uv$. Then $G - e$ is $(k-1)$-connected (see Exercise 5.22). Let W be a minimum $u - v$ separating set in $G - e$. Thus

$$|W| \geq \kappa(G - e) \geq k - 1.$$

By Theorem 5.16, $G - e$ contains at least $k - 1$ internally disjoint $u - v$ paths, and so G contains at least k internally disjoint $u - v$ paths.

For the converse, assume that G is a graph containing at least k internally disjoint $u - v$ paths for every pair u, v of distinct vertices of G. Let U be a minimum vertex-cut of G. Then $|U| = \kappa(G)$. Let x and y be vertices in distinct components of $G - U$. Thus U is an $x - y$ separating set of G. Since there are at least k internally disjoint $x - y$ paths in G, it follows by Theorem 5.16 that $\kappa(G) = |U| \geq k$. Therefore, G is k-connected. ∎

Whitney's theorem (Theorem 5.17) then gives us an alternative method for determining the connectivity of a graph G. Not only is $\kappa(G)$ the minimum number of vertices whose removal from G results in a disconnected or trivial graph but $\kappa(G)$ is the maximum positive integer k for which every two vertices u and v in G are connected by k internally disjoint $u - v$ paths in G.

In the preceding section we described for integers r and n with $2 \le r < n$ a class of graphs $H_{r,n}$ of order n called the Harary graphs. Each such graph contains a n-cycle $C = (v_1, v_2, \ldots, v_n, v_1)$. If $r = 2k$ is even, then $H_{r,n}$ is the kth power C^k of C. If $r = 2k + 1$ is odd and $n = 2\ell$ is even, then $H_{r,n}$ consists of $H_{r-1,n} = C^k$ together with the edges $v_i v_{i+\ell}$ for $1 \le i \le \ell$; while if $r = 2k + 1$ and $n = 2\ell + 1$ are both odd, then $H_{r,n}$ consists of $H_{r-1,n} = C^k$ together with the edges $v_i v_{i+\ell+1}$ for $1 \le i \le \ell$ and the additional edge $v_1 v_{1+\ell}$. We saw that $\delta(H_{r,n}) = r$ in each case and that $\lfloor 2m/n \rfloor = r$; thus $\kappa(H_{r,n}) \le r$. Theorem 5.15 states that the connectivity of $H_{r,n}$ is, in fact, r for all such r and n. To verify that $\kappa(H_{r,n}) = r$, we need only show that (1) the removal of fewer than r vertices from $H_{r,n}$ results in a nontrivial connected graph or (2) every two vertices u and v in $H_{r,n}$ are connected by r internally disjoint $u - v$ paths (see Exercise 5.38).

The following result indicates one way of constructing a k-connected graph from a given k-connected graph. With the aid of Whitney's theorem, Theorem 5.18 can then be used to provide a proof of Corollary 5.19. Proofs of both Theorem 5.18 and Corollary 5.19 are left as exercises.

Theorem 5.18 *Let G be a k-connected graph and let S be any set of k vertices. If a graph H is obtained from G by adding a new vertex w and joining w to the vertices of S, then H is also k-connected.*

We have mentioned that a collection $\{P_1, P_2, \ldots, P_k\}$ of $u - v$ paths are internally disjoint if every two distinct paths in the collection have only u and v in common, that is, each internal vertex in one of the paths lies on no other path in the collection. More generally, for $k + 1$ distinct vertices $u, v_1, v_2, \ldots, v_k$, a collection $\{P_1, P_2, \ldots, P_k\}$ of k paths, where P_i is a $u - v_i$ path $(1 \le i \le k)$, are **internally disjoint** if every two distinct paths in the collection have only u in common. For $2k$ distinct vertices $u_1, u_2, \ldots, u_k, v_1, v_2, \ldots, v_k$, a collection $\{P_1, P_2, \ldots, P_k\}$ of k paths, where P_i is a $u_i - v_i$ path $(1 \le i \le k)$, are **disjoint** if no two distinct paths in the collection have a vertex in common.

Corollary 5.19 *If G is a k-connected graph and $u, v_1, v_2, \ldots, v_k$ are $k + 1$ distinct vertices of G, then there exist internally disjoint $u - v_i$ paths $(1 \le i \le k)$ in G.*

We saw in Theorem 5.7 that if a graph G is a 2-connected graph (that is, a nonseparable graph of order at least 3), then every two vertices of G lie on a common cycle. Gabriel Dirac obtained a generalization of this result for k-connected graphs.

Theorem 5.20 *If G is a k-connected graph, $k \ge 2$, then every k vertices of G lie on a common cycle of G.*

Proof. Let $S = \{v_1, v_2, \ldots, v_k\}$ be a set of k vertices of G. We show that there exists a cycle in G containing every vertex of S. Among all cycles in G, let C

be one containing a maximum number ℓ of vertices of S. We claim that $\ell = k$. Assume, to the contrary, that $\ell < k$. Since G is k-connected, $k \geq 2$, it follows that G is 2-connected and so $2 \leq \ell < k$ by Theorem 5.7. We may assume that C contains the vertices $v_1, v_2, \ldots, v_\ell$ of S and that the vertices of S on C appear in the order $v_1, v_2, \ldots, v_\ell$ as we proceed cyclically about C.

Since $\ell < k$, there is a vertex $u \in S$ that does not belong to C. Furthermore, since $2 \leq \ell < k$, the graph G is ℓ-connected as well. Suppose first that the order of C is ℓ. Applying Corollary 5.19 to the vertices $u, v_1, v_2, \ldots, v_\ell$, we see that G contains internally disjoint $u - v_i$ paths P_i $(1 \leq i \leq \ell)$. Replacing the edge $v_1 v_2$ by P_1 and P_2 produces a cycle containing the vertices $u, v_1, v_2, \ldots, v_\ell$, which gives a contradiction.

Hence we may assume that C contains a vertex $v_0 \notin S$. Since $2 \leq \ell + 1 \leq k$, the graph G is $(\ell + 1)$-connected. Applying Corollary 5.19 to the vertices $u, v_0, v_1, v_2, \ldots, v_\ell$, we see that G contains internally disjoint $u - v_i$ paths P_i $(0 \leq i \leq \ell)$. Let v_i' $(0 \leq i \leq \ell)$ be the first vertex of P_i that belongs to C (possibly $v_i' = v_i$) and let P_i' be the $u - v_i'$ subpath of P_i. Since there are $\ell + 1$ paths P_i' and ℓ vertices of C that belong to S, there are distinct vertices v_r' and v_t', where $0 \leq r, t \leq \ell$, for which there is a $v_r' - v_t'$ path P' on C having no interior vertices belonging to S. Deleting the interior vertices of P' from C and adding the paths P_r' and P_t' produces a cycle containing the vertices $u, v_1, v_2, \ldots, v_\ell$, which is a contradiction. ∎

All of the results mentioned in this section concern connectivity. There are also edge-connectivity analogues of both Theorems 5.16 and 5.17. We state these next.

Theorem 5.21 *For distinct vertices u and v in a graph G, the minimum number of edges of G that separate u and v equals the maximum number of pairwise edge-disjoint $u - v$ paths in G.*

Theorem 5.22 *A nontrivial graph G is k-edge-connected if and only if G contains k pairwise edge-disjoint $u - v$ paths for each pair u, v of distinct vertices of G.*

Exercises for Section 5.4

5.33 Let G be a 5-connected graph and let u, v and w be three distinct vertices of G. Prove that G contains two cycles C and C' that have only u and v in common but neither of which contains w.

5.34 Prove Theorem 5.18: *Let G be a k-connected graph and let S be any set of k vertices. If a graph H is obtained from G by adding a new vertex w and joining w to the vertices of S, then H is also k-connected.*

5.35 Prove Corollary 5.19: *If G is a k-connected graph and u, v_1, v_2, ..., v_k are $k+1$ distinct vertices of G, then there exist internally disjoint $u - v_i$ paths $(1 \leq i \leq k)$ in G.*

5.36 Let G be a k-connected graph of order $n \geq 2k$ and let U and W be two disjoint sets of k vertices of G. Prove that there exist k disjoint paths connecting U and W.

5.37 Determine the connectivity and edge-connectivity of the n-cube Q_n.

5.38 Verify that $\kappa(H_{r,n}) = r$ by showing that every two vertices u and v in $H_{r,n}$ are connected by r internally disjoint $u - v$ paths for the following values r and n. (a) $r = 3$ and $n = 8$. (b) $r = 4$ and $n = 8$.

5.5   Exploration: Powers and Edge Labelings

We have seen for integers r and n with $2 \leq r < n$ that the Harary graphs $H_{r,n}$ have minimum degree r, order n and size m with $\kappa(G) = r = \lfloor 2m/n \rfloor$. These graphs were defined in terms of powers of cycles. We also mentioned that the square of every connected graph of order 3 or more is 2-connected. In fact, if k and n are integers with $2 \leq k < n$ and G is a connected graph of order n, then G^k is k-connected (see Exercise 5.39).

In particular, for every connected graph G of order $n \geq 3$ and every two distinct vertices u and v of G, there exist in G^2 two internally disjoint $u - v$ paths P and P'. Of course, each edge of P and P' belongs either to G or to $G^2 - E(G)$. Label each edge of G^2 that belongs to G with the label 1 and label each edge of G^2 not belonging to G with the label 2. In general, an edge uv of G^k is labeled i $(1 \leq i \leq k)$ if $d_G(u,v) = i$. Such a graph G^k is then called a **distance-labeled graph**. A path P in G^k is called **proper** if every two adjacent edges in P have different labels. Later we will see that this is related to the topic of graph colorings (discussed in Chapter 10). While for every connected graph G of diameter 2 or more, the graph G^2 contains a proper $u - v$ path for every two vertices u and v, the graph G^2 need not contain two internally disjoint proper $u - v$ paths. (See Exercises 5.40 and 5.41.)

Let G be a connected graph of diameter $d \geq 2$ and let k be an integer with $2 \leq k \leq d$. If G^k is a distance-labeled graph and e is an edge labeled j where $1 < j \leq k$, then e is necessarily adjacent to an edge labeled i for every integer i with $1 \leq i < j$ (see Exercise 5.42). This property possessed by distance-labeled graphs G^k suggests other concepts.

Let G be a connected graph. By a **proper edge labeling** of G we mean a labeling of the edges of G from the set $\{1, 2, \ldots, k\}$ for some positive integer k such that no two adjacent edges are labeled the same.

We now turn our attention to proper edge labelings of trees. A proper edge labeling of a tree T from the elements of the set $\{1, 2, \ldots, k\}$ is called a **Grundy labeling** if the labeling has the property that whenever an edge e is labeled j where $1 < j \le k$, then e is adjacent to an edge labeled i for every integer i with $1 \le i < j$. Grundy labelings are named for Patrick Michael Grundy (1917–1959), whose interests included combinatorial games. The maximum positive integer k for which a tree T has a Grundy labeling from the set $\{1, 2, \ldots, k\}$ is called the **Grundy index** $\Gamma'(T)$ of T. For every tree T, the Grundy index $\Gamma'(T)$ exists and $\Delta(T) \le \Gamma'(T) \le 2\Delta(T) - 1$. (See Exercise 5.43.) For example, $\Gamma'(T) = 4$ for the tree T of Figure 5.15.

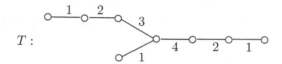

T:

Figure 5.15: A tree with Grundy index 4

By an **achromatic edge labeling** of a tree T we mean a proper edge labeling of T from the elements of the set $S = \{1, 2, \ldots, k\}$ for some positive integer k such that for every two distinct elements i and j of S, there exist adjacent edges labeled i and j. The maximum k for which T has an achromatic edge labeling is called the **achromatic index** of T and is denoted by $\psi'(T)$. For example, $\psi'(P_9) = 4$ (see Figure 5.16).

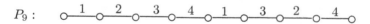

P_9:

Figure 5.16: An achromatic edge labeling of P_9

Exercises for Section 5.5

5.39 Prove that if G is a connected graph of order n and k is an integer with $2 \le k < n$, then G^k is k-connected.

5.40 Show for every connected graph G of diameter 2 or more and every two vertices u and v in G that G^2 contains a proper $u - v$ path but not necessarily two internally disjoint proper $u - v$ paths.

5.41 Show for every connected graph G of diameter 3 or more and every two vertices u and v in G that G^3 contains two internally disjoint proper $u - v$ paths but not necessarily three internally disjoint proper $u - v$ paths.

5.42 Let G be a connected graph of diameter $d \ge 2$ and let k be an integer with $2 \le k \le d$. Prove that if G^k is a distance-labeled graph and e is an edge labeled j where $1 < j \leq k$, then e is adjacent to an edge labeled i for every integer i with $1 \leq i < j$.

5.43 Show that the Grundy index $\Gamma'(T)$ exists for every tree T and that

$$\Delta(T) \leq \Gamma'(T) \leq 2\Delta(T) - 1.$$

5.44 Does there exist an integer $n < 9$ such that $\psi'(P_n) = 4$?

5.45 Show that $\Gamma'(T) \leq \psi'(T)$ for every nontrivial tree T.

5.46 Give an example of a nontrivial tree T such that $\Gamma'(T) \neq \psi'(T)$.

5.47 Show that if $\psi'(T) = k$ for some tree T, then $\sum_{v \in V(T)} \binom{\deg v}{2} \geq \binom{k}{2}$.

Chapter 6

Traversability

6.1 Eulerian Graphs

Figure 6.1(a) shows the layout of a housing development in a community, where mailboxes are placed along one side of each street (indicated by double lines in the diagram). Can a letter carrier make a round trip through the development and pass by each mailbox but once? Figure 6.1(b) shows that the answer to this question is yes.

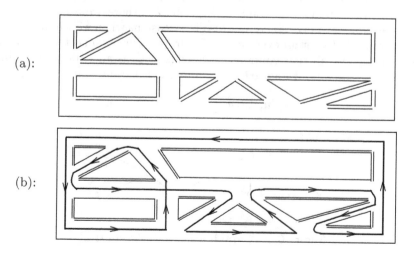

(a):

(b):

Figure 6.1: A housing development and a route of a letter carrier

The streets in the housing development can be represented quite naturally by the graph G of Figure 6.2, where the vertices represent the street intersections and the edges represent the streets. In terms of graph theory, the question for

the letter carrier can be rephrased as follows: Does the graph G of Figure 6.2
have a circuit that contains every edge of G? The letter carrier's route shown in
Figure 6.1(b) tells us that the circuit

$$C = (v_1, v_2, v_3, v_4, v_5, v_6, v_7, v_2, v_4, v_7, v_8, v_6, v_9, v_8, v_{11}, v_9, v_{10}, v_{11}, v_1)$$

in the graph G of Figure 6.2 has the desired property.

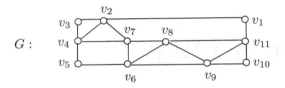

Figure 6.2: Modeling the streets of a housing development by a graph

This brings us to the main topic of this section. A circuit C in a graph G
is called an **Eulerian circuit** (pronounced oy-LEER-e-an) if C contains every
edge of G. Since no edge is repeated in a circuit, every edge appears exactly once
in an Eulerian circuit. Certainly, only graphs with one nontrivial component can
contain such a circuit. For this reason, we restrict ourselves to connected graphs
when investigating the question of whether a graph has an Eulerian circuit. A
connected graph that contains an Eulerian circuit is called an **Eulerian graph**.
In particular, the graph G of Figure 6.2 is an Eulerian graph.

An Eulerian circuit in a connected graph G is therefore a closed trail that
contains every edge of G. There will also be occasions when we will be interested
in *open* trails that contain every edge of a graph. For a connected graph G, we
refer to an open trail that contains every edge of G as an **Eulerian trail**. For
example, the graph G of Figure 6.3 contains the Eulerian trail

$$T = (u, s, t, u, v, w, y, u, w, x, z, w).$$

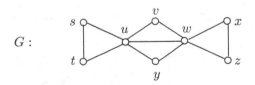

Figure 6.3: A graph with an Eulerian trail

To understand why the adjective "Eulerian" is used here, let us go back several years, indeed a few centuries, to 17th and 18th century Switzerland, when
thirteen members of the remarkable Bernoulli family became distinguished mathematicians. Among the most prominent were two brothers, Jaques and Jean, the
latter also known as John or Johann. Although the accomplishments of Johann

and the other Bernoullis were numerous, one of Johann's major accomplishments may have been to convince the father of a young Leonhard Euler (pronounced OY-ler) to have his son discontinue studying theology and study mathematics instead. Later Johann Bernoulli became the mathematical advisor of Euler. When one individual serves as the academic advisor (usually doctoral advisor) of another, the advisor is referred by some as the *academic father* (or *academic mother*) of the student. This provides a sense of "family" for teacher and student. Hence Johann Bernoulli could be called the academic father of Euler.

Euler was born in Basel, Switzerland on April 15, 1707. While in his 20s, he became ill and lost vision in one of his eyes. Later, he developed a cataract in his other eye and spent the last few years of his life totally blind. However, just as the magnificent composer Ludwig van Beethoven did much of his work while totally deaf, Euler did much of his mathematical research while totally blind. During his lifetime, more than 500 research papers and books of his were published. After his death (from a stroke) on September 18, 1783, another 400 were published. At that time and for a good many years afterwards, Euler had more publications than any other mathematician, only to be exceeded in the 20th century by Paul Erdős (pronounced AIR-dish), an academic descendant of Euler. We will visit Paul Erdős later.

While Euler made significant contributions to every area of mathematics that existed during his lifetime, it is a contribution he made to an area of mathematics that did not exist during his lifetime that is of primary interest to us here.

The city of Königsberg, located in Prussia, would play an interesting role in Euler's life and in the history of graph theory. The River Pregel flowed through Königsberg, separating it into four land areas. Seven bridges were built over the river that allowed the citizens of Königsberg to travel between these land areas. A map of Königsberg, showing the four land areas (labeled A, B, C, D), the location of the river and the seven bridges at that time are given in Figure 6.4.

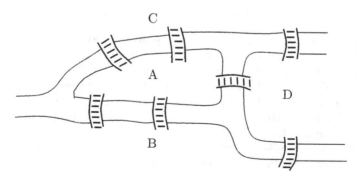

Figure 6.4: Königsberg in the early 18th century

The story goes that the citizens of Königsberg enjoyed going for walks throughout the city. Evidently, some citizens wondered whether it was possible to go for a walk in Königsberg and pass over each bridge exactly once. This became known as the **Königsberg Bridge Problem**. Evidently this problem remained unsolved for some time and became well known throughout the region. This problem eventually came to the attention of Euler (who was believed to be in St. Petersburg at the time). Although the subject of graph theory did not yet exist and certainly Euler's solution of this problem did not involve graphs, his solution had overtones of what would become graph theory. In particular, the situation in Königsberg can be represented by the multigraph in Figure 6.5.

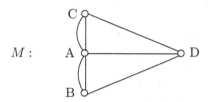

Figure 6.5: The Königsberg multigraph

When considering a walk W in some multigraph M, it is not enough to express W as a sequence of vertices. For example, if u, v are consecutive vertices in W and there are parallel edges joining u and v in M, then it must somehow be indicated which edge is being traversed, such as u, e, v.

In graph theory terms, the Königsberg Bridge Problem is: Does the multigraph M of Figure 6.5 contain an Eulerian circuit or an Eulerian trail? (Of course, Euler didn't use these terms.) Suppose that such a journey over the seven bridges of Königsberg was possible. Then it must begin at some land area and end at a land area (possibly the same one). Such a journey must therefore result in a trail of length 7 (one edge for each bridge) that encounters eight vertices (including repetition), namely,

$$T = (v_1, v_2, v_3, v_4, v_5, v_6, v_7, v_8).$$

Each of the vertices v_i $(1 \le i \le 8)$ represents a land area (A, B, C or D). Certainly each land area must appear in T. Observe that all four vertices of the multigraph M have odd degree. At least two vertices of M are neither the initial nor the terminal vertex of T. Thus each time that such a vertex is entered along T, it is exited. However, this implies that the vertex has even degree, which is impossible. Consequently, a walk that passes over every bridge in Königsberg exactly once is impossible as well.

Königsberg was founded in 1255 and was the capital of German East Prussia. The Prussian Royal Castle was located in Königsberg but it was destroyed during World War II, as was much of the city. Because of the outcome of the war, it was decided at the Potsdam Conference in 1945 that a region located between Poland and Lithuania, containing the city of Königsberg, should be made part of Russia. In 1946, Königsberg was renamed Kaliningrad after Mikhail Kalinin, the formal leader of the Soviet Union during 1919-1946. After the fall of the USSR, Lithuania and other former Soviet republics became independent and

Kaliningrad was no longer part of Russia. None of the attempts to change its name back to Königsberg have been successful.

Returning to the multigraph M of Figure 6.5, we are led to ask what characteristic it possesses that doesn't allow it to have an Eulerian circuit or an Eulerian trail. We have already mentioned that all four vertices of M have odd degree. This, it turns out, is the key observation. For a connected graph G to be Eulerian, it is both necessary and sufficient that each vertex of G has even degree.

Theorem 6.1 *A nontrivial connected graph G is Eulerian if and only if every vertex of G has even degree.*

Proof. Assume first that G is Eulerian. Then G contains an Eulerian circuit C. Suppose that C begins at the vertex u (and therefore ends at u). We show that every vertex of G is even. Let v be a vertex of G different from u. Since C neither begins nor ends at v, each time that v is encountered on C, two edges are accounted for (one to enter v and another to exit v). Thus v has even degree. Now to u. Since C begins at u, this accounts for one edge. Another edge is accounted for because C ends at u. If u is encountered at other times, two edges are accounted for. So u is even as well.

For the converse, assume that G is a nontrivial connected graph in which every vertex is even. We show that G contains an Eulerian circuit. Among all trails in G, let T be one of maximum length. Suppose that T is a $u - v$ trail. We claim that $u = v$. If not, then T ends at v. It is possible that v may have been encountered earlier in T. Each such encounter involves two edges of G, one to enter v and another to exit v. Since T ends at v, an odd number of edges at v has been encountered. But v has even degree. This means that there is at least one edge at v, say vw, that does not appear on T. But then T can be extended to w, contradicting the assumption that T has maximum length. Thus T is a $u - u$ trail, that is, $C = T$ is a $u - u$ circuit. If C contains all edges of G, then C is an Eulerian circuit and the proof is complete.

Suppose then that C does not contain all edges of G, that is, there are some edges of G that do not lie on C. Since G is connected, some edge $e = xy$ not on C is incident with a vertex x that is on C. Let $H = G - E(C)$, that is, H is the spanning subgraph of G obtained by deleting the edges of C. Every vertex of C is incident with an even number of edges on C. Since every vertex of G has even degree, every vertex of H has even degree. It is possible, however, that H is disconnected. On the other hand, H has at least one nontrivial component, namely, the component H_1 of H containing the edge xy. This means that H_1 is connected and every vertex of H_1 has even degree. Consider a trail of maximum length in H_1, beginning at x. As we just saw, this trail must also end at x and is an $x - x$ circuit C' of H_1.

Now if, in the circuit C, we were to attach C' when we arrive at x, we obtain a circuit C'' in G of greater length than C, which is a contradiction. This implies that C contains all edges of G and is an Eulerian circuit. ∎

Although Theorem 6.1 is credited to Leonhard Euler, as indeed it should, Euler's proof was incomplete. Euler failed to show that if every vertex of a connected graph G is even, then G is Eulerian. While Euler did not give a proof with the care and precision that is normally done today, it is quite likely that Euler felt this implication was clear. Nevertheless, it wasn't until 1873 that the missing portion of the proof was published. The proof was completed by Carl Hierholzer. Actually at that time Hierholzer had died but he had told a colleague what he had done and his colleague kindly wrote the paper for him and listed Hierholzer as its sole author.

With the aid of Theorem 6.1, it is now easy to characterize graphs possessing an Eulerian trail.

Corollary 6.2 *A connected graph G contains an Eulerian trail if and only if exactly two vertices of G have odd degree. Furthermore, each Eulerian trail of G begins at one of these odd vertices and ends at the other.*

Proof. Assume first that G contains an Eulerian trail T. Thus T is a $u - v$ trail for some distinct vertices u and v. We now construct a new connected graph H from G by adding a new vertex x of degree 2 and joining it to u and v. Then $C = (T, x, u)$ is an Eulerian circuit in H. By Theorem 6.1, every vertex of H is even and so only u and v have odd degrees in $G = H - x$.

For the converse, we proceed in a similar manner. Let G be a connected graph containing exactly two vertices u and v of odd degree. We show that G contains an Eulerian trail T, where T is either a $u - v$ trail or a $v - u$ trail. Add a new vertex x of degree 2 to G and join it to u and v, calling the resulting graph H. Therefore, H is a connected graph all of whose vertices are even. By Theorem 6.1, H is an Eulerian graph containing an Eulerian circuit C. Since it is irrelevant which vertex of C is the initial (and terminal) vertex, we assume that C is an $x - x$ circuit. Since x is incident only with the edges ux and vx, one of these is the first edge of C and the other is the final edge of C. Deleting x from C results in an Eulerian trail T of G that begins either at u or v and ends at the other. ∎

As a result of Theorem 6.1 and Corollary 6.2, it is now relatively easy to determine whether a graph contains an Eulerian circuit or an Eulerian trail. Furthermore, Theorem 6.1 and Corollary 6.2 both hold if "graph" is replaced by "multigraph."

Certainly, C_n, $n \geq 3$, is Eulerian and P_n, $n \geq 2$, contains an Eulerian trail. A complete graph is Eulerian if and only if $n \geq 3$ and n is odd; while $K_{s,t}$ is Eulerian if and only if s and t are both even. The graph $K_{2,t}$ contains an Eulerian trail if and only if t is odd. The n-cube Q_n is Eulerian if and only if $n \geq 2$ and n is even. We have seen that the n-cube Q_n, $n \geq 2$, is defined as the Cartesian product of the $(n-1)$-cube Q_{n-1} and K_2. This suggests a question.

Example 6.3 *Find a necessary and sufficient condition for the Cartesian product $G \times H$ of two nontrivial connected graphs G and H to be Eulerian.*

Solution. We can think of $G \times H$ as being constructed by replacing each vertex v of G by a copy H_v of H. Let x be a vertex of $G \times H$. Then x belongs to H_v for some vertex v of G. The vertex x is adjacent to its neighbors in H_v as well as to one vertex in H_u for every neighbor u of v in G. Thus

$$\deg_{G \times H} x = \deg_{H_v} x + \deg_G v.$$

Hence $\deg_{G \times H} x$ is even if and only if $\deg_{H_v} x$ and $\deg_G v$ are both even or both odd (that is, they are of the same parity).

If $\deg_{H_v} x$ is even, then $\deg_G v$ is even for every vertex v of G; while if $\deg_{H_v} x$ is odd, then $\deg_G v$ is odd for every vertex v of G. We have therefore arrived at the following:

Let G and H be nontrivial connected graphs. Then $G \times H$ is Eulerian if and only if both G and H are Eulerian or every vertex of G and H is odd. ◇

Exercises for Section 6.1

6.1 The diagram of Figure 6.6 shows the nine rooms on the second floor of a large house with doorways between various rooms. Is it possible to start in some room and go for a walk so that each doorway is passed through exactly once? How is this question related to graph theory?

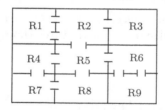

Figure 6.6: The diagram in Exercise 6.1

6.2 Let G_1 and G_2 be two Eulerian graphs with no vertex in common. Let $v_1 \in V(G_1)$ and $v_2 \in V(G_2)$. Let G be the graph obtained from $G_1 \cup G_2$ by adding the edge $v_1 v_2$. What can be said about G?

6.3 Let G_1, G_2 and G_3 be pairwise disjoint connected regular graphs and let $G = G_1 + (G_2 + G_3)$ be the graph obtained from G_1, G_2 and G_3 by adding edges between every two vertices belonging to two of G_1, G_2 and G_3. Prove that if G_1 and $\overline{G}_1$ are Eulerian but G_2 and G_3 are not Eulerian, then G is Eulerian.

6.4 Give an example of a graph G such that

(a) both G and $\overline{G}$ are Eulerian.

(b) G is Eulerian but $\overline{G}$ is not.

(c) neither G nor $\overline{G}$ is Eulerian and both G and $\overline{G}$ contain an Eulerian trail.

(d) neither G nor $\overline{G}$ is Eulerian, but G contains an Eulerian trail and $\overline{G}$ does not.

(e) G contains an Eulerian trail and an edge e such that $G - e$ is Eulerian.

6.5 Only one graph of order 5 has the property that the addition of any edge produces an Eulerian graph. What is it?

6.6 Let G be a connected regular graph that is not Eulerian. Prove that if $\overline{G}$ is connected, then $\overline{G}$ is Eulerian.

6.7 Let G be an r-regular graph of odd order n and let $F \cong \overline{G}$, where F and G have disjoint vertex sets. A graph H is constructed from G and F by adding two new vertices u and v and joining u and v to each other as well as to every vertex of G and F. Which of the following is true?

(a) H is Eulerian.

(b) H has an Eulerian trail.

(c) H has neither an Eulerian circuit nor an Eulerian trail.

6.8 (a) Show that every nontrivial connected graph G has a closed spanning walk that contains every edge of G exactly twice.

(b) Which nontrivial connected graphs G have closed spanning walks that contain every edge of G exactly three times?

6.2 Hamiltonian Graphs

Figure 6.7 shows a diagram of a modern art museum that is divided into 15 exhibition rooms. At the end of each day, a security officer enters the reception room by the front door and checks each exhibition room to make certain that everything is in order. It would be most efficient if the officer could visit each room only once and return to the reception room. Can this be done?

This question can be rephrased in terms of graphs. A graph G can be associated with this museum where the vertices of G are the exhibition rooms and two vertices (rooms) are joined by an edge if there is a doorway between the two rooms. This graph G is shown in Figure 6.8. The question above can now be asked as follows: Does the graph G of Figure 6.8 have a cycle that contains every vertex of G? The answer is yes; indeed,

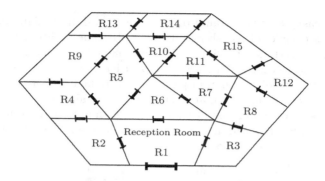

Figure 6.7: A digram of the exhibition rooms in a museum

$$C = \text{(R1, R2, R4, R9, R13, R14, R10, R5, R6, R7, R11, R15, R12, R8, R3, R1)}$$

is such a cycle. This brings us to the main topic of this section.

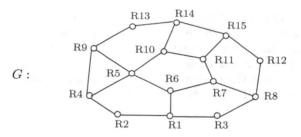

Figure 6.8: A graph that models the exhibition rooms
and doorways of the museum of Figure 6.7

A cycle in a graph G that contains every vertex of G is called a **Hamiltonian cycle** of G. Thus a Hamiltonian cycle of G is a spanning cycle of G. A **Hamiltonian graph** is a graph that contains a Hamiltonian cycle. Therefore, the graph G of Figure 6.8 is Hamiltonian. Certainly the graph C_n ($n \geq 3$) is Hamiltonian. Also, for $n \geq 3$, the complete graph K_n is a Hamiltonian graph.

A path in a graph G that contains every vertex of G is called a **Hamiltonian path** in G. If a graph contains a Hamiltonian cycle, then it contains a Hamiltonian path. In fact, removing any edge from a Hamiltonian cycle produces a Hamiltonian path. If a graph contains a Hamiltonian path, however, it need not contain a Hamiltonian cycle. For example, the graph P_n clearly contains a Hamiltonian path but P_n contains no cycles at all.

The graph $G = K_{3,3}$ of Figure 6.9 is a Hamiltonian graph. For example, $C = (u, x, v, y, w, z, u)$ is a Hamiltonian cycle of G. Since G is 3-regular and every Hamiltonian cycle contains two edges incident with each vertex of G, every Hamiltonian cycle of the graph G of Figure 6.9 fails to contain exactly one of the three edges incident with each vertex of G. By redrawing G in Figure 6.9, we can see more easily that G is Hamiltonian. Indeed, a Hamiltonian graph of order n consists of a cycle C of length n, with possibly some additional edges joining non-consecutive vertices of C. Since a Hamiltonian cycle C in a graph G of order $n \geq 3$ is a connected 2-regular subgraph of order n, every proper subgraph of C is a path or a (disjoint) union of paths. In particular, C contains no cycle of order less than n as a subgraph, and certainly C contains no subgraph having a vertex of degree 3 or more. Furthermore, since a Hamiltonian cycle C of G is a 2-regular subgraph, if G contains a vertex v of degree 2 in G, then both edges of G incident with v must lie on C.

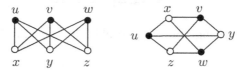

Figure 6.9: The Hamiltonian graph $K_{3,3}$

The graph G of Figure 6.10 is *not* a Hamiltonian graph. In order to see this, let's suppose that G *is* Hamiltonian. Then G contains a Hamiltonian cycle C. Since C contains the vertex t, which has degree 2, both tu and tz lie on C. By the same reasoning, xy and xz lie on C, as do vw and vz. However, this says that z is incident with three edges on C, which is impossible. Therefore, as we claimed, the graph G of Figure 6.10 is not Hamiltonian.

$G:$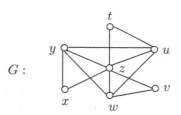

Figure 6.10: A non-Hamiltonian graph

The famous Petersen graph (shown in Figure 6.11) is also not Hamiltonian. The Petersen graph is a 3-regular graph of order 10. This graph can be considered as being constructed from two 5-cycles, namely the outer-cycle $C' = (u_1, u_2, u_3, u_4, u_5, u_1)$, the inner-cycle $C'' = (v_1, v_3, v_5, v_2, v_4, v_1)$ and the five edges $u_1v_1, u_2v_2, u_3v_3, u_4v_4, u_5v_5$. (By interchanging the labels of C' and C'', we see that it doesn't matter which 5-cycle is referred to as the outer-cycle or inner-cycle.) In fact, the length of a smallest cycle in the Petersen graph is 5. There are several ways to show that the Petersen graph is not Hamiltonian. We describe one of these.

Theorem 6.4 *The Petersen graph is non-Hamiltonian.*

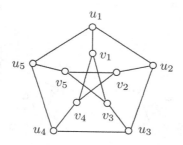

Figure 6.11: The Petersen graph: A non-Hamiltonian graph

Proof. Suppose that the Petersen graph, which we will denote by PG, is Hamiltonian. Then PG contains a Hamiltonian cycle C. This cycle contains ten edges. Two of the three edges incident with each vertex of PG necessarily belong to C. Certainly, C contains all, some or none of the five edges $u_i v_i$ $(1 \leq i \leq 5)$; so at least five edges of C belong either to C' or to C''. Therefore, either C' contains at least three edges of C or C'' contains at least three edges of C. Without loss of generality, assume that C' contains at least three edges of C. First, observe that all five edges of C' cannot belong to C since no cycle contains a smaller cycle as a subgraph. Suppose that C contains exactly four edges of C', say the edges $u_4 u_5, u_5 u_1, u_1 u_2, u_2 u_3$ (see Figure 6.12(a), where the dashed edges of PG cannot belong to C). However, the cycle C must then contain the edges $u_4 v_4, u_3 v_3$ as well as $v_1 v_3, v_1 v_4$ (see Figure 6.12(b)). But this implies that C contains an 8-cycle, which is impossible.

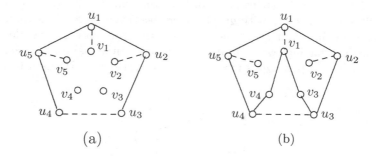

Figure 6.12: The cycle C contains exactly four edges of C'

One case remains then, namely that C contains exactly three edges of C'. There are two possibilities: (1) the three edges of C' on C are consecutive on C' or (2) these three edges are not consecutive on C'. These possibilities are shown in Figures 6.13(a) and 6.13(b), respectively. The situation in Figure 6.13(a) is impossible as $u_1 v_1$ is the only edge incident with u_1 that could lie on C. Likewise, the situation in Figure 6.13(b) is impossible since C would have to contain the smaller cycle $(u_4, v_4, v_1, v_3, u_3, u_4)$. Therefore, as claimed, the Petersen graph is not Hamiltonian.   ■

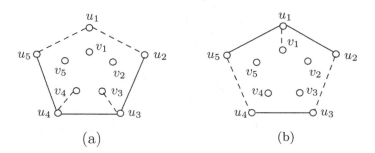

(a)   (b)

Figure 6.13: The cycle C contains exactly three edges of C'

The adjective "Hamiltonian" is named for the Irish mathematician William Rowan Hamilton (1805–1865). Although Hamilton's personal life was filled with sorrow, his professional life was marked by numerous accomplishments. His interests included poetry, optics, astronomy and mathematics (especially algebra). Indeed, as with poetry, Hamilton felt that mathematics too was artistic. Hamilton became acquainted with the distinguished poet William Wordsworth and the two had discussions about science and poetry. Wordsworth told Hamilton that his talents were with science, however, not poetry. One thing that worked against Hamilton was his inability to write well. Hamilton was knighted in 1835 for his accomplishments in physics and thus became *Sir* William Rowan Hamilton. It was Hamilton who discovered quaternions (the first noncommutative division algebra) and he spent much of his later years working on this topic.

Hamilton's connection with graph theory was even more slight than Euler's. This connection involved a well-known geometric figure. A regular dodecahedron is a polyhedron having twelve faces (all regular pentagons) and twenty corners with three edges of pentagons meeting at each corner (see Figure 6.14).

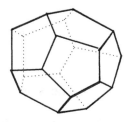

Figure 6.14: A regular dodecahedron

A regular dodecahedron has been known to be used as a desk calendar since a month can be displayed on each of its twelve faces. In 1857, Hamilton invented a game called the **Icosian Game**. It consisted of a regular dodecahedron made from wood. Each corner of the dodecahedron was marked with one of twenty cities beginning with different consonants. A goal of the game was to travel "Around the World," that is, to find a round trip that passed through each of the twenty cities exactly once. To make it easier to remember which city had already been visited, there was a peg at each corner so the journey could be described by a string that was wrapped around the peg at each city visited. This game proved to be awkward and a board game version of it was produced (see Figure 6.15).

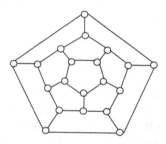

Figure 6.15: The graph of the dodecahedron

Hamilton sold the game to J. Jacques and Sons (well known even today as makers of quality chess sets) for 25 pounds and obtained a patent for it in London in 1859. The game was a commercial failure. In graph theory terms, the object of the game was to find a Hamiltonian cycle in the graph (of the dodecahedron) in Figure 6.15. Of course, this term did not exist at that time but it was Hamilton's game that gave rise to his name being used for these terms. Curiously, some two years before this, Thomas Penyngton Kirkman had asked in a paper of his whether there exists a cycle passing through every vertex of certain polyhedra. So Hamilton was not the first to deal with Hamiltonian graphs.

Let's see what we know about Hamiltonian graphs. If G is a Hamiltonian graph, then certainly G is connected. Since G contains a Hamiltonian cycle, G has no cut-vertices and has order at least 3; so G is 2-connected. This says that for every vertex v of G, the graph $G - v$ is connected. If we remove more vertices from G, then the number of components of the resulting graph cannot be too large. Recall that $k(G)$ denotes the number of components in a graph G.

Theorem 6.5 *If G is a Hamiltonian graph, then for every nonempty proper set S of vertices of G,*

$$k(G - S) \leq |S|.$$

Proof. Let S be a nonempty proper subset of $V(G)$. Suppose that $k(G - S) = k$ and that $G_1, G_2, \ldots, G_k$ are the components of $G - S$. Since G is Hamiltonian, G contains a Hamiltonian cycle C. Whenever C encounters a vertex of G_i for the last time $(1 \leq i \leq k)$, the next vertex of C must belong to S. This implies that S must contain at least k vertices, that is, $k = k(G - S) \leq |S|$. ∎

Theorem 6.5 gives a necessary condition for a graph to be Hamiltonian, that is, this theorem describes a property possessed by every Hamiltonian graph. The main benefit of a necessary condition lies in the contrapositive of the statement:

Let G be a graph. If $k(G-S) > |S|$ for some nonempty proper subset S of $V(G)$, then G is not Hamiltonian.

That is, Theorem 6.5 gives us a sufficient condition for a graph to be non-Hamiltonian. In particular, this gives us the obvious fact when a connected graph G contains a cut-vertex v (and where $S = \{v\}$) and $k(G - v) \geq 2 > 1 = |S|$:

If a graph G contains a cut-vertex, then G cannot be Hamiltonian.

Consider the graph G of Figure 6.16, where $S = \{u, v, w\}$. Then $G - S$ contains four components. However then, $k(G - S) = 4 > 3 = |S|$ and by Theorem 6.5, G is not Hamiltonian. This sufficient condition for a graph to be non-Hamiltonian is not necessary, however; that is, there exist non-Hamiltonian graphs G such that $k(G - S) \leq |S|$ for *every* nonempty proper subset S of $V(G)$ (see Exercise 6.9).

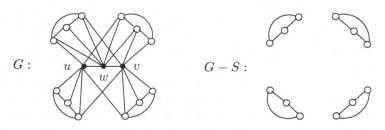

Figure 6.16: Illustrating Theorem 6.5

Although the definitions of Eulerian circuits and Hamiltonian cycles seem similar, these concepts are, in fact, quite different. While we have seen that determining whether a graph contains an Eulerian circuit is not difficult, determining whether a graph contains a Hamiltonian cycle can be exceedingly difficult. Indeed, while there is a very simple characterization of Eulerian graphs (Theorem 6.1), there is *no* such useful characterization of Hamiltonian graphs. This illustrates how making what appears to be a small change in a problem, namely from seeking an Eulerian circuit to seeking a Hamiltonian cycle, can change the problem from one that is easily solvable to one where no practical method of solution appears to exist.

As is ordinarily the case when there is no characterization of graphs possessing a certain property, one looks for sufficient conditions for a graph to have such a property. One sufficient condition for a graph to be Hamiltonian that is not difficult to apply was discovered by Oystein Ore.

Theorem 6.6 *Let G be a graph of order $n \geq 3$. If*

$$\deg u + \deg v \geq n$$

for each pair u, v of nonadjacent vertices of G, then G is Hamiltonian.

Proof. Assume, to the contrary, that there exists a non-Hamiltonian graph G of order $n \geq 3$ such that $\deg u + \deg v \geq n$ for each pair u, v of nonadjacent vertices of G. It may be the case that if we add certain edges to G, then the resulting graph is still not Hamiltonian. Of course, if we add all possible edges to G, we obtain K_n, which is obviously Hamiltonian. Add as many edges as possible to G so that the resulting graph H is not Hamiltonian. Therefore, adding any edge to H results in a Hamiltonian graph. Also, $\deg_H u + \deg_H v \geq n$ for every pair u, v of nonadjacent vertices of H.

Since H is not complete, H contains pairs of nonadjacent vertices. Let x and y be two nonadjacent vertices of H. Thus $H + xy$ is Hamiltonian. Furthermore, every Hamiltonian cycle of $H + xy$ must contain the edge xy. This means that H contains a Hamiltonian $x - y$ path $P = (x = x_1, x_2, \ldots, x_n = y)$. We claim that whenever $x_1 x_i$ is an edge of H, where $2 \leq i \leq n$, then $x_{i-1} x_n$ is not an edge of H (see Figure 6.17), for otherwise,

$$(x_1, x_i, x_{i+1}, \ldots, x_n, x_{i-1}, x_{i-2}, \ldots, x_1)$$

is a Hamiltonian cycle of H, which is impossible. Hence for each vertex in $\{x_2, x_3, \ldots, x_n\}$ that is adjacent to x_1, there is a vertex in $\{x_1, x_2, \ldots, x_{n-1}\}$ that is not adjacent to x_n. However, this means that $\deg x_n \leq (n-1) - \deg x_1$ and so

$$\deg_H x + \deg_H y \leq n - 1.$$

This is a contradiction. ∎

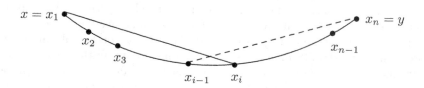

Figure 6.17: A step in the proof of Theorem 6.6

The bound given in Theorem 6.6 is sharp. For example, suppose that $n = 2k+1 \geq 3$ is an odd integer. Let G be the graph obtained by identifying a vertex in a copy of K_{k+1} and a vertex in another copy of K_{k+1}. So $G = K_1 + (2K_k)$. For $n = 7$, the graph G is shown in Figure 6.18. Certainly G is non-Hamiltonian since G contains a cut-vertex. If u and v are any two nonadjacent vertices of G, then $\deg u = \deg v = k$ and so $\deg u + \deg v = 2k = n - 1$. Hence the bound given in Theorem 6.6 cannot be lowered to produce the same conclusion.

G :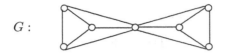

Figure 6.18: The graph $G = K_1 + (2K_3)$

There is a corollary to Theorem 6.6 due to Gabriel Dirac, whom we encountered earlier and which is even easier to apply. Dirac was a distinguished research mathematician whose stepfather Paul Dirac was well known for his work in quantum mechanics. In fact, Paul Dirac was awarded the Nobel Prize for physics in 1933. Gabriel Dirac was a professor at Aarhus Universitet in Denmark.

Corollary 6.7 *Let G be a graph of order $n \geq 3$. If $\deg v \geq n/2$ for each vertex v of G, then G is Hamiltonian.*

Proof. Certainly, if $G = K_n$, then G is Hamiltonian. We may therefore assume that G is not complete. Let u and v be two nonadjacent vertices of G. Thus

$$\deg u + \deg v \geq \frac{n}{2} + \frac{n}{2} = n.$$

By Theorem 6.6, G is Hamiltonian.   ∎

If we look at the proof of Theorem 6.6 more carefully, then we might see that it provides a proof of another result, due to J. Adrian Bondy and Vašek Chvátal.

Theorem 6.8 *Let u and v be nonadjacent vertices in a graph G of order n such that $\deg u + \deg v \geq n$. Then $G + uv$ is Hamiltonian if and only if G is Hamiltonian.*

Proof. If G is a Hamiltonian graph, then certainly $G + uv$ is Hamiltonian for any two nonadjacent vertices u and v of G. Thus we need only verify the converse.

Let $G + uv$ be a Hamiltonian graph for two nonadjacent vertices u and v of a graph G and assume, to the contrary, that G is not Hamiltonian. This implies that every Hamiltonian cycle in $G + uv$ must contain the edge uv and so G contains a Hamiltonian $u - v$ path. Since $\deg_G u + \deg_G v \geq n$, the proof of Theorem 6.6 tells us that G contains a Hamiltonian cycle. This is a contradiction.   ∎

The **closure** $C(G)$ of a graph G of order n is the graph obtained from G by recursively joining pairs of nonadjacent vertices whose degree sum is at least n (in the resulting graph at each stage) until no such pair remains. Figure 6.19 illustrates how to obtain the closure of a graph G, while Figure 6.20 shows three other graphs and their closures. (The graph G_3 of Figure 6.20 is the graph of Figure 6.19.) In the graph G_4 of Figure 6.20, the vertices u and v are the only two nonadjacent vertices whose degree sum is at least 10 (the order of G_4).

Repeated applications of Theorem 6.8 give us the following result.

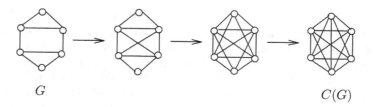

G $C(G)$

Figure 6.19: Constructing the closure of a graph

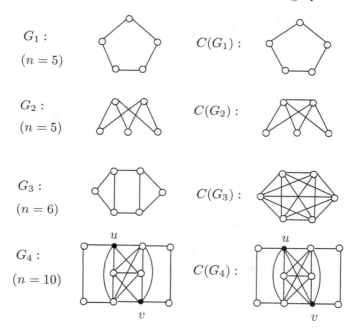

Figure 6.20: Graphs and their closures

Theorem 6.9 *A graph is Hamiltonian if and only if its closure is Hamiltonian.*

A simple but useful consequence of Theorem 6.9 is stated next.

Corollary 6.10 *If G is a graph of order at least 3 such that $C(G)$ is complete, then G is Hamiltonian.*

We saw in Corollary 6.7 (Dirac's theorem) that if the degree of every vertex is at least $n/2$ in a graph G of order $n \geq 3$, then G is Hamiltonian. Moreover, Ore's theorem (Theorem 6.6) requires a weaker condition to be satisfied for a graph to be Hamiltonian, as only the degree sum of every two nonadjacent vertices must be at least n. Actually, Bondy and Chvátal's Closure Theorem (Theorem 6.8) implies this as well. Indeed, Theorem 6.8 suggests that many more graphs are

Hamiltonian. Yet, it is not clear what conditions on the degrees of the vertices of a graph G of order $n \geq 3$ must be satisfied in order for Theorem 6.8 to guarantee that G is Hamiltonian. An example of such a result (which preceded Theorem 6.8 chronologically) is due to Lajos Pósa. The following theorem is even more remarkable because Pósa was barely a teenager when he discovered this result. Despite having a promising future in research, Pósa went on to devote himself to the mathematics education of the everyday student in his native Hungary.

Theorem 6.11 *Let G be a graph of order $n \geq 3$. If for every integer j with $1 \leq j < \frac{n}{2}$, the number of vertices of G with degree at most j is less than j, then G is Hamiltonian.*

Proof. We show that $C(G)$ is complete. Assume, to the contrary, that this is not the case. Among all pairs of nonadjacent vertices in $C(G)$, let u, w be a pair for which $\deg_{C(G)} u + \deg_{C(G)} w$ is maximum. Necessarily, $\deg_{C(G)} u + \deg_{C(G)} w \leq n - 1$. We may also assume that $\deg_{C(G)} u \leq \deg_{C(G)} w$. Let $\deg_{C(G)} u = k$. Thus $k \leq \frac{n-1}{2}$ and so

$$\deg_{C(G)} w \leq n - k - 1. \tag{6.1}$$

Let W be the set of all vertices distinct from w that are not adjacent to w. Therefore, $u \in W$. Observe that if $v \in W$, then $\deg_{C(G)} v \leq k$, for otherwise

$$\deg_{C(G)} v + \deg_{C(G)} w > \deg_{C(G)} u + \deg_{C(G)} w,$$

contradicting the defining property of the pair u, w. Therefore, the degree of every vertex of W is at most k. So by hypothesis, $|W| \leq k - 1$. Hence

$$\deg_{C(G)} w \geq (n - 1) - (k - 1) = n - k,$$

which contradicts (6.1).   ∎

For $j = 1$, Theorem 6.11 says that G has no vertex of degree 1. For $j = 2$, the graph G is allowed to have a vertex of degree 2. For $j = 3$, the graph G is allowed to have a vertex of degree 2 *and* a vertex of degree 3 *or* two vertices of degree 3.

Exercises for Section 6.2

6.9 We have seen that the graph G of Figure 6.10 is not Hamiltonian. Show that $k(G - S) \leq |S|$ for every nonempty proper subset S of $V(G)$. What does this say about Theorem 6.5?

6.10 Let G be a 6-regular graph of order 10 and let $u, v \in V(G)$. Prove that G, $G - v$ and $G - u - v$ are all Hamiltonian.

6.11 Prove that $\overline{C}_n$ is Hamiltonian for $n \geq 5$.

6.12 Let G be a 3-regular graph of order 12 and H a 4-regular graph of order 11.

   (a) Is $G + H$ Eulerian?

   (b) Is $G + H$ Hamiltonian?

6.13 Give an example of a graph G that is

   (a) Eulerian but not Hamiltonian. (Explain why G is not Hamiltonian.)

   (b) Hamiltonian but not Eulerian. (Explain why G is not Eulerian.)

   (c) Hamiltonian and has an Eulerian trail but is not Eulerian.

   (d) neither Eulerian nor Hamiltonian, but has an Eulerian trail.

6.14 Give an example of a graph with the following properties or explain why no such example exists:

   (a) a 2-connected (that is, connected, order at least 3 and no cut-vertices) Eulerian graph that is not Hamiltonian.

   (b) a Hamiltonian graph G that is not Eulerian but whose complement $\overline{G}$ is Eulerian.

6.15 The **subdivision graph** of a graph G is that graph obtained from G by deleting *every* edge uv of G and replacing it by a vertex w of degree 2 that is joined to u and v. Is it true that if the subdivision graph of a graph G is Hamiltonian, then G is Eulerian?

6.16 Let G be a connected r-regular graph of even order n such that $\overline{G}$ is also connected. Show that

   (a) either G or $\overline{G}$ is Eulerian.

   (b) either G or $\overline{G}$ is Hamiltonian.

6.17 For a graph G of order $n \geq 3$, the graph $G(3)$ is obtained from G by adding a new vertex v_S for each 3-element subset S of $V(G)$ and joining v_S to each vertex in S. Find all such graphs G for which $G(3)$ is Hamiltonian.

6.18 Show that the bound in Corollary 6.7 is sharp.

6.19 Let G_1 and G_2 be two graphs of order $n \geq 3$, each of which satisfies the hypothesis of Dirac's theorem (Corollary 6.7) on Hamiltonian graphs. A graph G is constructed from $G_1 \cup G_2$ by adding edges between G_1 and G_2 such that every vertex of G_1 is joined to at least half the vertices of G_2 in such a way that every vertex of G_2 is joined to at least half the vertices of G_1. Prove that G is Hamiltonian.

6.20 Let G be a graph of order $n \geq 3$ having the property that for each $v \in V(G)$, there is a Hamiltonian path with initial vertex v. Show that G is 2-connected (connected, order at least 3 and no cut-vertices) but not necessarily Hamiltonian.

6.21 Let G be a graph of order $n \geq 3$ such that $\deg u + \deg v \geq n - 1$ for every two nonadjacent vertices u and v. Prove that G must contain a Hamiltonian path.

6.22 (a) Does there exist a graph G of order 10 and size 28 that is not Hamiltonian?

  (b) Does there exist a graph H of order 10 and size 28 that is not Hamiltonian, where 8 of the 10 vertices have the following degrees: 5, 5, 5, 5, 5, 6, 6, 6?

6.23 (a) Does there exist a graph G of order $n = 2k \geq 6$ and size $m = k^2 + k - 2$ that is not Hamiltonian?

  (b) Does there exist a graph H of order $n = 2k \geq 6$ and size $m = k^2 + k - 2$ that is not Hamiltonian, where k vertices of H have degree k and $k - 2$ vertices of H have degree $k + 1$?

6.24 (a) A connected graph G of order $n = 2k + 1$ has $k + 1$ vertices of degree 2, no two of which are adjacent, while the remaining k vertices have degree 3 or more. Show that G is not Hamiltonian.

  (b) Give an example of a Hamiltonian graph H of order $n = 2k$ for some $k \geq 2$, where k vertices have degree 2, no two vertices of which are adjacent, while the remaining vertices have degree 3 or more.

6.3   Exploration: Hamiltonian Walks

While certainly not every connected graph of order at least 3 contains a Hamiltonian cycle, every connected graph does contain a closed spanning walk. Indeed, if every edge of a connected graph G is replaced by two parallel edges, then the resulting multigraph M is Eulerian (see Figure 6.21). Since an Eulerian circuit in M gives rise to a closed spanning walk in G in which each edge of G appears twice, it follows that a connected graph of size m has a closed spanning walk of length $2m$ in G.

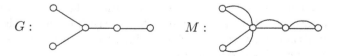

Figure 6.21: Closed spanning walks in graphs

A **Hamiltonian walk** in a connected graph G is a closed spanning walk of minimum length in G. From our earlier observation, every connected graph G

of size m contains a Hamiltonian walk and the length of such a walk is at most $2m$. The length of a Hamiltonian walk in G is denoted by $h(G)$. Therefore, for a connected graph G of order $n \geq 3$, it follows that $h(G) = n$ if and only if G is Hamiltonian. The concept of a Hamiltonian walk was introduced by Seymour Goodman and Stephen Hedetniemi in 1973.

Although it is often difficult to determine whether a graph G is Hamiltonian, we have seen that if G satisfies any of a number of sufficient conditions, then G is Hamiltonian. However, none of these conditions is necessary and so G can be Hamiltonian without satisfying any of these conditions. In such a case, our only option may be to construct a Hamiltonian cycle in G. So the problem is reduced to finding a way to list all of the vertices of G in a cyclic sequence $(v_1, v_2, \ldots, v_n, v_1)$ so that every two consecutive vertices in the sequence are adjacent. Another way to say this is to list the vertices of G in a cyclic sequence $(v_1, v_2, \ldots, v_n, v_1)$ such that $d(v_i, v_{i+1}) = 1$ for $1 \leq i \leq n-1$ and $d(v_n, v_1) = 1$. If we also write v_1 as v_{n+1}, then the cyclic sequence $(v_1, v_2, \ldots, v_n, v_{n+1} = v_1)$ is a Hamiltonian cycle if and only if

$$\sum_{i=1}^{n} d(v_i, v_{i+1}) = n.$$

Looking at Hamiltonian cycles in this manner suggests another concept. For a connected graph G of order $n \geq 3$ and a cyclic ordering

$$s : v_1, v_2, \cdots, v_n, v_{n+1} = v_1$$

of $V(G)$, define the number $d(s)$ by

$$d(s) = \sum_{i=1}^{n} d(v_i, v_{i+1}).$$

Therefore, $d(s) \geq n$ for each cyclic ordering s of $V(G)$. Moreover, G is Hamiltonian if and only if there exists a cyclic ordering s' of $V(G)$ such that $d(s') = n$. The **Hamiltonian number** $h^*(G)$ of G is defined by

$$h^*(G) = \min \{d(s)\},$$

where the minimum is taken over all cyclic orderings s of $V(G)$. For the graph $G = K_{2,3}$ of Figure 6.22, $d(s_1) = 8$ and $d(s_2) = 6$ for the cyclic orderings

$$s_1 : v_1, v_2, v_3, v_4, v_5, v_1 \text{ and } s_2 : v_1, v_3, v_2, v_4, v_5, v_1$$

of $V(G)$. Since G is a non-Hamiltonian graph of order 5 and $d(s_2) = 6$, it follows that $h^*(G) = 6$.

We are about to see that there is an alternative way to define the length $h(G)$ of a Hamiltonian walk in G. Denote the length of a walk W by $L(W)$.

Theorem 6.12 *For every connected graph G,*

$$h^*(G) = h(G).$$

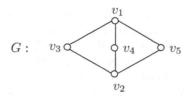

$$G: \quad v_3$$

Figure 6.22: A graph G with $h^*(G) = 6$

Proof. First, we show that $h(G) \le h^*(G)$. Let $s : v_1, v_2, \cdots, v_n, v_{n+1} = v_1$ be a cyclic ordering of $V(G)$ for which $d(s) = h^*(G)$. For each integer i with $1 \le i \le n$, let P_i be a $v_i - v_{i+1}$ geodesic in G. Thus $L(P_i) = d(v_i, v_{i+1})$. The union of the paths P_i form a closed walk W in G containing all vertices of G. Therefore,

$$h(G) \le L(W) = \sum_{i=1}^{n} L(P_i) = \sum_{i=1}^{n} d(v_i, v_{i+1}) = d(s) = h^*(G).$$

Next, we show that $h^*(G) \le h(G)$. Let W be a Hamiltonian walk in G with $L(W) = h(G)$. Suppose that $W = (x_1, x_2, \ldots, x_N, x_1)$, where then $N \ge n$. Define $v_1 = x_1$ and $v_2 = x_2$. For $3 \le i \le n$, define v_i to be x_{j_i}, where j_i is the smallest positive integer such that $x_{j_i} \notin \{v_1, v_2, \ldots, v_{i-1}\}$. Then $s :$ $v_1, v_2, \ldots, v_n, v_{n+1} = v_1$ is a cyclic ordering of $V(G)$. For each i with $1 \le i \le n$, let Q_i be the $v_i - v_{i+1}$ subwalk of W and so $d(v_i, v_{i+1}) \le L(Q_i)$. Since

$$h^*(G) \le \sum_{i=1}^{n} d(v_i, v_{i+1}) \le \sum_{i=1}^{n} L(Q_i) \le L(W) = h(G),$$

we have the desired result.   ∎

As a consequence of Theorem 6.12, we can denote the Hamiltonian number of a graph G by $h(G)$, which is then the length of Hamiltonian walk in G.

For the graph $G = K_{2,3}$ of Figure 6.22 and the cyclic orderings $s_1 : v_1, v_2,$ v_3, v_4, v_5, v_1 and $s_2 : v_1, v_3, v_2, v_4, v_5, v_1$ of $V(G)$, we saw that $d(s_1) = 8$ and $d(s_2) = 6$. Actually, it is not difficult to show that $d(s)$ is either 8 or 6 for *every* cyclic ordering s of $V(G)$. This suggests another parameter of a connected graph. The **upper Hamiltonian number** $h^+(G)$ of a connected graph G is defined as

$$h^+(G) = \max \{d(s)\},$$

where the maximum is taken over all cyclic orderings s of $V(G)$. Therefore, $h^+(G) = 8$ for the graph G of of Figure 6.22.

As another example, we consider the Petersen graph. Label the vertices of the Petersen graph PG as shown in Figure 6.23. Since PG is a non-Hamiltonian graph of order 10, $h(PG) \ge 11$. On the other hand, let $s : x_1, x_2, \ldots, x_{11} = x_1$ be any cyclic ordering of the vertices of PG. Since diam$(PG) = 2$, it follows that

$d(x_i, x_{i+1}) \leq 2$ for $1 \leq i \leq 10$. Hence $d(s) \leq 2 \cdot 10 = 20$ and so $h^+(PG) \leq 20$. Therefore,

$$11 \leq h(PG) \leq h^+(PG) \leq 20.$$

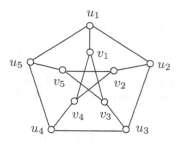

Figure 6.23: The Petersen graph

In fact, $h(PG) = 11$ and $h^+(PG) = 20$. Moreover, consider the sequences s_i ($1 \leq i \leq 10$):

$$s_1 : u_1, u_2, u_3, u_4, u_5, v_5, v_2, v_4, v_3, v_1, u_1$$
$$s_2 : u_1, u_2, u_3, u_4, u_5, v_5, v_2, v_3, v_4, v_1, u_1$$
$$s_3 : u_1, u_2, u_3, u_5, u_4, v_4, v_2, v_3, v_5, v_1, u_1$$
$$s_4 : u_1, u_3, u_5, u_2, u_4, v_4, v_2, v_5, v_3, v_1, u_1$$
$$s_5 : u_1, u_3, u_5, u_2, u_4, v_3, v_5, v_2, v_4, v_1, u_1$$
$$s_6 : u_1, u_3, u_5, u_2, u_4, v_5, v_2, v_4, v_3, v_1, u_1$$
$$s_7 : u_1, u_3, u_5, u_2, u_4, v_3, v_5, v_4, v_2, v_1, u_1$$
$$s_8 : u_1, u_3, u_5, u_2, v_2, u_4, v_3, v_4, v_5, v_1, u_1$$
$$s_9 : u_1, u_3, u_5, u_2, u_4, v_2, v_3, v_4, v_5, v_1, u_1$$
$$s_{10} : u_1, u_3, u_5, u_2, u_4, v_1, v_2, v_3, v_4, v_5, u_1$$

Since $d(s_i) = 10 + i$ for $1 \leq i \leq 10$, it follows that for each integer k with $11 \leq k \leq 20$, there exists a cyclic ordering s of $V(PG)$ such that $d(s) = k$.

Exercises for Section 6.3

6.25 For $G_1 = K_n$ and $G_2 = K_{s,t}$, where $n \geq 3$ and $1 \leq s \leq t$, find $h(G_i)$ and $h^+(G_i)$ for $i = 1, 2$.

6.26 Give an example of a graph G of order $n \geq 3$ such that $h(G) = n + 1$. Verify that your example is correct.

6.27 Give an example of a graph G of order $n \geq 3$ such that $h(G) = 2n - 2$. Verify that your example is correct.

6.28 Let C_n be a cycle of order $n \geq 3$.

(a) What is $h^+(C_n)$ if n is even?

(b) What can you say about $h^+(C_n)$ if n is odd?

6.29 (a) Prove that if G is a connected graph of order n and diameter $d \geq 2$, then $h^+(G) \leq nd$.

(b) Is the upper bound in (a) is sharp?

6.30 Determine all graphs G for which $h(G) = h^+(G)$.

6.31 Instead of considering cyclic sequences of the vertices of a graph G (and studying the Hamiltonian and upper Hamiltonian numbers), consider linear sequences of the vertices of G.

6.33 Ask and answer a question of your own concerning Hamiltonian numbers and/or upper Hamiltonian numbers.

6.4   Excursion: Early Books of Graph Theory

Theorem 6.1, published in 1736, is considered to be the beginning of graph theory. It wasn't until 1936, however, when the first textbook (in German) was written on graph theory by Dénes König. König was born in Budapest, Hungary on September 21, 1884. König was interested in mathematics at an early age, no doubt influenced by his father who was a well-known mathematics professor. Indeed, he even published a paper as a high school student.

König spent nine semesters doing university work, the first four in Budapest and the last five in Göttingen. He attended lectures by Hermann Minkowski on analysis situs, which is what topology was called in its early days. The fact that Minkowski was interested in graph theory played a big role in the mathematics König decided to work on. König received his doctorate in 1907 and wrote his dissertation in geometry. He acquired a faculty position at the Technische Hochschule in Budapest that year and remained a member of the faculty there until his death. Among the courses König taught was graph theory, although the name "graph theory" never appeared in the catalogue at the university until 1927. It fell under the heading of analysis situs prior to that time. In 1935 König became a full professor. König was well known for his enthusiastic lectures, although his lectures were not always well attended. Because of him, however, a number of dedicated students were introduced to this new area of mathematics. Indeed, under his influence, Hungarian researchers turned to this field, including Paul Erdős, Tibor Gallai and Paul Turán.

Although an excellent mathematician, König's main accomplishment is probably the popularization and recognition of graph theory. Because of his efforts, graph theory grew from being a collection of isolated results in a branch of recreational mathematics to a recognized new area of the mathematical sciences. Although he was belittled by some mathematicians, he was not discouraged. He believed in the future of graph theory. Indeed, König would often begin a lecture on graph theory by saying:

> *Graph theory is one of the most interesting of mathematical disciplines.*

In 1936, König's book *Theorie der endlichen und unendlichen Graphen* (the very first book ever written solely on graph theory) was published in Leipzig although Oswald Veblen had discussed graph theory in his 1922 book *Analysis Situs*, the first book written on topology. König worked on his book with great care for many years. His book awakened the interest of many young mathematicians in graph theory, although its impact was only felt after World War II. In 1944, after the occupation of Hungary by the Nazis, König worked to assist persecuted mathematicians. Rather than be persecuted himself, he committed suicide on October 19, 1944. In 1950 König's book was reprinted in the United States. For over 20 years, König's work was the only book on graph theory, until 1958 when *Théorie des Graphes et Ses Applications* by the French mathematician Claude Berge was published.

Claude Berge was born on June 5, 1926. He was the one individual to spread the word of graph theory throughout France. Despite König's book, prior to the 1950s many mathematicians thought little of combinatorics and graph theory. Most French mathematicians tended to resist graph theory and preferred the term "network" instead. However, because of Berge's efforts, all this changed.

When graph theory was introduced to Berge, it was a subject that was unknown in France. It was Berge's intent to make some sense of this new field. At first, he worked on graph theory just for himself, but in 1958 he published his book on graph theory (others would come later). He found the subject interesting and alive with many applications. However, it wasn't only the French to whom Berge introduced graph theory. He traveled widely and lectured on the subject. Many of Berge's works were translated into other languages. In 1993 Berge was awarded the Euler Medal by the Institute of Combinatorics and Its Applications.

Berge made contributions to other areas of mathematics, including game theory and topology. He introduced an alternative to the Nash equilibrium (named after John Nash whose life was chronicled in the 2002 academy award winning movie *A Beautiful Mind*) called the Berge equilibrium. Berge also coined the term *hypergraph*.

Berge had many interests besides mathematics. He had a special affinity for Chinese works of art and was a skillful chess player. Berge loved to write. In 1994 he authored a mathematical murder mystery, titled *Who Killed the Duke of Densmore?* in which the detective investigates the murder of the Duke of Densmore and uses graphs (actually interval graphs, which we do not discuss here) to find the murderer. Berge died on June 30, 2002.

The next major book on graph theory, titled *Theory of Graphs*, was published in 1962 by the American Mathematical Society, four years after Berge's book. Its author, Oystein Ore (pronounced OR-ah) was born in Oslo, Norway in 1899. Ore attended the University of Oslo, from which he graduated in 1922. He received his Ph.D. in 1924. After this, he spent time in Paris and Gottingen and then returned to the University of Oslo. In 1927 he went to the United States where he became a professor of mathematics at Yale University. Ore wrote over a hundred papers and a number of books.

Ore was well known for his work in algebra and number theory before he started working in graph theory. He had much to do with introducing graph theory to the English-speaking world with his 1962 book. Ore once wrote, in reference to Euler's solution to the Königberg Bridge Problem, that

> *The theory of graphs is one of the few fields of mathematics with a definite birth date.*

Ore died during the summer of 1968, only a few months before Ore was to attend and be the principal speaker at the first of nine quadrennial graph theory conferences to be held at Western Michigan University.

Although some books dealing with special topics in graph theory and applications of graph theory were published during 1959-1969, only two other major books on graph *theory* were published before 1970, both in 1969. One of these was the 1969 book *Teoriia Konechnykn Grafov* (*Theory of Finite Graphs*) by Alexander A. Zykov, who has had the greatest influence on the development of and interest in graph theory in Russia. Born in Kiev in 1922, Zykov was introduced to graph theory during 1943-44 while he was a student at Gor'kiy State University. He was in Novosibirsk during 1959-69 and organized a graph theory seminar at the Mathematical Institute of the Siberian branch of the Academy Sciences in the USSR. Since 1969 Zykov has been at Odessa State Polytechnic University in Ukraine, where he organized the Odessa seminar in discrete mathematics.

The second book, titled *Graph Theory*, was written by Frank Harary (1921-2005). Harary received his Ph.D. in 1948 from the University of California in Berkeley and became a faculty member at the University of Michigan, where he stayed until 1986. He then became a faculty member at New Mexico State University.

While working with social psychologists at the University of Michigan, he independently discovered graph theory. Harary spent much of his life traveling and lecturing on graph theory, thereby introducing this subject to many mathematicians around the world. Along the way, he acquired numerous co-authors, which led to hundreds of publications on all aspects of the subject. Harary, known for his lucid writing style and his enthusiasm for the subject, was often referred to as the *Ambassador of Graph Theory*.

At one time Alexander Zykov gave Frank Harary a framed set of three pictures of lions, namely (1) a lion sleeping, (2) a lion awakening and (3) a lion roaring to represent (1) graph theory yesterday, (2) graph theory today and (3) graph theory tomorrow. Although meant as a comical gift, Zykov indicated that if one were to define "yesterday" as "before 1936," "today" as "1936-1970" and "tomorrow" as "after 1970," then there is a certain amount of truth to this representation of graph theory.

William T. Tutte (1917–2002), a mathematician who made numerous significant contributions to graph theory, was a shy man with a clever sense of humor. Tutte loved to write poetry, often under the pseudonym of Blanche Descartes. As Descartes, he reflected on graph theory in his 1969 poem titled "The Expanding Unicurse":

Some citizens of Königsberg
Were walking on the strand
Beside the river Pregel
With its seven bridges spanned.

"O Euler, come and walk with us,"
Those burghers did beseech.
"We'll roam the seven bridges o'er,
And pass but once by each."

"It can't be done," thus Euler cried.
"Here comes the Q. E. D.
Your islands are but vertices,
And four have odd degree."

From Königsberg to König's book,
So runs the graphic tale,
And still it grows more colorful,
In Michigan and Yale.

The mention of Michigan and Yale in the last line of the poem refers to the universities of Frank Harary and Oystein Ore, respectively. We will encounter William Tutte again.

Chapter 7

Digraphs

7.1 Strong Digraphs

We saw in Section 6.1 that the street system of a town can be naturally represented by a graph. In this case, the vertices of the graph are the street intersections in the town, while the edges of the graph are the street segments between intersections. The street systems of two towns A and B are shown in Figure 7.1 together with the graphs G_A and G_B that model them.

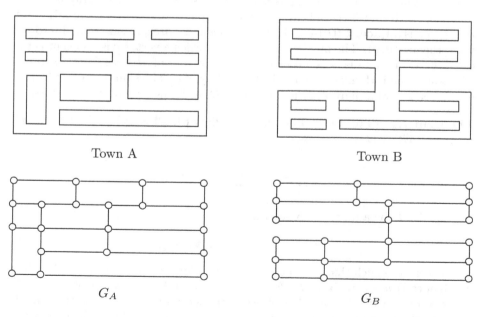

Figure 7.1: Two towns and two graphs modeling them

Both graphs G_A and G_B of Figure 7.1 have the important property that they are connected, meaning that it is possible to travel between any two locations in both Town A and Town B. (Of course, this is a characteristic one would expect of any town.) The graph G_B has a bridge, however, while G_A does not. In fact, it may be the case that the street segment in Town B that gives rise to the bridge in G_B is a road that goes over a bridge in the town. Of course, a major disadvantage of having such a street in Town B is that if it should ever become necessary to close that street, then traveling between some pairs of locations in Town B is impossible. We have no such difficulties in Town A, however. The fact that we can travel between any two street intersections in Town A even after one of its streets may be closed allows us to do something else in Town A, as we are about to discover. Before continuing with this discussion, however, it is convenient to revisit the concept of a digraph (directed graph).

Recall that a **digraph** D consists of a finite nonempty set V of objects called **vertices** and a set E of ordered pairs of distinct vertices. Each element of E is an **arc** or a **directed edge**. If a digraph D has the property that for each pair u, v of distinct vertices of D, at most one of (u, v) and (v, u) is an arc of D, then D is an **oriented graph**. An oriented graph can also be obtained by assigning a direction to (that is, orienting) each edge of a graph G. The digraph D is then referred to as an **orientation** of G. A digraph H is called a **subdigraph** of a digraph D if $V(H) \subseteq V(D)$ and $E(H) \subseteq E(D)$.

A digraph D is **symmetric** if whenever (u, v) is an arc of D, then (v, u) is an arc of D as well. We will rarely be interested in symmetric digraphs, however, since studying symmetric digraphs is really the same as studying graphs.

Also, recall that if (u, v) is an arc of a digraph, then u is said to be **adjacent to** v and v is **adjacent from** u. The vertices u and v are also said to be **incident with** the arc (u, v). The number of vertices *to which* a vertex v is adjacent is the **outdegree** of v and is denoted by od v. The number of vertices *from which* v is adjacent is the **indegree** of v and is denoted by id v. The sum of the outdegrees of the vertices of a digraph D is the size of D, as is the sum of its indegrees.

Theorem 7.1 (The First Theorem of Digraph Theory) *If D is a digraph of size m with $V(D) = \{v_1, v_2, \ldots, v_n\}$, then*

$$\sum_{i=1}^{n} \operatorname{od} v_i = \sum_{i=1}^{n} \operatorname{id} v_i = m.$$

Now let D be a digraph. A sequence

$$W = (u = u_0, u_1, \ldots, u_k = v) \tag{7.1}$$

of vertices of D such that u_i is adjacent to u_{i+1} for all i ($0 \leq i \leq k-1$) is called a (**directed**) $u - v$ **walk** in D. Each arc (u_i, u_{i+1}), $0 \leq i \leq k-1$, is said to lie on or belong to W. The number of occurrences of arcs on a walk is the **length** of the walk. So the length of the walk W in (7.1) is k. A walk in which no arc is repeated is a (**directed**) **trail**, while a walk in which no vertex is repeated is a (**directed**) **path.** A $u - v$ walk is **closed** if $u = v$ and is **open** if $u \neq v$. A closed trail of length at least 2 is a (**directed**) **circuit**; a closed walk of length at least 2 in which no vertex is repeated except for the initial and terminal vertices is a (**directed**) **cycle**. Consequently, whenever we refer to any kind of a walk in a digraph, we mean a *directed* walk, that is, we always proceed in the direction of the arrows. As with graphs, the subdigraph of a digraph consisting of the vertices and arcs of a path, cycle, trail or circuit is referred to by the same term.

To illustrate these concepts, consider the digraph D of Figure 7.2. Since (t, w) and (w, t) are both arcs of D, the digraph D is not an oriented graph. First, $W = (y, w, v, x, y, w, t)$ is a $y - t$ walk of length 6. The arc (y, w) occurs twice on W, so W is not a trail. However, $T = (y, w, t, w, v)$ is a $y - v$ trail. Since the vertex w is repeated in T, it is not a path. Also, $C = (v, t, w, t, u, v)$ is a circuit that is not a cycle, while $C' = (v, x, y, w, v)$ is a cycle of length 4. The cycle $C'' = (t, w, t)$ has length 2.

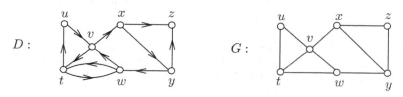

Figure 7.2: Walks in a digraph

The **underlying graph** of a digraph D is obtained by removing all directions from the arcs of D and replacing any resulting pair of parallel edges by a single edge. Equivalently, the underlying graph of a digraph D is obtained by replacing each arc (u, v) or pair $(u, v), (v, u)$ of arcs by the edge uv. So the graph G of Figure 7.2 is the underlying graph of the digraph D of that figure. Also, if D is an orientation of a graph G, then G is the underlying graph of D.

While a graph is either connected or it's not, for a digraph there is another alternative. A digraph D is **connected** (sometimes called **weakly connected**) if the underlying graph of D is connected. In particular, the digraph D of Figure 7.2 is connected. A digraph D is **strong** (or **strongly connected**) if D contains both a $u - v$ path and a $v - u$ path for every pair u, v of distinct vertices of D. The digraph D of Figure 7.2 is not strong since there is no $z - y$ path in D. Indeed, there is no path from z to any other vertex of D.

Distance is defined in digraphs as well as in graphs. Let u and v be vertices in a digraph D. The **directed distance** or, more simply, the **distance** $\vec{d}(u, v)$ **from** u **to** v is the length of a shortest $u - v$ path in D. A $u - v$ path of length $\vec{d}(u, v)$ is a $u - v$ **geodesic**. Once again, let us emphasize that the paths we are discussing here are *directed* paths. In order for $\vec{d}(u, v)$ to be defined for every pair u, v of vertices of D, the digraph D must be strong.

If D is a strong digraph, then necessarily every vertex of D has positive outdegree and indegree. This is only a necessary condition for a digraph to be strong, however, not a sufficient condition. Every vertex in the digraph D of

Figure 7.3 has positive outdegree and indegree; yet, there is no $u - x$ path, for example.

$D:$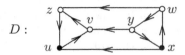

Figure 7.3: A digraph that is not strong

While every $u - v$ path in a digraph D is a $u - v$ walk, we have seen that the converse is not true. On the other hand, the presence of a $u - v$ walk in D implies the existence of a $u - v$ path in D. The statement and proof are nearly identical to the corresponding result for graphs (Theorem 1.6).

Theorem 7.2 *If a digraph D contains a $u - v$ walk of length ℓ, then D contains a $u - v$ path of length at most ℓ.*

Proof. Among all $u - v$ walks in D, let W be one of minimum length. Suppose that $W = (u = u_0, u_1, \ldots, u_k = v)$. Then $k \le \ell$. If the vertices $u_0, u_1, u_2, \ldots, u_k$ are distinct, then W is a $u - v$ path and the proof is complete. Otherwise, there are vertices u_i and u_j such that $u_i = u_j$, where $1 \le i < j \le k$. If we delete $u_{i+1}, u_{i+2}, \ldots, u_j$ from W, then we obtain a $u - v$ walk

$$W' = (u = u_0, u_1, \ldots, u_i, u_{j+1}, u_{j+2}, \ldots, u_k = v)$$

whose length is less than k, which is impossible. Thus W is a $u - v$ path of length $k \le \ell$. ∎

The following result provides a necessary and sufficient condition for a digraph to be strong.

Theorem 7.3 *A digraph D is strong if and only if D contains a closed spanning walk.*

Proof. Since every trivial digraph is strong, we may assume that D is nontrivial. First, let D be a digraph that contains a closed spanning walk $W = (w_0, w_1, \ldots, w_k = w_0)$. Let u and v be any two distinct vertices of D. Then $u = w_i$ and $v = w_j$ for integers i and j with $0 \le i < j \le k$. Since $W' = (u = w_i, w_{i+1}, \ldots, w_j = v)$ is a $u - v$ walk and $W'' = (v = w_j, w_{j+1}, \ldots, w_k = w_0, w_1, \ldots, w_i = u)$ is a $v - u$ walk, it follows by Theorem 7.2 that D contains both a $u - v$ path and a $v - u$ path.

Now we verify the converse. Let D be a strong digraph and suppose that $V(D) = \{v_1, v_2, \ldots, v_n\}$. Since D is strong, D contains a $v_i - v_{i+1}$ path P_i for $i = 1, 2, \ldots, n-1$. Let P_n be a $v_n - v_1$ path. For $1 \le i \le n-1$, let P_i' be the path obtained by deleting the final vertex of P_i. Then

$$W = (P_1', P_2', \ldots, P_{n-1}', P_n)$$

is a closed spanning walk in D. ∎

By Theorem 7.3, a digraph that contains a spanning circuit is strong. There is one type of spanning circuit that is of added interest to us. An **Eulerian circuit** in a (strong) digraph D is a circuit containing every arc of D. An **Eulerian digraph** is a digraph containing an Eulerian circuit. The digraph D of Figure 7.4 is Eulerian and $C = (u, v, w, y, z, x, y, x, w, u)$ is an Eulerian circuit.

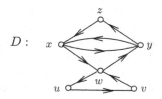

Figure 7.4: An Eulerian digraph

Just as Eulerian graphs are easy to characterize, so too are Eulerian digraphs. Indeed, the proof is similar to the proof of the characterization of Eulerian graphs (Theorem 6.1).

Theorem 7.4 *A nontrivial connected digraph D is Eulerian if and only if* $\operatorname{od} v = \operatorname{id} v$ *for every vertex v of D.*

Proof. First, let D be an Eulerian digraph. Then D contains an Eulerian circuit C. Let v be a vertex of C. Assume first that v is not the initial vertex of C (and so v is not the terminal vertex either). Whenever v is encountered on C, an arc is used to enter v and another is used to exit v. This contributes 1 to both the indegree and outdegree of v. If v is encountered k times on C, then $\operatorname{od} v = \operatorname{id} v = k$. If v is the initial vertex of C, then an arc is used to exit v. The final arc of C enters v. Any other occurrences of v on C contribute 1 to both the indegree and outdegree of v and so $\operatorname{od} v = \operatorname{id} v$ in this case as well.

For the converse, let D be a nontrivial connected digraph for which $\operatorname{od} v = \operatorname{id} v$ for every vertex v of D. For a vertex u of D, let T be a trail of maximum length with initial vertex u. Suppose that T is a $u-v$ trail. Assume first that $u \ne v$ and that v is encountered k times on T, where $k \ge 1$. Then T contains k arcs directed towards v and $k-1$ arcs directed away from v. However, since $\operatorname{od} v = \operatorname{id} v$, there is an arc directed away from v that does not belong to T. This means, however, that T can be extended to a longer trail with initial vertex u. Since this is impossible, $u = v$ and T is a circuit C in D. Consequently, D contains circuits and C is a circuit of maximum length in D.

We claim that C contains all of the arcs of D and that C is therefore an Eulerian circuit. Assume, to the contrary, that C does not contain all of the arcs of D. Since D is connected, there is a vertex w on C that is incident with arcs not on C. Let $D' = D - E(C)$ be the spanning subdigraph of D whose arcs are those not belonging to C. Since $\operatorname{od}_D v = \operatorname{id}_D v$ and $\operatorname{od}_C v = \operatorname{id}_C v$ for every vertex v on C, it follows that $\text{od}_{D'} v = \text{id}_{D'} v$ for every vertex of D'. Let T' be a trail of maximum length in D' with initial vertex w. As before, T' is a circuit C' in D''. If we attach C' to C at w, then we produce a circuit C'' in D containing more arcs than C, which is impossible. Hence C is an Eulerian circuit.   ∎

We mentioned earlier that in Town A (Figure 7.1) it is possible to close down any street segment in the town and, afterwards, still be able to travel between any two street intersections; while in Town B this is not possible because of the existence of a bridge in the graph G_B that models the street system of this town. We also mentioned that this characteristic of Town A allows something else to be done there. Suppose that the town commissioners in Town A, in their infinite wisdom, decide that it would be convenient (for whatever reason) to convert all the streets in the town to one-way streets. Can this be done? The answer to this question is of course yes but this is not the question that should be asked. Is it possible to convert all the streets in Town A to one-way streets so that, afterwards, it is possible to drive (legally) from any place in Town A to any other place. The answer to *this question* is also yes and one way to accomplish this is shown in Figure 7.5. This new street system is modeled by the digraph D_A in Figure 7.5. You have probably noticed that the question that we have just asked can be rephrased as follows: Does there exist a strong orientation of the graph G_A of Figure 7.1? Of course, we now know that the answer to this question is yes. The more general question is: Which graphs have strong orientations? The graphs G_A and G_B of Figure 7.1 provide the clue.

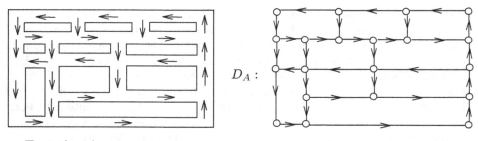

Town A with one-way streets

Figure 7.5: A digraph modeling the one-way streets system

Theorem 7.5 *A nontrivial connected graph G has a strong orientation if and only if G contains no bridges.*

Proof. Suppose first that G is a nontrivial connected graph that contains a bridge, say $e = uv$. Let D be any orientation of G. Then either (u, v) or (v, u) is an arc of D, say (u, v). Surely, D contains a $u - v$ path. We claim that there is no $v - u$ path in D; for if D contains a $v - u$ path P, then P can be considered as a $v - u$ path P' in G that does not contain uv in G. However, the path P' in G together with the edge uv produce a cycle in G that contains e, which is impossible since e is a bridge. Thus, as claimed, D contains no $v - u$ path and so D is not strong.

To verify the converse, let G be a connected graph that contains no bridges. We show that G has a strong orientation. Since G contains no bridges, G has a cycle C. If we direct the edges of C to produce a directed cycle C', then for every two vertices x and y on C', there is both an $x - y$ path and a $y - x$ path on C'. Thus, it is certainly possible to direct some of the edges of G to obtain a digraph D' so that, afterwards, there is a set U of vertices of D' where there is both an $x - y$ path and a $y - x$ path for every two vertices x and y of U. If an edge of G that joins two vertices of U has not been assigned a direction, then we may assign any direction and obtain the same conclusion.

Consequently, there is a set S of vertices of G of largest cardinality and an orientation D of the edges of G joining two vertices of S such that, afterwards, for every two distinct vertices x and y in S there is both an $x - y$ path and a $y - x$ path in D. If $S = V(G)$, then the proof is complete. Assume, however, that $S \neq V(G)$. Since G is connected, there must be a vertex $u \in S$ and a vertex $v \notin S$ such that $uv \in E(G)$. Since uv is not bridge, uv lies on a cycle

$$C'' = (u, v = v_1, v_2, \ldots, v_s = u)$$

of G. Of course, $u \in S$ but u may not be the first vertex of C'' following v that belongs to S. Let $w = v_t$ $(t \leq s)$ be the first such vertex. Now direct the edges $uv, vv_2, \ldots, v_{t-1}v_t$ as $(u, v), (v, v_2), \ldots, (v_{t-1}, v_t)$ (see Figure 7.6) and let P be the (directed) $v - v_t$ path produced. If any other edge joins a vertex of $T = \{v_1, v_2, \ldots, v_{t-1}\}$ and a vertex of $S \cup T$, then direct this edge arbitrarily. Let D' be the resulting digraph.

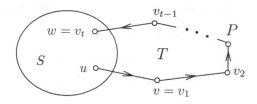

Figure 7.6: Producing a strong orientation

It then follows that for each pair x, y of vertices of $S \cup T$, there is both an $x - y$ path and a $y - x$ path in D'. This contradicts S as being a proper subset of $V(G)$ of largest cardinality for which there exist both an $x - y$ path and a $y - x$ path for each pair $x, y \in S$ in an orientation of G. ∎

It was mentioned in Chapter 5 that a graph G is 2-edge-connected if G is a nontrivial connected graph that contains no bridges. Hence Theorem 7.5 can be restated as follows:

> *A nontrivial connected graph G has a strong orientation if and only if G is 2-edge-connected.*

Theorem 7.5 is due to Herbert E. Robbins (1915-2001). The paper in which this theorem appears, titled "A theorem on graphs, with an application to a problem of traffic control," was published in 1939 in the *American Mathematical Monthly*, only a year after he received his Ph.D. from Harvard University in topology, under the direction of Hassler Whitney. This was only Robbins' second publication of what was to become a long and impressive list. Also, in 1939 at age 24, Robbins began work on the classic book *What Is Mathematics?* with Richard Courant. This book has been classified by Robbins as more of a literary work than a scientific work. A few years later Robbins became interested in and devoted his research to statistical analysis, in which he made major contributions. He spent many years as a Professor of Mathematical Statistics at Columbia University. In 1958 he had a doctoral student, Herbert Wilf, who also has made major contributions to combinatorics and graph theory.

Exercises for Section 7.1

7.1 (a) Prove that if D is an oriented graph of order 4 such that $D - v$ is strong for every vertex v of D, then D is strong.

   (b) Show that no oriented graph D of order 4 has the property that $D - v$ is strong for every vertex v of D.

7.2 Prove that a graph G has an Eulerian orientation if and only if G is Eulerian.

7.3 Show that each of the graphs G_1 and G_2 in Figure 7.7 is orientable by assigning a direction to each edge so that the resulting digraph is strong.

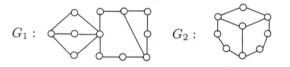

Figure 7.7: The graphs of Exercise 7.3

7.4 The **converse** $\vec{D}$ of a digraph D is obtained from D by reversing the direction of every arc of D. Show that a digraph D is strong if and only if its converse $\vec{D}$ is strong.

7.5 Prove that a nontrivial digraph D is strong if and only if for every edge-cut S of the underlying graph G of D separating $V(G - S)$ into two sets A and B, there is an arc in D directed from A to B and an arc in D directed from B to A.

7.6 Does there exist a nontrivial digraph D in which no two vertices of D have the same outdegree but every two vertices of D have the same indegree?

7.2 Tournaments

It is difficult to know just how far back competitions go. There have been competitions between two individuals (tennis, chess, bridge, jousting) and competitions between two teams (soccer, basketball, baseball). There have even been competitions between frogs, as Mark Twain wrote of Dan'l Webster in *The Celebrated Jumping Frog of Calaveras County*. In some competitions, there is a single match between two individuals or two teams and the victor in the match decides the outcome of the competition. In other competitions, often called tournaments, several individuals (or teams) are involved and there is a formula to decide who plays whom. Losing a match causes that individual or team to be eliminated and the tournament continues with those individuals who have won the earlier matches. Other tournaments are "double elimination," where a player or team is allowed to lose one match but is eliminated when a second loss occurs.

Other tournaments are "round robin tournaments," where each team plays every other team exactly once in the competition and the outcome of the tournament is decided only after all these matches have been played. For example, suppose that a round robin tournament involves eight teams (denoted by 1, 2, ..., 8). Then every team must play each of the other seven teams once. In the first "round," there are then four matches, each involving a pair of teams. There are seven rounds in this round robin tournament. Figure 7.8 shows how such a schedule might look. If only seven teams were involved, then in any round robin tournament only three matches can take place in a round with one team not playing (this team receives a "bye"). In this case, we can replace each occurrence of 8 in Figure 7.8 with "bye." We will see in Section 8.2 how such schedules can be constructed.

Round 1	Round 2	Round 3	Round 4	Round 5	Round 6	Round 7
1–2	1–3	1–4	1–5	1–6	1–7	1–8
3–4	2–8	2–6	2–3	2–7	2–5	2–4
5–6	4–5	3–7	6–8	3–5	3–8	3–6
7–8	6–7	5–8	4–7	4–8	4–6	5–7

Figure 7.8: A round robin tournament with eight teams

Round robin tournaments give rise quite naturally to a class of digraphs, not so coincidentally called tournaments. A **tournament** is an orientation of a complete graph. Therefore, a **tournament** can be defined as a digraph such that for every pair u, v of distinct vertices, exactly one of (u, v) and (v, u) is an arc. A tournament T then models a round robin tournament. The vertices of T are the teams in the round robin tournament and (u, v) is an arc in T if team u defeats team v. (Ties are not permitted.)

Recall that two digraphs D and D' are **isomorphic**, written $D \cong D'$, if there exists a bijective function $\phi : V(D) \to V(D')$ such that $(u, v) \in E(D)$ if and only if $(\phi(u), \phi(v)) \in E(D')$. Such a function ϕ is called an **isomorphism**. There is only one tournament of order 1 and only one tournament of order 2 (up to isomorphism). There are two tournaments of order 3 and four tournaments of order 4. The tournaments of order 4 or less are shown in Figure 7.9. There are also 12 tournaments of order 5. Based on this information, it probably comes as a great surprise to learn that there are over 154 billion tournaments of order 12.

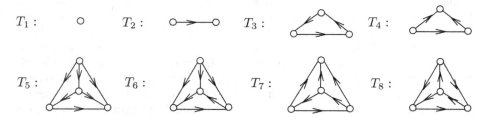

Figure 7.9: Tournaments of order 4 or less

A tournament T is **transitive** if whenever (u, v) and (v, w) are arcs of T, then (u, w) is also an arc of T. The tournaments T_1, T_2, T_4 and T_5 of Figure 7.9 are transitive. In fact, for every positive integer n, there is a unique transitive tournament of order n (again, up to isomorphism). If T is a transitive tournament of order n and i is an integer with $0 \le i \le n - 1$, there is a vertex v_i in T such that od $v_i = i$. Transitive tournaments have a property that no other tournaments have.

Theorem 7.6 *A tournament T is transitive if and only if T has no cycles.*

Proof. Let T be a transitive tournament and assume, to the contrary, that T contains a cycle $C = (v_1, v_2, \ldots, v_k, v_1)$. Since (v_1, v_2) and (v_k, v_1) are arcs of T, there are vertices on C to which v_1 is adjacent and vertices on C from which v_1 is adjacent. Hence there must be a vertex v_i $(2 \le i \le k - 1)$ such that (v_1, v_i) and (v_{i+1}, v_1) are arcs of T. Since (v_1, v_i) and (v_i, v_{i+1}) are arcs of a transitive tournament, (v_1, v_{i+1}) is an arc of T. This is a contradiction.

For the converse, assume that T is a tournament that contains no cycles. Let (u, v) and (v, w) be two arcs of T. Since T contains no cycles, (w, u) is not an arc of T, implying that (u, w) *is* an arc of T and so T is transitive. ∎

As we have seen, if T is a transitive tournament of order n, then there is a unique vertex u of T having outdegree $n - 1$, which is certainly the largest outdegree of any vertex of T. Therefore, u is adjacent to all other vertices of T and so $\vec{d}(u, v) \le 1$ for every vertex v of T. Nearly the same thing is true for any vertex of maximum outdegree in a tournament (transitive or not).

Theorem 7.7 *If u is a vertex of maximum outdegree in a tournament T, then $\vec{d}(u, v) \le 2$ for every vertex v of T.*

Proof. Suppose that od $u = k$ and let $v_1, v_2, \ldots, v_k$ be the k vertices of T that are adjacent from u. If there are no other vertices of T, then $\vec{d}(u, v) \le 1$ for every vertex v of T.

Assume then that there are some vertices of T adjacent to u, say $w_1, w_2, \ldots, w_\ell$ (see Figure 7.10). We have already noted that $\vec{d}(u, v_i) = 1$ for $1 \leq i \leq k$. We show that $\vec{d}(u, w_j) = 2$ for each vertex w_j with $1 \leq j \leq \ell$. If some vertex v_i $(1 \leq i \leq k)$ is adjacent to w_j, then certainly $\vec{d}(u, w_j) = 2$. On the other hand, if this is not the case, then w_j is adjacent to all of the vertices $v_1, v_2, \ldots, v_k$. Since w_j is also adjacent to u, it follows that $\operatorname{od} w_j \geq k + 1 > k = \operatorname{od} u$, which is impossible. ∎

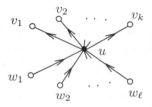

Figure 7.10: A step in the proof of Theorem 7.7

Suppose that we have a collection of teams involved in a round robin tournament. As we have seen, the results of the matches can be modeled by a tournament T (a digraph). The outdegree of a vertex in T is then the number of matches won by this team. Let A be a team that has won the most matches. According to Theorem 7.7, if B is any other team, then either (1) A defeated B or (2) A defeated a team that defeated B.

As with graphs, a path P in a digraph D is a **Hamiltonian path** of D if P contains all vertices of D. A cycle C in D is a **Hamiltonian cycle** if C contains every vertex of D. If D has a Hamiltonian cycle, then D is a **Hamiltonian digraph**. We now describe a property possessed by all tournaments that was first observed by László Rédei.

Theorem 7.8 *Every tournament contains a Hamiltonian path.*

Proof. Let P be a path of greatest length in a tournament T, say

$$P = (v_1, \, v_2, \, \ldots, \, v_k).$$

If P contains every vertex of T, then P is a Hamiltonian path. Suppose then that P is not a Hamiltonian path. Thus there exists a vertex v of T that is not on P (see Figure 7.11). Neither (v, v_1) nor (v_k, v) is an arc of P, for otherwise T contains a path whose length exceeds the length of P. Thus (v_1, v) and (v, v_k) are arcs of T. This implies, however, that there must be a vertex v_i, $1 \leq i \leq k - 1$, such that v is adjacent from v_i and v_{i+1} is adjacent from v. However then, $P' = (v_1, v_2, \ldots, v_i, v, v_{i+1}, \ldots, v_k)$ is a path whose length is greater than that of P, which is impossible. ∎

According to Theorem 7.8 then, if we have any collection of teams that have participated in a round robin tournament, then the teams can be ordered, say

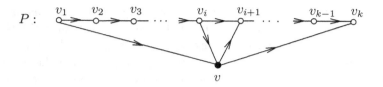

Figure 7.11: A step in the proof of Theorem 7.8

as A_1, A_2, ..., A_n, such that team A_1 has defeated A_2, team A_2 has defeated A_3 and so on. This doesn't necessarily mean that A_1 is the best team and A_n is the worst team, however. For example, in the strong tournament T of order 3 shown in Figure 7.12, there is no logical ordering of the three teams. For example, (A_1, A_2, A_3) and (A_2, A_3, A_1) as well as (A_3, A_1, A_2) are Hamiltonian paths. Indeed, the only time that there is a clear ordering of the teams is when the resulting digraph (tournament) is transitive.

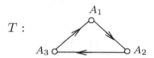

Figure 7.12: Hamiltonian paths in a tournament

If T is a tournament that is not transitive, then not only does T contain a Hamiltonian path but, by Theorem 7.6, T contains cycles. If T is strong, then even more can be said.

Theorem 7.9 *Every vertex in a nontrivial strong tournament belongs to a triangle.*

Proof. Let v be a vertex in a nontrivial strong tournament T. Since T is strong, $\text{od}\,v > 0$ and $\text{id}\,v > 0$. Let U be the set of vertices to which v is adjacent, and let W be the set of vertices from which v is adjacent (see Figure 7.13). Thus $U \neq \emptyset$ and $W \neq \emptyset$. Since T is strong, there is a $v - w$ path for each $w \in W$. Such a path necessarily contains an arc (u, w) for some $u \in U$ and some $w \in W$ and so v lies on the triangle (v, u, w, v).   ■

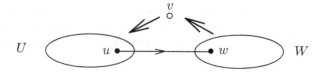

Figure 7.13: A step in the proof of Theorem 7.9

While every tournament contains a Hamiltonian path, certainly not every tournament contains a Hamiltonian cycle. If a tournament T contains a Hamiltonian cycle, then (by Theorem 7.3) T is strong. What may be surprising is that the converse is true as well.

Theorem 7.10 *A nontrivial tournament T is Hamiltonian if and only if T is strong.*

Proof. We have already seen that every Hamiltonian tournament is strong. For the converse, assume that T is a nontrivial strong tournament. Thus T contains cycles. Let C be a cycle of maximum length in T. If C contains all of the vertices of T, then C is a Hamiltonian cycle. So assume that C is not Hamiltonian, say

$$C = (v_1, v_2, \ldots, v_k, v_1),$$

where $3 \le k < n$. If T contains a vertex v that is adjacent to some vertex of C and adjacent from some vertex of C, then there must be a vertex v_i of C that is adjacent to v such that v_{i+1} is adjacent from v. In this case,

$$C' = (v_1, v_2, \ldots, v_i, v, v_{i+1}, \ldots, v_k, v_1)$$

is a cycle whose length is greater than that of C, producing a contradiction. Hence, every vertex of T that is not on C is either adjacent to every vertex of C or adjacent from every vertex of C. Since T is strong, there must be vertices of each type.

Let U be the set of all vertices of T that are not on C and such that each vertex of C is adjacent to every vertex of U and let W be the set of those vertices of T that are not on C such that every vertex of W is adjacent to each vertex of C (see Figure 7.14). Then $U \ne \emptyset$ and $W \ne \emptyset$.

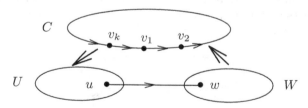

Figure 7.14: A step in the proof of Theorem 7.10

Since T is strong, there is a path from every vertex of C to every vertex of W. Since no vertex of C is adjacent to any vertex of W, there must be a vertex $u \in U$ that is adjacent to a vertex $w \in W$. However then,

$$C'' = (v_1, v_2, \ldots, v_k, u, w, v_1)$$

is a cycle whose length is greater than that of C, a contradiction. ∎

Theorem 7.10 is due to Paul Camion. The results of Rédei and Camion are the fundamental theorems on traversability in tournaments. There is only one strong tournament T of order 4. Since every vertex in a strong tournament belongs to a triangle, there is a vertex v in T such that $T - v$ is also strong. This statement is true for strong tournaments of every order greater than 4 as well. The proof of the following result uses the same approach as the proof of the preceding result.

Theorem 7.11 *If T is a strong tournament of order $n \geq 4$, then there exists a vertex v of T such that $T - v$ is a strong tournament.*

Proof. Since the result is true for $n = 4$, we can assume that $n \geq 5$. Assume, to the contrary, that the theorem is false. Then there exists a strong tournament of order $n \geq 5$ such that for every vertex v of T, the tournament $T - v$ is not strong. By Theorem 7.10, this implies that T contains no cycle of length $n - 1$. Let C be a cycle of greatest length in T that is not a Hamiltonian cycle. Suppose that

$$C = (v_1, v_2, \ldots, v_k, v_1),$$

where $3 \leq k \leq n - 2$. If there exists a vertex x not on C that is adjacent to some vertices of C and adjacent from some vertices of C, then there is some vertex v_i on C such that (v_i, x) and (x, v_{i+1}) are arcs of T. However, then,

$$C' = (v_1, v_2, \ldots, v_i, x, v_{i+1}, \ldots, v_k, v_1)$$

is a cycle of length $k + 1$, which is a contradiction.

This implies that every vertex of T that is not on C is either adjacent to all vertices of C or is adjacent from all vertices of C. Let U be the set of vertices of T that are not on C and that are adjacent from all vertices of C and let W be the set of vertices of T that are not on C and that are adjacent to all vertices of C. Then $U \neq \emptyset$ and $W \neq \emptyset$ (see Figure 7.15).

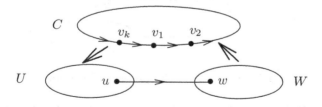

Figure 7.15: A step in the proof of Theorem 7.11

Since T is strong, there exist vertices $u \in U$ and $w \in W$ such that (u, w) is an arc of T. However, then,

$$C'' = (v_1, v_2, \ldots, v_{k-1}, u, w, v_1)$$

is a cycle of length $k + 1$, which is impossible.   ∎

As a consequence of Theorem 7.11, every strong tournament of order $n \geq 3$ contains an induced strong tournament of order k for every integer k with $3 \leq k \leq n$.

Exercises for Section 7.2

7.7 Prove that there is only one tournament T of order n, where $3 \leq n \leq 5$, such that T and $T - (u, v) + (v, u)$ are strong for every arc (u, v) of T.

7.8 If every vertex of some tournament of order n has the same outdegree x, then what is x?

7.9 Prove that a tournament T is transitive if and only if every two vertices of T have distinct outdegrees.

7.10 Prove that if u and v are vertices of a tournament such that $\vec{d}(u, v) = k$, then id $u \geq k - 1$.

7.11 Let T be a tournament of order $n \geq 3$ with $V(T) = \{v_1, v_2, \ldots, v_n\}$. Prove that if od $v_i >$ id v_i for $1 \leq i \leq n - 1$, then T is not strong.

7.12 Prove or disprove:

(a) If every vertex of a tournament T belongs to a cycle in T, then T is strong.

(b) For every pair u, v of vertices in a strong tournament T, there exists either a Hamiltonian $u - v$ path or a Hamiltonian $v - u$ path.

(c) If (u, v) is an arc of a strong tournament T, then (u, v) lies on a Hamiltonian cycle of T.

7.13 Let u and v be distinct vertices in a tournament such that $\vec{d}(u, v)$ and $\vec{d}(v, u)$ are defined. Show that $\vec{d}(u, v) \neq \vec{d}(v, u)$.

7.14 (a) Show that if an odd number of teams play in a round robin tournament, then it is possible for all teams to tie for first place.

(b) Show that if an even number of teams play in a round robin tournament, then it is not possible for all teams to tie for first place.

7.15 Prove that if T is a strong tournament of order $n \geq 3$, then T contains a cycle of length k for every integer k with $3 \leq k \leq n$.

7.3   Excursion: Decision–Making

In the United States presidential election of 2000, George W. Bush narrowly defeated Al Gore. Although Gore received a higher popular vote total than Bush, the winner of the election was Bush because his electoral vote total was higher than that of Gore.

During a typical year, there are numerous occasions when decisions are made by voting. Whether it's electing a president, a prime minister, a senator, a governor, a mayor or a student representative on a committee, decisions must be made as to which individuals will hold these positions. Furthermore, a procedure must be in place to determine how this decision will be made. The answer may seem simple. The decision is made by voting. However, this is not as simple as it may first appear. If there are several candidates for a certain position, then there is a variety of ways of deciding the outcome of an election. It would seem that it is easy to decide the outcome of a two-person election and in general this is true, with the aforementioned 2000 United States presidential election being a possible exception (even though there were more than two candidates for president). Making a decision among several choices is not restricted to governmental or college elections, however.

Example 7.12 *Al, his wife Barbara and their three children Cassie, Donna and Edwin have discussed which new car they should purchase and have agreed that the choice should be made from a General Motors car (GM), a Honda (H), a Chrysler (C), a Toyota (T) and a Ford (F). Al and Barbara also agreed that this should be a family decision and that each family member would have an equal voice in the decision.*

Actually, Al's preferences coincide exactly with the order of cars listed above. That is, Al prefers a General Motors car to a Honda, a Honda to a Chrysler and so on. Al's preferences are given in the tournament shown in Figure 7.16. For example, the directed edge from C to F indicates that Al prefers a Chrysler to a Ford. The tournament in Figure 7.16 is called the **tournament of paired comparisons** for Al's preferences as it indicates his preferred choices for each pair of cars.

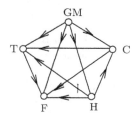

Figure 7.16: Al's tournament of paired comparisons

The tournaments of paired comparisons for all family members are given in Figure 7.17. All of these tournaments are transitive, as expected. For example, since Barbara prefers a Honda over a Ford and a Ford over a Toyota, one would expect that she prefers a Honda over a Toyota, which, in fact, she does. For the decision as to which car should be purchased, Figure 7.18 shows the single tournament of paired comparisons for the entire family. For example, all three children prefer a Ford over a General Motors car, while both parents prefer the General Motors car to a Ford. Since the majority of the family members prefer a Ford to a General Motors car, the family prefers a Ford to a General Motors car and so there is an arc from F to GM. All other arcs in this tournament are obtained in the same manner. Now that we have all the information, the question is: Which car should be purchased? This question doesn't appear to have an easy answer. At least, it doesn't seem to have an obvious answer. The problem is that even though every tournament in Figure 7.17 is transitive, the tournament in Figure 7.18 constructed from these five tournaments is not transitive. ◇

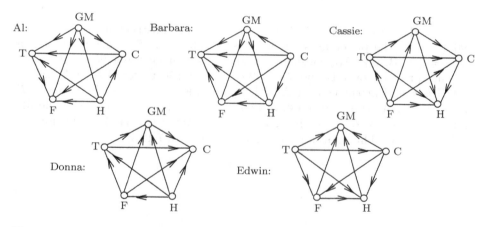

Figure 7.17: The tournaments of paired comparisons for all five family members

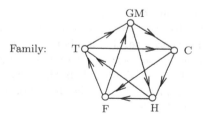

Figure 7.18: The family tournament of paired comparisons

Let's turn to another example.

Example 7.13 *Suppose that a college is having an election for student president and this year there are three candidates: Atkins, Bennett and Chapman.*

In order to have the full input of the students, each student is asked to cast his/her vote by making one of the choices listed below:

○	○	○	○	○	○
Atkins	Atkins	Bennett	Bennett	Chapman	Chapman
Bennett	Chapman	Chapman	Atkins	Bennett	Atkins
Chapman	Bennett	Atkins	Chapman	Atkins	Bennett

For example, checking the box in the third column in the list would mean that the first choice of the person voting is Bennett (B), the second choice is Chapman (C) and the third choice is Atkins (A). The voting takes place and here is the outcome:

100	500	75	425	50	350
A	A	B	B	C	C
B	C	A	C	A	B
C	B	C	A	B	A

What do we do with this information? Let's construct a tournament T that provides the preferences among these candidates. The vertex set of T is $V(T) = \{A, B, C\}$. We consider A and B first. Since $100 + 500 + 50 = 650$ students prefer A over B and $75 + 425 + 350 = 850$ prefer B over A, we see that B is the clear choice over A and the directed edge (B, A) is drawn in T. Similarly, (C, A) is a directed edge in T since 825 prefer C over A, while 675 prefer A over C. Also, (C, B) is an arc since 900 prefer C over B, while only 600 prefer B over C. This tournament is shown in Figure 7.19.

$T:$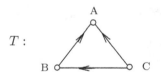

Figure 7.19: The tournament of paired comparisons for the college election

Looking at the tournament T in Figure 7.19, we see that not only is Chapman preferred over the other two candidates, each of Bennett and Chapman is preferred over Atkins. Furthermore, the number of voters with these preferences is quite one-sided. So the decision appears to be very clear. Or is it? After all, how often do you see a voting decision made in this way? It is common to count only the votes for the candidates who are the first choice of the voters. If this is done, then the outcome of the election is

A - 600 votes   B - 500 votes   C - 400 votes.

And the winner is: Atkins. On the other hand, there are often primary elections to determine the top two vote-getters to face off in a general election. For example, in the election above, Chapman received the least votes and would be eliminated from appearing on the ballot in the general election. That is, in the general election, it would be Atkins versus Bennett. But we have already seen that between these two candidates, Bennett would receive more votes than Atkins and would win quite easily. Therefore, by the two most common ways of deciding an election, the winning candidate is not the preferred candidate. ◇

Exercises for Section 7.3

7.16 The preferences of 98 voters for three candidates are shown below.

number of voters	1st choice	2nd choice	3rd choice
18	A	B	C
17	A	C	B
13	B	A	C
16	B	C	A
16	C	A	B
18	C	B	A

(a) If the candidate who is the first choice of most voters wins, then who would win?

(b) Draw the tournament of paired comparisons. Indicate how you decided to draw the three arcs. According to this tournament, which candidate should win?

7.17 This year there are four candidates for the president of the student council at a local college: Archer (A), Benson (B), Chase (C) and Dawkins (D). Each student is asked to order his/her preferences among the four candidates by voting for one of the 4! = 24 ordered lists of candidates. A total of 408 students vote in the election and the outcome is as follows:

12	11	28	10	27	26	11	10	25	9	24	29
A	A	A	A	A	A	B	B	B	B	B	B
B	B	C	C	D	D	A	A	C	C	D	D
C	D	B	D	B	C	C	D	A	D	A	C
D	C	D	B	C	B	D	C	D	A	C	A

10	9	22	12	21	20	11	8	21	7	20	25
C	C	C	C	C	C	D	D	D	D	D	D
A	A	B	B	D	D	A	A	B	B	D	C
B	D	A	D	A	B	B	C	A	C	A	B
D	B	D	A	B	A	C	B	C	A	B	A

Of course, the question now is: Who won the election?

(a) Determine the winner of the election by counting only the first choice of each voter.

(b) Determine the winner of the election by eliminating the candidate who received the smallest number of votes in (a) and then recounting the votes of the three remaining candidates?

(c) Determine the winner of the election by eliminating the two candidates who received the smallest number of votes in (a) and then
recounting the votes of the two remaining candidates?

(d) Determine the winner of the election by constructing the tournament
of paired comparisons of the four candidates.

(e) Who *should* win the election?

7.18 Return to Example 7.12 of the family trying to decide which of five cars
to purchase. Edwin has another idea to make the decision. Start with the
Honda (H) and Ford (F) and determine which of these is the preferred car
of the family. Then do the same thing for the Toyota (T) and the General
Motors car (GM). Then compare the Chrysler (C) and the preferred car
between H and F. The preferred car here is compared against the car
preferred between T and GM. Which car does the family prefer using this
method? Is this a good method?

7.4   Exploration: Wine Bottle Problems

There are games and problems in which success is attained by proceeding through
a sequence of steps. That is, in the process of playing the game or attempting to
solve the problem, an individual may find himself or herself at one of a number
of states and from that state it is possible to move to certain other states by
a single (allowable) step. Such a situation can be modeled by a graph whose
vertices are the states and where two states A and B are adjacent if it is possible
to proceed from A to B by a single step. This is under the assumption that
moving from A to B is reversible by a single reverse step. If, on the other hand,
there are states A and B such that it is possible to proceed from A to B by single
step but not so from B to A, then this situation is more appropriately modeled
by a digraph rather than a graph. We now look at a class of problems that can
be modeled by digraphs.

Example 7.14 *Three wine bottles A, B and C have capacities of 1, 3 and 4
liters, respectively. These bottles are not graduated, however. That is, there are
no markings on the bottles. So, looking at a single bottle, it would be impossible
to know exactly how much wine is in it, unless, of course, the bottle was full or
empty. The largest bottle is filled with wine and the other two containers are
empty. By a* **pouring,** *we mean that the contents of some bottle X containing
wine is poured into a bottle Y until either bottle Y is filled or bottle X is empty.
We wish to divide the wine into two equal portions by pouring successively from
one bottle to another. The problem then is: Can we obtain 2 liters of wine in the
largest bottle and 2 liters in the medium-size bottle and if so, what is the fewest
possible number of pourings needed to accomplish this?*

At any particular time, suppose that bottle A contains a liters of wine, B has b liters of wine and C has c liters of wine. Thus $a + b + c = 4$ and initially $a = b = 0$. Indeed, knowing only a and b tells us how much wine is in all three bottles. To help us answer this question, we construct a digraph D such that

$$V(D) = \{(a, b) : a \in \{0, 1\}, b \in \{0, 1, 2, 3\}\},$$

where (a_1, b_1) is adjacent *to* (a_2, b_2) if we can proceed from (a_1, b_1) to (a_2, b_2) by a single pouring. The answer to the question is therefore the distance from the vertex $(0, 0)$ to the vertex $(0, 2)$ in D. The digraph D is shown in Figure 7.20.

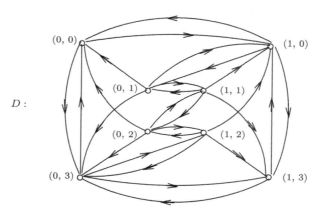

Figure 7.20: The digraph of Example 7.14

Observe that while some steps are reversible, others are not. For example, there is an arc from $(0, 1)$ to $(0, 0)$ but not from $(0, 0)$ to $(0, 1)$. That is, if bottle A is empty and bottle B contains exactly 1 liter of wine, then the contents of bottle B can be poured into bottle C; while if bottle C contains 4 liters of wine, then it is not possible to pour exactly 1 liter of the contents of bottle C into bottle B. Looking at the digraph D of Figure 7.20, we see that the distance from $(0, 0)$ to $(0, 2)$ is 3 and a geodesic is the path

$$((0, 0), (0, 3), (1, 2), (0, 2)).$$

Consequently, beginning with the bottle C filled with wine, it is possible to divide the wine into two equal portions by performing three pourings, but no fewer. Notice that there are paths from $(0, 0)$ to $(0, 2)$ of greater length as well.

Exercises for Section 7.4

7.19 Three wine bottles A, B and C have capacities of 3, 5 and 8 liters, respectively. What is the smallest number of pourings needed to produce

(a) two bottles, each containing 4 liters of wine?

(b) two bottles, one of which contains 2 liters of wine and the other 6 liters of wine?

(c) two bottles, one of which contains 1 liter of wine and the other 7 liters of wine?

7.20 Create a problem of your own (as in Exercise 7.19) by selecting different capacities of three wine bottles.

Chapter 8

Matchings and Factorization

8.1 Matchings

A mathematics department at a university has acquired a collection of 12 different mathematics books on a variety of subjects to be presented to students who have performed well on a competitive mathematics exam (one book to each successful student). Of course, there would be a problem if more than 12 students qualified for these books. It turns out, however, that this is not a problem as only 10 students did well enough on the exam to receive books. Nevertheless, another possible difficulty has arisen. Some of the students already have copies of some books and there are some books that certain students have no need for. The question is this: Is there a way of distributing 10 of the 12 books to the 10 students so that each student receives a book that he or she would like to have? The answer to this problem may be *no* even though there are more books than students. For example, there may be three or more books that no student wants. Also, perhaps there are four students only interested in the same three books, in which case it would be impossible to distribute four books to these four students.

It may already be clear that this situation can be modeled by a graph G whose vertices are the students, say $S_1, S_2, \ldots, S_{10}$ and the books, say $B_1, B_2, \ldots, B_{12}$, where two vertices of G are adjacent if one of these vertices is a student and the other is a book that this student would like to have. Certainly then, G is a bipartite graph with partite sets $U = \{S_1, S_2, \ldots, S_{10}\}$ and $W = \{B_1, B_2, \ldots, B_{12}\}$. For example, if student S_1 would like to have any of the books B_2, B_3, B_5, B_7, then the graph G contains the subgraph shown in Figure 8.1. What we are seeking then is a set A of 10 edges in the graph G (where G is only partially drawn in Figure 8.1), no two of which are adjacent. If such a set A exists, then each vertex S_i ($1 \leq i \leq 10$) is incident with exactly one edge in A.

There is a related mathematical question here. Let U and W be two sets such that $|U| = 10$ and $|W| = 12$. Does there exist a one-to-one function $f : U \rightarrow W$?

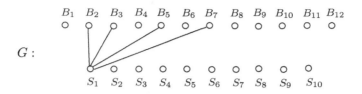

G :

Figure 8.1: A subgraph of a bipartite graph

If this is all there is to the question, then the answer is *yes*. However, what if the image of each element of U cannot be just *any* element of W? The image of each element of U is required to be an element of some prescribed subset of W. Consequently, what we are asking is that if we know the sets of possible images of the elements of U, is there a one-to-one function $f : U \to W$ that satisfies these conditions?

This discussion leads us to some new concepts. A set of edges in a graph is **independent** if no two edges in the set are adjacent. By a **matching** in a graph G, we mean an independent set of edges in G. Thus the problem we were discussing asks whether a particular graph contains a certain matching. Since many problems of this type involve bipartite graphs, as does the problem we were discussing, we first consider these concepts for bipartite graphs only.

Let G be a bipartite graph with partite sets U and W, where $r = |U| \le |W|$. A matching in G is therefore a set $M = \{e_1, e_2, \ldots, e_k\}$ of edges, where $e_i = u_i w_i$ for $1 \le i \le k$ such that $u_1, u_2, \ldots, u_k$ are k distinct vertices of U and $w_1, w_2, \ldots, w_k$ are k distinct vertices of W. We say that M **matches** the set $\{u_1, u_2, \ldots, u_k\}$ to the set $\{w_1, w_2, \ldots, w_k\}$. Necessarily, for any matching of k edges, we must have $k \le r$. The term "matching" is used since the edges of M match or pair off k elements of U with k elements of W. The question in which we are interested can now be phrased as follows: Does G contain a matching of cardinality r? Before continuing with this discussion, let's consider two examples.

Example 8.1 *As a result of doing well on an exam, six students Ashley (A), Bruce (B), Charles (C), Duane (D), Elke (E) and Faith (F) have earned the right to receive a complimentary textbook in either algebra (a), calculus (c), differential equations (d), geometry (g), history of mathematics (h), programming (p) or topology (t). There is only one book on each of these subjects. The preferences of the students are*

  A: d, h, t;   B: g, p, t;   C: a, g, h;   D: h, p, t;   E: a, c, d;   F: c, d, p.

Can each of the students receive a book he or she likes?

Solution. This situation can be modeled by the bipartite graph G of Figure 8.2(a) having partite sets $U = \{A, B, C, D, E, F\}$ and $W = \{a, c, d, g, h, p, t\}$. We are asking if G contains a matching with six edges. Such a matching does

exist, as shown in Figure 8.2(b). From the matching shown in Figure 8.2(b), we
see how six of the seven books can be paired off with the six students.   ◇

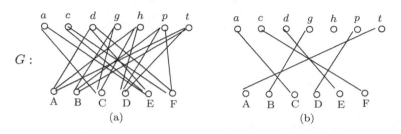

Figure 8.2: A matching in a bipartite graph

Example 8.2 *Seven seniors Ben (B), Don (D), Felix (F), June (J), Kim
(K), Lyle (L) and Maria (M) are looking for positions after they graduate.
The University Placement Office has posted open positions for an accountant
(a), consultant (c), editor (e), programmer (p), reporter (r), secretary (s) and
teacher (t). Each of the seven students has applied for some of these positions:*

B: c, e;	*D: a, c, p, s, t;*	*F: c, r;*	*J: c, e, r;*
K: a, e, p, s;	*L: e, r;*	*M: p, r, s, t.*	

Is it possible for each student to be hired for a job for which he or she has applied?

Solution. This situation can be modeled by the bipartite graph G of Figure 8.3,
where one partite set $U = \{B, D, F, J, K, L, M\}$ is the set of students and the
other partite set $W = \{a, c, e, p, r, s, t\}$ is the set of positions. A vertex $u \in U$ is
joined to a vertex $w \in W$ if u has applied for position w.

The answer to this question is *no* as Ben, Felix, June and Lyle have only
applied for some or all the positions of consultant, editor and reporter. So not
all of these four students can be hired for the jobs for which they have applied.
Consequently, not all seven students can be hired for the seven positions. What
we have observed for the bipartite graph G of Figure 8.3 is that there is no
matching with seven edges. What we gave for an explanation is that there is
a subset $X = \{B, F, J, L\}$ of U containing four vertices whose neighbors belong
to a set $\{c, e, r\}$ of only three vertices. As we are about to see, this is the key
reason why this or any bipartite graph with partite sets U and W such that
$r = |U| \le |W|$ does not contain a matching with r edges.   ◇

Let G be a bipartite graph with partite sets U and W such that $|U| \le |W|$.
For a nonempty set X of U, the **neighborhood** $N(X)$ of X is the union of the
neighborhoods $N(x)$, where $x \in X$. Equivalently, $N(X)$ consists of all those
vertices of W that are the neighbors of one or more vertices in X. The graph
G is said to satisfy **Hall's condition** if $|N(X)| \ge |X|$ for *every* nonempty
subset X of U. This condition is named for Philip Hall, whom we will visit

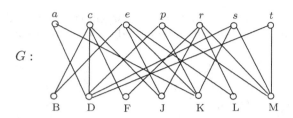

Figure 8.3: A graph modeling the situation in Example 8.2

shortly. The partite set $U = \{B, D, F, J, K, L, M\}$ in the bipartite graph G of Example 8.2 (shown in Figure 8.3) does not satisfy Hall's condition since the subset $X = \{B, F, J, L\}$ of U has the property that $|N(X)| < |X|$. It turns out, however, that the bipartite graph G of Example 8.1 does satisfy Hall's condition.

Theorem 8.3 *Let G be a bipartite graph with partite sets U and W such that $r = |U| \le |W|$. Then G contains a matching of cardinality r if and only if G satisfies Hall's condition.*

Proof. If Hall's condition is not satisfied, then there is some subset S of U such that $|S| > |N(S)|$. Since S cannot be matched to a subset of W, it follows that U cannot be matched to a subset of W.

The converse is verified by the Strong Principle of Mathematical Induction. We proceed by induction on the cardinality of U. Suppose first that Hall's condition is satisfied and $|U| = 1$. Since $|N(U)| \ge |U| = 1$, there is a vertex in W adjacent to the vertex in U and so U can be matched to a subset of W. Assume, for an integer $k \ge 2$, that if G_1 is any bipartite graph with partite sets U_1 and W_1, where

$$|U_1| \le |W_1| \text{ and } 1 \le |U_1| < k,$$

that satisfies Hall's condition, then U_1 can be matched to a subset of W_1. Let G be a bipartite graph with partite sets U and W, where $k = |U| \le |W|$, such that Hall's condition is satisfied. We show that U can be matched to a subset of W. We consider two cases.

Case 1. For every subset S of U such that $1 \le |S| < |U|$, it follows that $|N(S)| > |S|$. Let $u \in U$. By assumption, u is adjacent to two or more vertices of W. Let w be a vertex adjacent to u. Now let H be the bipartite subgraph of G with partite sets $U - \{u\}$ and $W - \{w\}$. For each subset S of $U - \{u\}$, $|N(S)| \ge |S|$ in H. By the induction hypothesis, $U - \{u\}$ can be matched to a subset of $W - \{w\}$. This matching together with the edge uw shows that U can be matched to a subset of W.

Case 2. There exists a proper subset X of U such that $|N(X)| = |X|$. Let F be the bipartite subgraph of G with partite sets X and $N(X)$. Since Hall's condition is satisfied in F, it follows by the induction hypothesis that X can be matched to a subset of $N(X)$. Indeed, since $|N(X)| = |X|$, the set X can be matched to $N(X)$. Let M' be such a matching.

Next, consider the bipartite subgraph H of G with partite sets $U - X$ and $W - N(X)$. Let S be a subset of $U - X$ and let

$$S' = N(S) \cap (W - N(X)).$$

We show that $|S| \leq |S'|$. By assumption, $|N(X \cup S)| \geq |X \cup S|$. Hence

$$|N(X)| + |S'| = |N(X \cup S)| \geq |X| + |S|.$$

Since $|N(X)| = |X|$, it follows that $|S'| \geq |S|$. Thus Hall's condition is satisfied in H and so there is a matching M'' from $U - X$ to $W - N(X)$. Therefore, $M' \cup M''$ is a matching from U to W in G. ∎

Theorem 8.3 is also due to Philip Hall, who was a well-known algebraist. Hall was born on April 11, 1904 in Hempstead, London, England and grew to love mathematics as a young student. His interest in mathematics was greatly influenced by the mathematics teachers he had, who were not only fine mathematicians, they were enthusiastic mathematicians. Hall excelled in English as well. Although neither outgoing nor athletic, Hall was popular as a student. He went on to King's College Cambridge where he was encouraged to study the work of William Burnside and became interested in group theory. Hall received his B.A. in 1925. Only after a great deal of thought did he decide to pursue an academic career.

Hall obtained a fellowship at King's College in 1927. He corresponded with Burnside who was very helpful to Hall, although the two never met. Later in 1927 Hall obtained an important result in group theory, generalizing the Sylow theorems for finite solvable groups, which is now often called Hall's theorem. This theorem was published in 1928. Even though his fellowship was renewed in 1930, a second renewal appeared unlikely due to his lack of mathematical activity for three years. He then published a paper in 1932 on groups of prime power order, perhaps his best known work. In 1933 he was then appointed a Lecturer at Cambridge. In 1935 his theorem on matchings (Theorem 8.3) was published, although it was not stated in terms of graph theory.

Except for a period during World War II when he worked for a Foreign Office at Bletchley Park, he remained at Cambridge from 1933 to 1967. Hall spent much of his life making important contributions to algebra and is considered one of the great mathematicians of the 20th century. A man of high intellectual standards and sound judgment, Hall cared greatly for his students and his students cared greatly for him. Hall was an elegant writer but when it became necessary for him to criticize the writing of his students, he found gentle ways to suggest improvements. Even after his students left upon completing their degrees, he remained in contact with them and encouraged them. He died on December 30, 1982.

As we mentioned, Theorem 8.3 was not stated in terms of graphs. Let $S_1, S_2, \ldots, S_n$ be nonempty finite sets. Then this collection of sets has a **system**

of distinct representatives if there exist n distinct elements $x_1, x_2, \ldots, x_n$ such that $x_i \in S_i$ for $1 \leq i \leq n$. Of course, in order for the sets $S_1, S_2, \ldots, S_n$ to have a system of distinct representatives, $|S_1 \cup S_2 \cup \ldots \cup S_n| \geq n$.

For example, consider the sets $S_1, S_2, \ldots, S_7$, where

$$S_1 = \{1, 2, 3\} \quad S_2 = \{2, 4, 6\} \quad S_3 = \{3, 4, 5\} \quad S_4 = \{1, 4, 7\}$$
$$S_5 = \{1, 5, 6\} \quad S_6 = \{3, 6, 7\} \quad S_7 = \{2, 5, 7\}.$$

Then this collection of sets has a system of distinct representatives. In particular, $1, 2, \ldots, 7$ (that is, $i \in S_i$ for $i = 1, 2, \ldots, 7$) is a system of distinct representatives. On the other hand, the sets $S_1', S_2', \ldots, S_6'$, where

$$S_1' = \{1, 3, 5, 6\} \quad S_2' = \{3, 4\} \quad S_3' = \{4, 5\}$$
$$S_4' = \{3, 4, 5\} \quad S_5' = \{1, 2, 4, 6\} \quad S_6' = \{3, 5\},$$

do not have a system of distinct representatives as $S_2' \cup S_3' \cup S_4' \cup S_6' = \{3, 4, 5\}$, so distinct representatives do not exist for the sets S_2', S_3', S_4', S_6'.

These examples may very well suggest that what is needed for a collection of sets to have a system of distinct representatives is exactly what is required in a bipartite graph G to have one partite set U matched to a subset of the other partite set W of G.

Theorem 8.4 *A collection $\{S_1, S_2, \ldots S_n\}$ of nonempty finite sets has a system of distinct representatives if and only if for each integer k with $1 \leq k \leq n$, the union of any k of these sets contains at least k elements.*

Proof. Suppose that $\{S_1, S_2, \ldots, S_n\}$ has a system of distinct representatives. Then, necessarily, for each integer k with $1 \leq k \leq n$, the union of any k of these sets contains at least k elements. So only the converse needs to be verified.

Let $\{S_1, S_2, \ldots, S_n\}$ be a collection of n sets such that for each integer k with $1 \leq k \leq n$, the union of any k of these sets contains at least k elements. We construct a bipartite graph G with partite sets $U = \{S_1, S_2, \ldots, S_n\}$ and $W = S_1 \cup S_2 \cup \cdots \cup S_n$, where a vertex S_i $(1 \leq i \leq n)$ in U is adjacent to a vertex w in W if $w \in S_i$. Let X be any subset of U, where $|X| = k$ with $1 \leq k \leq n$. Since the union of any k sets contains at least k elements, $|N(X)| \geq |X|$. Therefore, G satisfies Hall's condition. By Theorem 8.3, G contains a matching of cardinality n, which pairs off the sets $S_1, S_2, \ldots, S_n$ with n distinct elements in $S_1 \cup S_2 \cup \ldots \cup S_n$, producing a system of distinct representatives for these sets.   ∎

Hall actually proved Theorem 8.4 on sets which has Theorem 8.3 as an equivalent formulation. It was in fact Dénes König who recognized Hall's theorem as a theorem in graph theory. Theorem 8.3 is sometimes stated in "more friendly" terms and goes by the name of the **Marriage Theorem**.

Theorem 8.5 (The Marriage Theorem) *In a collection of r women and r men, a total of r marriages between acquainted couples is possible if and only if for each integer k with $1 \leq k \leq r$, every subset of k women is collectively acquainted with at least k men.*

In Theorem 8.3 our interest was not only in matchings but matchings of maximum cardinality. Such a matching is called a **maximum matching**. At this point, we no longer assume that we are dealing with bipartite graphs only. If G is a graph of order n, then the cardinality of a maximum matching cannot exceed $\lfloor n/2 \rfloor$. That is, if G is a graph of (odd) order $2\ell + 1$, then no matching contains more than ℓ edges; while if G has (even) order $2k$, then no matching contains more than k edges. If a graph G of order $2k$ has a matching M of cardinality k, then this (necessarily maximum) matching M is called a **perfect matching** as M matches *every* vertex of G to some vertex of G. For example, every nonempty regular bipartite graph contains a perfect matching, a result obtained by Dénes König.

Theorem 8.6 *Every r-regular bipartite graph ($r \geq 1$) has a perfect matching.*

Proof. Let G be an r-regular bipartite graph with partite sets U and W. Necessarily $|U| = |W|$. Let X be a nonempty subset of U. Suppose that $|X| = k \geq 1$. Since every vertex of X has degree r in G, there are kr edges of G incident with vertices of X. Furthermore, since each vertex of W is incident with at most r of these kr edges, every vertex in $N(X)$ is incident with at most r edges and so $|N(X)| \geq k = |X|$. By Theorem 8.3, G has a perfect matching. ∎

There is a parameter directly associated with matchings (and maximum matchings). The **edge independence number** $\alpha'(G)$ of a graph G is the maximum cardinality of an independent set of edges. Therefore, if M is a maximum matching in G, then $\alpha'(G) = |M|$. Furthermore, a graph G of order n has a perfect matching if and only if n is even and $\alpha'(G) = n/2$. For an integer $n \geq 3$ and integers r and s with $1 \leq r \leq s$,

$$\alpha'(C_n) = \alpha'(K_n) = \lfloor n/2 \rfloor \text{ and } \alpha'(K_{r,s}) = r.$$

There is another parameter closely related to the edge independence number. A vertex and an incident edge are said to **cover** each other. An **edge cover** of a graph G without isolated vertices is a set of edges of G that covers all vertices of G. The **edge covering number** $\beta'(G)$ of a graph G is the minimum cardinality of an edge cover of G. An edge cover of G of cardinality $\beta'(G)$ is a **minimum edge cover** of G. Thus $\beta'(G)$ is defined if and only if G has no isolated vertices. For an integer $n \geq 3$ and integers r and s with $1 \leq r \leq s$,

$$\beta'(C_n) = \beta'(K_n) = \lceil n/2 \rceil \text{ and } \beta'(K_{r,s}) = s.$$

Therefore,

$$\alpha'(C_n) + \beta'(C_n) = \alpha'(K_n) + \beta'(K_n) = n,$$

while

$$\alpha'(K_{r,s}) + \beta'(K_{r,s}) = r + s.$$

These illustrate the following theorem.

Theorem 8.7 *For every graph G of order n containing no isolated vertices,*

$$\alpha'(G) + \beta'(G) = n.$$

Proof. First, suppose that $\alpha'(G) = k$. Then a maximum matching of G consists of k edges, which then cover $2k$ vertices. The remaining $n - 2k$ vertices of G can be covered by $n - 2k$ edges. Thus $\beta'(G) \leq k + (n - 2k) = n - k$. Hence

$$\alpha'(G) + \beta'(G) \leq k + (n - k) = n.$$

It remains only to show that $\alpha'(G) + \beta'(G) \geq n$.

Let X be a minimum edge cover of G. Hence $|X| = \ell = \beta'(G)$. Consider the subgraph $F = G[X]$ induced by X. We begin with an observation: F contains no trail T of length 3. If F did contain a trail T of length 3 and e is the middle edge of T, then $X - \{e\}$ also covers all vertices of G, which is impossible. Therefore, F contains no cycles and no paths of length 3 or more, implying that every component of F is a star.

Since a forest of order n and size $n - k$ contains k components and the size of F is $\ell = n - (n - \ell)$, it follows that F contains $n - \ell$ nontrivial components. Selecting one edge from each of these $n - \ell$ components produces a matching of cardinality $n - \ell$, that is, $\alpha'(G) \geq n - \ell$. Therefore,

$$\alpha'(G) + \beta'(G) \geq (n - \ell) + \ell = n.$$

Consequently, $\alpha'(G) + \beta'(G) = n$.   ∎

Theorem 8.7 is due to Tibor Gallai, known as Tibor Grünwald in his early years. Gallai, a Hungarian born in 1912, was a winner of national mathematics competitions, along with Paul Erdős and Paul Turán and became life-long friends of both. As a consequence of his accomplishments, Gallai was admitted to Pázmány University in Budapest. He was one of a group of enthusiastic students in the 1930s in Budapest that included Paul Erdős, Paul Turán, George Szekeres and Esther Klein. Some of these students attended the graph theory course given by Dénes König, who was a professor at the Technical University of Budapest. This was to have a profound effect on Gallai's mathematical interests. Gallai helped König with his graph theory book and König mentioned some of Gallai's results in the book and used other ideas of Gallai. Many of Gallai's contributions to graph theory were to prove fundamental to the subject and aided in the rapid development of graph theory and combinatorics. For example, he was among the first to recognize the importance of so-called min-max theorems.

Gallai was an exceptionally modest individual and rarely made public appearances or attended conferences. In fact, much of his work only became known through the efforts of his students. While Gallai was quick to praise the work of others, he often underestimated the merits of his own contributions, even though he had important results in many areas of graph theory. Consequently, he was notoriously slow to publish his own results. Several of his results went unpublished, later to be independently rediscovered (and published) by others. Gallai died in 1992.

Independence of vertices is an equally important topic in graph theory. A set of vertices in a graph is **independent** if no two vertices in the set are adjacent. The **vertex independence number** (or the **independence number**) $\alpha(G)$ of a graph G is the maximum cardinality of an independent set of vertices in G. An independent set in G of cardinality $\alpha(G)$ is called a **maximum independent set**.

There is an analogous covering concept for vertices. A **vertex cover** in a graph G is a set of vertices that covers all edges of G. The minimum number of vertices in a vertex cover of G is the **vertex covering number** $\beta(G)$ of G. A vertex cover of cardinality $\beta(G)$ is a **minimum vertex cover** in G. For an integer $n \geq 3$ and integers r and s with $1 \leq r \leq s$,

$$\alpha(C_n) = \lfloor n/2 \rfloor, \ \alpha(K_n) = 1 \text{ and } \alpha(K_{r,s}) = s;$$

while

$$\beta(C_n) = \lceil n/2 \rceil, \ \beta(K_n) = n - 1 \text{ and } \beta(K_{r,s}) = r.$$

Here too observe that

$$\alpha(C_n) + \beta(C_n) = \alpha(K_n) + \beta(K_n) = n \text{ and } \alpha(K_{r,s}) + \beta(K_{r,s}) = r + s.$$

There is an analogue to Theorem 8.7 for vertices, also due to Gallai. The proof is similar to the proof of Theorem 8.7 and is left as an exercise. The results in Theorems 8.7 and 8.8 are often referred to as the **Gallai Identities.**

Theorem 8.8 *For every graph G of order n containing no isolated vertices,*

$$\alpha(G) + \beta(G) = n.$$

The independence and covering concepts that we have just discussed for a graph G are summarized below.

$\alpha(G)$	vertex independence number	maximum number of vertices, no two of which are adjacent
$\beta(G)$	vertex covering number	minimum number of vertices that cover all edges of G
$\alpha'(G)$	edge independence number	maximum number of edges, no two of which are adjacent
$\beta'(G)$	edge covering number	minimum number of edges that cover all vertices of G

Example 8.9 *Determine the values of $\alpha(G)$, $\beta(G)$, $\alpha'(G)$ and $\beta'(G)$ for the graph $G = K_1 + 2K_3$ of Figure 8.4.*

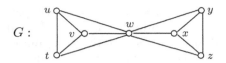

Figure 8.4: The graph G in Example 8.9

Solution. Since the order of G is 7, it follows that $\alpha'(G) \leq \lfloor 7/2 \rfloor = 3$. Because $\{tu, vw, yz\}$ is an independent set of three edges, $\alpha'(G) = 3$. By Theorem 8.7, $\beta'(G) = 4$. Note that $\{tu, vw, wx, yz\}$ is a minimum edge cover for G.

The vertex w is adjacent to all other vertices of G. Furthermore, $G[\{t, u, v\}] = K_3$ and $G[\{x, y, z\}] = K_3$. Thus $\alpha(G) = 2$. One example of a maximum independent set is $\{t, z\}$. By Theorem 8.8, $\beta(G) = 5$. One example of a minimum vertex cover of G is $\{t, u, w, y, z\}$. $\diamond$

Exercises for Section 8.1

8.1 Let G be the bipartite graph with partite sets $U = \{u_0, u_1, \ldots, u_6\}$ and $W = \{w_0, w_1, \ldots, w_6\}$, where the elements of U are the statements listed below, and $w_i = i$ for $0 \leq i \leq 6$. Vertices u_i and w_j ($0 \leq i, j \leq 6$) are adjacent if the integer w_j is a correct response to the statement u_i.

u_0: The size of a nontrivial complete graph.

u_1: The number of distinct $u - v$ paths for vertices u and v in a tree.

u_2: The number of Hamiltonian cycles in a transitive tournament.

u_3: The number of bridges in a tree of order 6.

u_4: The value of r for a nonempty r-regular graph of order 7.

u_5: The maximum degree of a tree of order 5.

u_6: The maximum number of cut-vertices among all graphs of order 5.

(a) Draw the graph G.

(b) Does G contain a perfect matching? If no, explain why not; if yes, draw a perfect matching and indicate what this means in this case.

8.2 There are positions open in seven different divisions of a major company: advertising (a), business (b), computing (c), design (d), experimentation (e), finance (f) and guest relations (g). Six people are applying for some of these positions, namely:

Alvin (A) : a, c, f;   Beverly (B): a, b, c, d, e, g;
Connie (C): c, f;   Donald (D): b, c, d, e, f, g;
Edward (E): a, c, f;   Frances (F): a, f.

(a) Represent this situation by a bipartite graph.

(b) Is it possible to hire all six applicants for six different positions?

8.3 Figure 8.5 shows two bipartite graphs G_1 and G_2, each with partite sets $U = \{v, w, x, y, z\}$ and $W = \{a, b, c, d, e\}$. In each case, can U be matched to W?

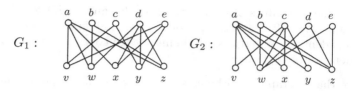

Figure 8.5: The graphs G_1 and G_2 in Exercise 8.3

8.4 A connected bipartite graph G has partite sets U and W, where $|U| = |W| = k \geq 2$. Prove that if every two vertices of U have distinct degrees in G, then G contains a perfect matching.

8.5 Prove that every tree has at most one perfect matching.

8.6 (a) Prove that every connected graph of order 4 that is not $K_{1,3}$ has a perfect matching.

(b) Let G be a connected graph of even order. Prove that if G contains no induced subgraph isomorphic to $K_{1,3}$, then G has a perfect matching.

8.7 Give an example of a connected non-Hamiltonian graph that contains two disjoint perfect matchings.

8.8 Show that the Petersen graph does not contain two disjoint perfect matchings. (Recall that the smallest cycle in the Petersen graph has length 5.)

8.9 For each integer i with $1 \leq i \leq 4$, give an example of a connected graph G_i of *smallest order* such that $\alpha(G_i) + \alpha'(G_i) = 5$ and $\alpha(G_i) = i$.

8.10 Show, for every connected graph G of order 6 with four independent vertices, that either $\alpha(G) = 5$ or $\alpha'(G) \geq 2$.

8.11 Give an example of an infinite class of graphs G for which $\alpha(G) = \beta'(G)$.

8.12 Prove or disprove:

(a) Every vertex cover of a graph contains a minimum vertex cover.

(b) Every independent set of edges in a graph is contained in a maximum independent set of edges.

H :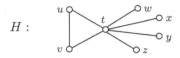

Figure 8.6: The graph G in Exercise 8.13

8.13 Determine the values of $\alpha(H)$, $\beta(H)$, $\alpha'(H)$ and $\beta'(H)$ for the graph H of Figure 8.6. Give an example of a minimum vertex cover, a maximum independent set of vertices, a minimum edge cover and a maximum independent set of edges of H.

8.14 Prove that a graph G without isolated vertices has a perfect matching if and only if $\alpha'(G) = \beta'(G)$.

8.15 Two vertex-disjoint graphs G_1 and G_2 have orders n_1 and n_2, respectively, where (1) $n_2 > n_1$, (2) n_1 and n_2 are of the same parity and (3) $\alpha'(G_2) \geq (n_2 - n_1)/2$. Let $G = G_1 + G_2$. What is $\beta'(G)$?

8.16 Prove that if G is a graph of order n, maximum degree Δ and having no isolated vertices, then $\beta(G) \geq \frac{n}{\Delta+1}$.

8.2   Factorization

We have mentioned that a matching M in a graph G of order n is a **perfect matching** if n is even and $|M| = n/2$. The subgraph $F = G[M]$ induced by M is therefore a 1-regular spanning subgraph of G. A 1-regular spanning subgraph of a graph G is also called a **1-factor** of G. Consequently, the edge set of a 1-factor of a graph is a perfect matching of the graph. So a graph G has a 1-factor if and only if G has a perfect matching.

For even integers $n \geq 4$, the graphs C_n and K_n have 1-factors, while for positive integers r and s, the complete bipartite graph $K_{r,s}$ has a 1-factor if and only if $r = s$. The Petersen graph PG (see Figure 8.7) also has a 1-factor, for example, $F = PG[X]$, where $X = \{u_i v_i : 1 \leq i \leq 5\}$ is a 1-factor of the Petersen graph. Of course, the Petersen graph is a 3-regular graph. Many other 3-regular graphs have 1-factors. Indeed all of the graphs in Figure 8.7 have 1-factors.

Not every 3-regular graph contains a 1-factor, however. For example, the 3-regular graph H of order 16 shown in Figure 8.8 does not contain a 1-factor. This brings up a question: Which graphs contain 1-factors? Certainly, only graphs of even order can contain a 1-factor. If G is a Hamiltonian graph of even order, then G contains a 1-factor. By taking every other edge in a Hamiltonian cycle, a 1-factor is obtained. Indeed, a Hamiltonian graph of even order contains two disjoint perfect matchings.

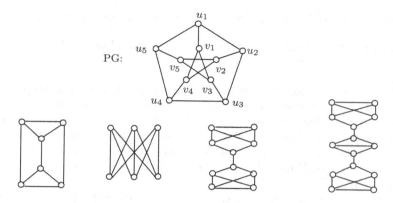

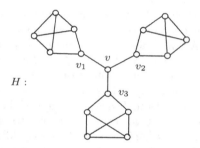

Figure 8.7: 3-regular graphs containing 1-factors

Figure 8.8: A 3-regular graph containing no 1-factor

If G is a Hamiltonian graph of even order, then $k(G - S) \leq |S|$ for every nonempty proper subset S of $V(G)$, where, recall, $k(G - S)$ denotes the number of components of $G - S$. This is a consequence of Theorem 6.5. We have seen that the converse of this theorem is not true. For example, $k(PG - S) \leq |S|$ for every nonempty proper subset S of the vertex set of the Petersen graph PG but the Petersen graph is not Hamiltonian. Yet the Petersen graph does contain a 1-factor.

We have already noted that the graph H of Figure 8.8 does not contain a 1-factor. If it did contain a 1-factor F, then exactly one edge of F is incident with the vertex v. Since $H - v$ consists of three components of odd order, two of these components must contain a 1-factor, which, of course, is impossible. This implies that if G is a graph of even order containing a nonempty proper subset S of $V(G)$ such that $G - S$ has more than $|S|$ components of odd order, then G cannot contain a 1-factor. It turns out that this observation is a critical one. A component of a graph is **odd** or **even** according to whether its order is odd or even. We write $k_o(G)$ for the number of odd components of a graph G. In particular, if G is a Hamiltonian graph of even order n (and thus G contains a 1-factor), then $k_o(G - S) \leq |S|$ for every proper subset S of $V(G)$. The following theorem provides a characterization of graphs containing a 1-factor.

Theorem 8.10 *A graph G contains a 1-factor if and only if $k_o(G - S) \leq |S|$
for every proper subset S of $V(G)$.*

Proof. Assume first that G contains a 1-factor F. Let S be a proper subset
of $V(G)$. If $G - S$ has no odd components, then $k_o(G - S) = 0$ and certainly
$k_o(G - S) \leq |S|$. Suppose that $k_o(G - S) = k \geq 1$ and let $G_1, G_2, \ldots, G_k$ be
the odd components of $G - S$. (There may also be even components of $G - S$.)
Since G contains the 1-factor F and the order of each subgraph G_i ($1 \leq i \leq k$)
is odd, some edge of F must be incident to both a vertex of G_i and a vertex of
S and so $k_o(G - S) \leq |S|$.

For the converse, assume that $k_o(G - S) \leq |S|$ for every proper subset S of
$V(G)$. In particular, for $S = \emptyset$, we have $k_o(G - S) = k_o(G) = 0$, that is, every
component of G is even and so G has even order. We now show by induction
that every graph G of even order with this property has a 1-factor. There is
only one graph of order 2 having only even components, namely K_2, which, of
course, has a 1-factor. Assume, for an even integer $n \geq 4$, that all graphs H of
even order less than n for which $k_o(H - S) \leq |S|$ for every proper subset S of
$V(H)$ have a 1-factor. Let G be a graph of order n satisfying $k_o(G - S) \leq |S|$
for every proper subset S of $V(G)$. Thus every component of G has even order.

First, we make an observation. Since every nontrivial component of G contains a vertex that is not a cut-vertex (Corollary 5.6), there are subsets R of
$V(G)$ for which $k_o(G - R) = |R|$. (For example, we could choose $R = \{v\}$, where
v is not a cut-vertex of G.) Among all such sets, let S be one of maximum
cardinality and let $G_1, G_2, \ldots, G_k$ be the k odd components of $G - S$. Thus
$k = |S| \geq 1$.

Observe that $G_1, G_2, \ldots, G_k$ are the only components of $G - S$, for otherwise
$G - S$ has an even component G_0 containing a vertex u_0 that is not a cut-vertex.
Then for the set $S_0 = S \cup \{u_0\}$ of cardinality $k + 1$,

$$k_o(G - S_0) = |S_0| = k + 1,$$

which is impossible. Therefore, as claimed, the odd components $G_1, G_2, \ldots, G_k$
are, in fact, the only components of $G - S$.

Now, for each integer i with $1 \leq i \leq k$, let S_i be the set of vertices of S that
are adjacent to at least one vertex in G_i. Since G has only even components,
each set S_i is nonempty. We claim next that for each integer ℓ with $1 \leq \ell \leq k$,
the union of any ℓ of the sets $S_1, S_2, \ldots, S_k$ contains at least ℓ vertices. Assume,
to the contrary, that there exists an integer j such that the union T of j of the
sets $S_1, S_2, \ldots, S_k$ has fewer than j elements. Without loss of generality, we may
assume that $T = S_1 \cup S_2 \cup \cdots \cup S_j$ and $|T| < j$. Then

$$k_o(G - T) \geq j > |T|,$$

which is impossible. Thus, as claimed, for each integer ℓ with $1 \leq \ell \leq k$, the
union of any ℓ of the sets $S_1, S_2, \ldots, S_k$ contains at least ℓ vertices.

By Theorem 8.4, there exists a set $\{v_1, v_2, \ldots, v_k\}$ of k distinct vertices such that $v_i \in S_i$ for $1 \leq i \leq k$. Since every graph G_i $(1 \leq i \leq k)$ contains a vertex u_i for which $u_i v_i \in E(G)$, it follows that $\{u_i v_i : 1 \leq i \leq k\}$ is a matching of G (see Figure 8.9).

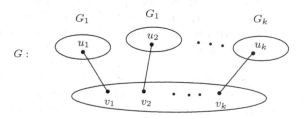

Figure 8.9: A step in the proof of Theorem 8.10

Next, we show that if G_i $(1 \leq i \leq k)$ is nontrivial, then $G_i - u_i$ has a 1-factor. Let W be a proper subset of $V(G_i - u_i)$. We claim that

$$k_o(G_i - u_i - W) \leq |W|.$$

Assume, to the contrary, that $k_o(G_i - u_i - W) > |W|$. Since $G_i - u_i$ has even order, $k_o(G_i - u_i - W)$ and $|W|$ are either both even or both odd. Hence $k_o(G_i - u_i - W) \geq |W| + 2$. Let $S' = S \cup W \cup \{u_i\}$. Then

$$
\begin{aligned}
|S'| &\geq k_o(G - S') = k_o(G - S) + k_o(G_i - u_i - W) - 1 \\
&\geq |S| + (|W| + 2) - 1 = |S| + |W| + 1 = |S'|,
\end{aligned}
$$

which implies that $k_o(G - S') = |S'|$, contradicting our choice of S. Therefore, $k_o(G_i - u_i - W) \leq |W|$, as claimed.

By the induction hypothesis, if G_i $(1 \leq i \leq k)$ is nontrivial, then $G_i - u_i$ has a 1-factor. The collection of 1-factors of $G_i - u_i$ for all nontrivial graphs G_i $(1 \leq i \leq k)$ and the edges in $\{u_i v_i : 1 \leq i \leq k\}$ produce a 1-factor of G.   ∎

Theorem 8.10 is due to William Thomas Tutte. Tutte was born on May 14, 1917 in Newmarket, England. Tutte's father was a gardener and his mother a caretaker. Although the family moved about as his father attempted to obtain a more stable position, his father eventually became a gardener at a hotel. The family lived in a flint cottage in the little village of Cheveley. Tutte went to the village school from ages 6 to 11. Because of his performance on a competitive exam, he won a scholarship at age 10 to attend a school in Cambridge. His parents felt that the distance to school was too great and so Tutte was kept at home. A year later, however, he took the exam again and this time he was permitted to go to school in Cambridge despite the lengthy commute each day.

In 1935 he entered Trinity College, Cambridge where he majored in chemistry and went on to do graduate work in chemistry. In fact, he had two publications in chemistry. Despite his major, chemistry was not Tutte's first love. His primary interest was mathematics. He was active in the Trinity Mathematical Society, where he met three undergraduates (all mathematics majors): Cedric Smith, R. Leonard Brooks, Arthur Stone, who became life-long friends of Tutte. The four students wrote a paper using electrical networks to solve a geometric problem, which became a standard reference in the field.

Tutte's academic career was then interrupted by World War II. At the invitation of his tutor in January 1941, Tutte went to Bletchley Park, the organization of code-breakers in Great Britain. In October 1941, Tutte encountered the first of a set of machine-coded messages from Berlin named Fish. While the Bletchley code-breakers, among whom Alan Turing was the most prominent, had success in deciphering naval and air force versions of Enigma codes, they did not have success with the army version. Because of this, they turned to Fish, which was used only by the army. The Fish code was used for high level communications between Berlin and the field commanders. Using only samples of messages, Tutte was able to discover the structure of machines that generated these codes. Tony Sale, who first described the work of Tutte and his colleagues, called this the "greatest intellectual feat of the whole war."

After World War II, Tutte returned to Cambridge, this time as a graduate student in mathematics. While a graduate student, he published some work that he had started earlier: a characterization of graphs containing 1-factors (Theorem 8.10). In his doctoral thesis, Tutte revitalized a subject that is now known as *matroid theory*, an area of mathematics (generally credited as being introduced by Hassler Whitney) that grew out of abstractions of different combinatorial objects (including graphs and matrices).

Tutte completed his Ph.D. at the University of Cambridge in 1948, with minimal assistance from his advisor Shaun Wylie. Tutte was invited to join the faculty at the University of Toronto at the invitation of H. S. M. (Donald) Coxeter (1907-2003), one of the great geometers of the 20th century. While in Toronto, Tutte became one of the preeminent researchers in the field of combinatorics. In 1957 the University of Waterloo in Canada was founded. Five years later Ralph G. Stanton (1923-2010), the Chair of the Department of Mathematics there and well known for his work in combinatorial designs, was successful in getting Tutte to join the Department. Tutte's presence at Waterloo was a major factor in the establishment of the reputation of the University. Although Tutte retired in 1984, he continued to work in graph theory and related areas of discrete mathematics. In his 1998 book *Graph Theory As I Have Known It*, Tutte described how he arrived at many of his fundamental results. He died on May 2, 2002.

Figure 8.7 shows several 3-regular graphs containing 1-factors, while Figure 8.8 shows a 3-regular graph that does not contain a 1-factor. Even though the graph of Figure 8.8 has bridges, there is no 3-regular graph without bridges that does not contain a 1-factor. This is a result of Julius Petersen. In fact, it is often called Petersen's theorem. Although this theorem preceded Theorem 8.10 historically, a simpler proof can be given with the aid of Theorem 8.10.

Theorem 8.11 (Petersen's Theorem) *Every 3-regular bridgeless graph contains a 1-factor.*

Proof. Let G be a 3-regular bridgeless graph and let S be a subset of $V(G)$ of cardinality $k \geq 1$. We show that the number $k_o(G-S)$ of odd components of $G - S$ is at most $|S|$. Since this is certainly the case if $G - S$ has no odd components, we may assume that $G - S$ has $\ell \geq 1$ odd components $G_1, G_2, \ldots, G_\ell$. Let X_i $(1 \leq i \leq \ell)$ denote the set of edges joining the vertices of S and the vertices of G_i. Since every vertex of each graph G_i has degree 3 in G and the sum of the degrees of the vertices in the graph G_i is even, $|X_i|$ is odd. Because G is bridgeless, $|X_i| \neq 1$ for each i $(1 \leq i \leq \ell)$ and so $|X_i| \geq 3$. Therefore, there are at least 3ℓ edges joining the vertices of S and the vertices of $G - S$. However, since $|S| = k$ and every vertex of S has degree 3 in G, at most $3k$ edges join the vertices of S and the vertices of $G - S$. Therefore,

$$3k_o(G - S) = 3\ell \leq 3k = 3|S|$$

and so $k_o(G - S) \leq |S|$. By Theorem 8.10, G has a 1-factor. ∎

Actually, Petersen proved a somewhat stronger result (whose proof is left as an exercise).

Theorem 8.12 *Every 3-regular graph with at most two bridges contains a 1-factor.*

In view of the graph H of Figure 8.8, the number of bridges in the statement of Theorem 8.12 cannot be increased to 3 and obtain the same conclusion.

If a 3-regular graph G contains a 1-factor F_1 and the edges of F_1 are removed from G (that is, the edges of a perfect matching are removed from G), then a 2-regular graph H results. If, in turn, H contains a 1-factor F_2 and the edges of F_2 are removed from H, the resulting graph F_3 is itself a 1-factor. If this were to occur, then G contains three pairwise edge-disjoint 1-factors. This brings us to our next concept.

A graph G is said to be **1-factorable** if there exist 1-factors $F_1, F_2, \ldots, F_r$ of G such that $\{E(F_1), E(F_2), \ldots, E(F_r)\}$ is a partition of $E(G)$. We then say that G is **factored** into the 1-factors $F_1, F_2, \ldots, F_r$, which form a **1-factorization** of G. Consequently, every edge of G belongs to exactly one of these 1-factors. Since each set $E(F_i)$, $1 \leq i \leq r$, is a perfect matching, every vertex v of G is incident with exactly one edge in each of the 1-factors $F_1, F_2, \ldots, F_r$, that is, $\deg_G v = r$, which implies that G is r-regular. Therefore, every 1-factorable graph is regular. However, the converse is not true. In fact, the 3-regular graph H of Figure 8.8 doesn't contain a single 1-factor. On the other hand, Theorem 8.11 only guarantees that a 3-regular graph has a 1-factor when it is bridgeless.

In 1884 Peter Tait wrote that he had shown every 3-regular graph is 1-factorable but this result was "not true without limitation." Petersen interpreted Tait's vague remark to mean that every 3-regular bridgeless graph is 1-factorable. However, in 1898 Petersen showed that even if a 3-regular graph is bridgeless, this doesn't mean that the graph is 1-factorable. He did this by giving an example of a 3-regular bridgeless graph that is not 1-factorable: the Petersen graph (shown in Figure 8.7).

Theorem 8.13 *The Petersen graph is not 1-factorable.*

Proof. Assume, to the contrary, that the Petersen graph PG is 1-factorable. Thus PG can be factored into three 1-factors F_1, F_2, F_3. Hence the spanning subgraph H of PG with $E(H) = E(F_1) \cup E(F_2)$ is 2-regular and so H is either a single cycle or a union of two or more cycles. Since PG is not Hamiltonian, H cannot be a single cycle and is therefore the union of two or more cycles. On the other hand, since the length of a smallest cycle in PG is 5, it follows that $H = 2C_5$. This is impossible, however, since $2C_5$ does not contain a 1-factor. ∎

Probably the best known 1-factorable graphs are the complete graphs of even order.

Theorem 8.14 *For each positive ineteger k, the complete graph K_{2k} is 1-factorable.*

Proof. Since the result is true for $k = 1$ and $k = 2$, we assume that $k \geq 3$. Let $G = K_{2k}$, where $V(G) = \{v_0, v_1, v_2, \ldots, v_{2k-1}\}$. Let $v_1, v_2, \ldots, v_{2k-1}$ be the vertices of a regular $(2k - 1)$-gon and place v_0 in the center of the $(2k - 1)$-gon. Draw each edge of G as a straight line segment. Let F_1 be the 1-factor of G consisting of the edge v_0v_1 and all edges of G perpendicular to v_0v_1, namely v_2v_{2k-1}, v_3v_{2k-2}, $\ldots$, v_kv_{k+1}. In general, for $1 \leq i \leq 2k - 1$, let F_i be the 1-factor of G consisting of the edge v_0v_i and all edges of G perpendicular to v_0v_i. Then G has a factorization into the 1-factors $F_1, F_2, \ldots, F_{2k-1}$.   ∎

In the 1-factorization of K_{2k} described in the proof of Theorem 8.14, the 1-factor F_2 can be obtained by rotating the 1-factor F_1 clockwise through an angle of $2\pi/(2k - 1)$ radians. By rotating F_1 through this angle twice, F_3 is obtained and so on. For this reason, the 1-factorization described in the proof is called a **cyclic factorization**. This 1-factorization is illustrated in Figure 8.10 for the case $k = 3$.

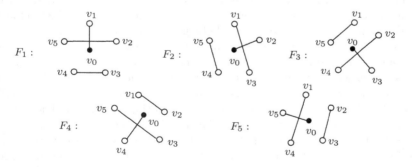

Figure 8.10: A cyclic 1-factorization of K_6

With the aid of Theorem 8.6, we can describe another class of 1-factorable graphs.

Theorem 8.15 *Every r-regular bipartite graph, $r \geq 1$, is 1-factorable.*

Proof. Let G be an r-regular bipartite graph, where $r \geq 1$. By Theorem 8.6, G contains a perfect matching M_1. Hence $G - M_1$ is $(r-1)$-regular. If $r \geq 2$, then $G - M_1$ contains a perfect matching M_2. Continuing in this manner and applying Theorem 8.6 r times, we see that $E(G)$ can be partitioned into perfect matchings, which gives rise to a 1-factorization of G. ∎

A **2-factor** in a graph G is a spanning 2-regular subgraph of G. Every component of a 2-factor is therefore a cycle. A graph G is said to be **2-factorable** if there exist 2-factors $F_1, F_2, \ldots, F_k$ such that $\{E(F_1), E(F_2), \ldots, E(F_k)\}$ is a partition of $E(G)$. Every 2-factorable graph is necessarily $2k$-regular then. That is, if G is a 2-factorable graph, then G is r-regular for some positive even integer r. In what might be considered an unexpected result, Petersen showed that the converse of this statement is true as well.

Theorem 8.16 *A graph G is 2-factorable if and only if G is r-regular for some positive even integer r.*

Proof. We have already observed that every 2-factorable graph is r-regular for some positive even integer r. Therefore, we need only establish the converse. Let G be an r-regular graph, where $r = 2k$ and $k \geq 1$. Without loss of generality, we may assume that G is connected. By Theorem 6.1, G is Eulerian and therefore contains an Eulerian circuit C. (Of course, a vertex of G can appear more than once in C. In fact, each vertex of G appears exactly k times in C.)

Let $V(G) = \{v_1, v_2, \ldots, v_n\}$. We construct a bipartite graph H with partite sets

$$U = \{u_1, u_2, \ldots, u_n\} \text{ and } W = \{w_1, w_2, \ldots, w_n\},$$

where the vertices u_i and w_j $(1 \leq i, j \leq n)$ are adjacent in H if v_j immediately follows v_i on C. Since every vertex of G appears exactly k times in C, the graph H is k-regular. By Theorem 8.15, H is 1-factorable and so H can be factored into k 1-factors $F'_1, F'_2, \ldots, F'_k$.

Next, we show each 1-factor F'_i $(1 \leq i \leq k)$ of H corresponds to a 2-factor F_i of G. Consider the 1-factor F'_1, for example. Since F'_1 is a perfect matching of H, it follows that $E(F'_1)$ is an independent set of k edges of H, say

$$E(F'_1) = \{u_1 w_{i_1}, u_2 w_{i_2}, \ldots, u_n w_{i_n}\},$$

where the integers $i_1, i_2, \ldots, i_n$ are the integers $1, 2, \ldots, n$ in some order and $i_j \neq j$ for each j $(1 \leq j \leq n)$. Suppose that $i_t = 1$. Then the 1-factor F'_1 gives rise to a cycle $C^{(1)} = (v_1, v_{i_1}, \ldots, v_t, v_{i_t} = v_1)$. If $C^{(1)}$ has length n, then the Hamiltonian cycle $C^{(1)}$ of G is a 2-factor of G. If the length of $C^{(1)}$ is less than n, then there is a vertex v_ℓ of G that is not on $C^{(1)}$. Suppose that $i_s = \ell$. This gives rise to a second cycle $C^{(2)} = (v_\ell, v_{i_\ell}, \ldots, v_{i_s} = v_\ell)$. Continuing in this manner, we obtain a collection of pairwise vertex-disjoint cycles that contain each vertex of G once, producing a 2-factor F_1 of G. In general then, the 1-factorization of H into 1-factors $F_1', F_2', \ldots, F_k'$ produces a 2-factorization of G into 2-factors $F_1, F_2, \ldots, F_k$, as desired. ∎

To illustrate the proof of Theorem 8.16, consider the 4-regular graph $G = K_{2,2,2}$ shown in Figure 8.11(a). One Eulerian circuit of G is

$$C = (v_1, v_2, v_3, v_1, v_4, v_5, v_6, v_4, v_2, v_5, v_3, v_6, v_1).$$

Since $V(G) = \{v_1, v_2, , \ldots, v_6\}$, we construct a bipartite graph H (shown in Figure 8.11(b)) with partite sets $U = \{u_1, u_2, \ldots, u_6\}$ and $W = \{w_1, w_2, \ldots, w_6\}$. Since $v_1 v_2$ and $v_2 v_3$ are edges of C, the edges $u_1 w_2$ and $u_2 w_3$ belong to H. Figure 8.11(c) shows a possible 1-factorization of H into the two 1-factors F_1' and F_2', which gives rise to the 2-factorization of G shown in Figure 8.11(d).

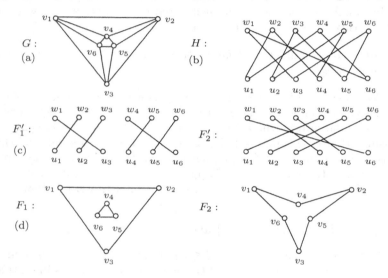

Figure 8.11: Constructing a 2-factorization of a 4-regular graph G

Since the complete graph K_{2k+1}, where $k \geq 1$, is $2k$-regular, K_{2k+1} is 2-factorable. Some 2-factorizations of K_3, K_5 and K_7 are shown in Figure 8.12. Observe that not only does there exist a 2-factorization of each of K_3, K_5 and K_7 but there exists a 2-factorization of each of these graphs in which each 2-factor is a Hamiltonian cycle. This is obvious in the cases of K_3 and K_5. A **Hamiltonian factorization** of a graph G is a 2-factorization of G in which each 2-factor is a Hamiltonian cycle. A graph G is **Hamiltonian-factorable** if there exists a Hamiltonian factorization of G. We now show that for every odd integer $n \geq 3$, the complete graph K_n is Hamiltonian-factorable. Before presenting a proof of this fact, we make a few comments about complete graphs that will be useful in this proof and in some future proofs.

For a graph $G = K_n$, where $n \geq 4$, with $V(G) = \{v_1, v_2, \ldots, v_n\}$, certainly G contains the Hamiltonian cycle $C = (v_1, v_2, \ldots, v_n, v_1)$. For every two distinct

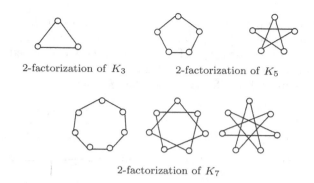

Figure 8.12: Some 2-factorizations of K_3, K_5 and K_7

vertices v_i and v_j of G (and on C), the distance $d_C(v_i, v_j)$ between v_i and v_j on the cycle C is one of the integers $1, 2, \ldots, n/2$ if n is even and is one of the integers $1, 2, \ldots, (n-1)/2$ if n is odd. In the construction of K_n, it is at times useful to label the edge $v_i v_j$ with the number $d_C(v_i, v_j)$. Hence if $n \geq 3$ is odd, then exactly n edges are labeled k for each integer k with $1 \leq k \leq (n-1)/2$. On the other hand, if $n \geq 4$ is even, then exactly n edges are labeled k for each integer k with $1 \leq k \leq (n-2)/2$, while $n/2$ edges are labeled $n/2$. This is illustrated in Figure 8.13 for K_4 and K_5. For K_6, only the edges labeled 1 or 3 are shown. The edges not drawn in K_6 are labeled 2. Sometimes it is convenient to enlarge a given complete graph K_n to K_{n+1} by adding a new vertex to K_n and joining that vertex to all vertices of K_n. When this occurs, we label each new edge in this construction with the integer 0.

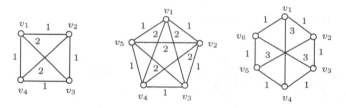

Figure 8.13: Labeling the edges of a complete graph

Theorem 8.17 *For every integer $k \geq 1$, the complete graph K_{2k+1} is Hamiltonian-factorable.*

Proof. Since the theorem is true for $1 \leq k \leq 3$, we may assume that $k \geq 4$. Let $G = K_{2k+1}$ and $H = K_{2k}$, where $V(H) = \{v_1, v_2, \ldots, v_{2k}\}$ and let $V(G) = V(H) \cup \{v_0\}$. Let the vertices of H be the vertices of a regular $2k$-gon and let the edges of H be straight line segments (see Figure 8.14(a) for the case $k = 4$). A Hamiltonian cycle C of G is now constructed from a Hamiltonian path P of H

that begins with $v_1, v_2, v_{2k}, v_3, v_{2k-1}$ and then continues in this zig-zag pattern until arriving at the vertex v_{k+1}. The path P is a $v_1 - v_{k+1}$ path where edges are those parallel to $v_1 v_2$ or $v_2 v_{2k}$. The cycle C in G is completed by placing v_0 at some convenient location within the regular $2k$-gon and joining v_0 to both v_1 and v_{k+1} (see Figure 8.14(b) for the case when $k = 4$).

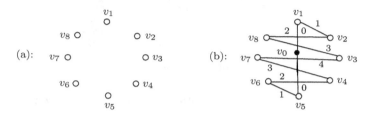

Figure 8.14: Forming a Hamiltonian cycle in K_9

Observe that the Hamiltonian cycle C of G just constructed consists of two edges labeled with each of the integers $0, 1, \ldots, k-1$ and one edge labeled k. By rotating the Hamiltonian path P of H clockwise through an angle of π/k radians, a new Hamiltonian path P' of H is constructed that is edge-disjoint with P'. The path P' is a $v_2 - v_{k+2}$ path. By joining v_0 to v_2 and v_{k+2}, a new Hamiltonian cycle C' of G is obtained that is edge-disjoint with C. We continue this until k Hamiltonian cycles of G are obtained producing a Hamiltonian factorization of G. (This is illustrated for the complete graph K_9 ($k = 4$) in Figure 8.15.) In general, for each i with $1 \le i \le k$, the ith Hamiltonian cycle of G is

$$(v_0, v_i, v_{i+1}, v_{i-1}, v_{i+2}, v_{i-2}, \cdots, v_{k+i+1}, v_{k+i-1}, v_{k+i}, v_0),$$

where the subscripts are expressed modulo $2k$.   ∎

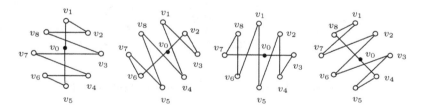

Figure 8.15: A Hamiltonian factorization of K_9

More generally, a spanning subgraph F of a graph G is called a **factor** of G. The graph G is said to be **factorable** into the factors $F_1, F_2, \ldots, F_k$ if $\{E(F_1), E(F_2), \ldots, E(F_k)\}$ is a partition of $E(G)$. If each factor F_i is isomorphic to some graph F, then G is F-**factorable**. For a graph G of order n, G is $(n/2)K_2$-factorable if and only if G is 1-factorable, while G is C_n-factorable if and only if G is Hamiltonian-factorable.

We have seen that K_9 is Hamiltonian-factorable (into four Hamiltonian cycles). Of course, we saw earlier (by Theorem 8.16) that K_9 is 2-factorable. That K_9 is factorable into Hamiltonian cycles is only one type of 2-factorization of K_9. For example, K_9 can be factored into copies of the graph $3K_3$.

Example 8.18 *The graph K_9 is also $3K_3$-factorable.*

Solution. As in Figure 8.16, we place the vertices v_1, v_2, ..., v_8 cyclically as the vertices of a regular 8-gon and place v_0 in some convenient location within the 8-gon. A factor $F_1 = 3K_3$ of K_9 is shown in Figure 8.16, where two edges are labeled 0, 1, 2 and 3, as described in Figure 8.16 and one edge is labeled 4. By rotating F_1 clockwise through an angle of $\pi/4$ radians three times, three new factors F_2, F_3 and F_4 are produced, each of which is $3K_3$. This produces a $3K_3$-factorization of K_9. ◇

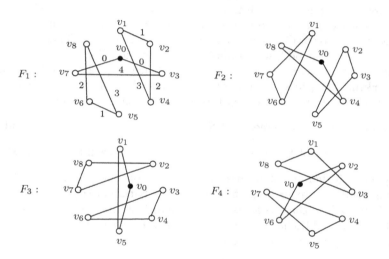

Figure 8.16: A $3K_3$-factorization of K_9

The following problem was posed by Thomas Kirkman in 1850 and has since become known as **Kirkman's Schoolgirl Problem**:

> *A school mistress has fifteen schoolgirls whom she wishes to take on a daily walk. The girls are to walk in five rows of three girls each. It is required that no two girls should walk in the same row more than once a week. Can this be done?*

If we think about Kirkman's Schoolgirl Problem a bit, we see that the question can be rephrased as follows: Is there a $5K_3$-factorization of K_{15}? If we label the vertices of K_{15} by the schoolgirls, numbered $1, 2, \ldots, 15$ say, then we see that a solution is given in the next table.

Day 1	Day 2	Day 3	Day 4	Day 5	Day 6	Day 7
{1, 2, 3}	{1, 4, 5}	{1, 6, 7}	{1, 8, 9}	{1, 10, 11}	{1, 12, 13}	{1, 14, 15}
{4, 8, 12}	{2, 9, 11}	{2, 8, 10}	{3, 5, 7}	{2, 12, 15}	{2, 5, 6}	{2, 4, 7}
{5, 10, 14}	{3, 13, 15}	{3, 12, 14}	{2, 13, 14}	{3, 4, 6}	{3, 9, 10}	{3, 8, 11}
{6, 9, 15}	{6, 8, 14}	{4, 9, 13}	{4, 10, 15}	{5, 8, 13}	{4, 11, 14}	{5, 9, 12}
{7, 11, 13}	{7, 10, 12}	{5, 11, 15}	{6, 11, 12}	{7, 9, 14}	{7, 8, 15}	{6, 10, 13}

Although there is a $5K_3$-factorization of K_{15}, it turns out that there is no cyclic $5K_3$-factorization of K_{15} (which makes such a factorization more difficult to construct). Since we saw in Figure 8.16 that K_9 is $3K_3$-factorable, the following is true:

Nine schoolgirls can take four daily walks in three rows of three girls each so that no two girls walk in the same row twice.

Suppose that the complete graph $G = K_n$ of order n has a 2-factorization in which every component of each 2-factor is a triangle. Then G is tK_3-factorable for some positive integer t and so $n = 3t$. Since the degree of every vertex of G is 2 in each 2-factor of G, every vertex has even degree in G. Therefore, $n - 1$ is even and so $n = 3t$ is odd, which implies that t is odd. Therefore, $t = 2k + 1$ for some nonnegative integer k and $n = 6k + 3$. Hence G is $(2k + 1)K_3$-factorable.

A **Kirkman triple system** of order n is a set S of cardinality n, a collection T of 3-element subsets of S, called **triples**, and a partition $\mathcal{P}$ of T such that

(1) every two distinct elements of S belong to a unique triple in T and

(2) every element of S belongs to a unique triple in each element of $\mathcal{P}$.

Consequently, if there is a Kirkman triple system of order n, then $n = 6k + 3$ for some nonnegative integer k. In fact, there is a Kirkman triple system of order $6k + 3$ if and only if there is a $(2k + 1)K_3$-factorization of K_{6k+3}. In 1971 Dijen Ray-Chaudhuri and Richard Wilson established the existence of a Kirkman triple system for every nonnegative integer k.

Theorem 8.19 *A Kirkman triple system of order $n \geq 3$ exists if and only if $n \equiv 3 \pmod 6$.*

Although it is impossible for K_{2k}, $k \geq 2$, to be Hamiltonian-factorable since K_{2k} is $(2k - 1)$-regular, K_{2k} is very close to being Hamiltonian-factorable.

Theorem 8.20 *For every integer $k \geq 1$, the complete graph K_{2k} can be factored into $k - 1$ Hamiltonian cycles and a 1-factor.*

Proof. Since the result is true for $k = 1$ and $k = 2$, we assume that $k \geq 3$. Let $G = K_{2k}$, where $V(G) = \{v_0, v_1, \ldots, v_{2k-1}\}$. Let $v_1, v_2, \ldots, v_{2k-1}$ be the vertices of a regular $(2k-1)$-gon and place v_0 in the center of the $(2k-1)$-gon. Join each two vertices of G by a straight line segment. Let G_1 be the spanning subgraph of G whose edges consist of (1) v_0v_1 and v_0v_{k+1}, (2) all edges parallel to v_0v_1 and (3) all edges parallel to v_0v_{k+1}. Then $G_1 = C_{2k}$. For $1 \leq i \leq k - 1$, let G_i

be the spanning subgraph of G whose edges consist of (1) v_0v_i and v_0v_{k+i}, (2) all edges parallel to v_0v_i and (3) all edges parallel to v_0v_{k+i}. Then $G_i = C_{2k}$ for each i ($1 \le i \le k-1$) and every edge of G belongs to some subgraph G_i ($1 \le i \le k-1$) except for the edges $v_1v_{2k-1}, v_2v_{2k-2}, \ldots, v_{k-1}v_{k+1}$ and v_0v_k, which are the edges of a 1-factor G_k of G. Thus G can be factored into $G_1, G_2, \ldots, G_k$. (See Figure 8.17 for the case $k = 4$.) ■

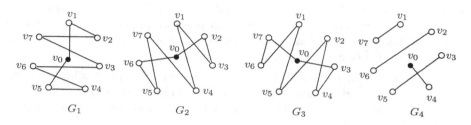

Figure 8.17: A factorization of K_8

Exercises for Section 8.2

8.17 Determine which of the cubic graphs G_1, G_2 and G_3 in Figure 8.18

 (a) has a 1-factor,

 (b) is 1-factorable.

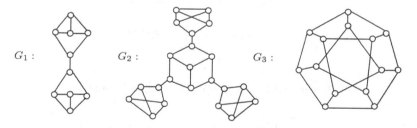

Figure 8.18: The graphs in Exercise 8.17

8.18 Give an example of a 5-regular graph that contains no 1-factor.

8.19 Nine members of a book club meet for dinner each week (4 times a month) to discuss the book they have read the preceding week. How can the nine people be seated at a circular dinner table for each of the four meetings during a month so that every two members sit next to each other exactly once during the month?

8.20 Use the technique employed in the proof of Theorem 8.11 to prove Theorem 8.12: *Every 3-regular graph with at most two bridges contains a 1-factor.*

8.21 Use Tutte's characterization of graphs with 1-factors (Theorem 8.10) to show that $K_{3,5}$ does not have a 1-factor.

8.22 (a) Show that Petersen's theorem (Theorem 8.11) can be extended somewhat by proving that if G is a bridgeless graph, every vertex of which has degree 3 or 5 and such that G has at most two vertices of degree 5, then G has a 1-factor.

(b) Show that the result in (a) cannot be extended further by giving an example of a bridgeless graph G containing exactly three vertices of degree 5 such that all remaining vertices of G have degree 3 but G has no 1-factor.

8.23 Prove that if the bridges of a 3-regular graph lie on a single path, then G has a 1-factor.

8.24 Show that every 3-regular bridgeless graph contains a 2-factor.

8.25 Show that $C_n \times K_2$ is 1-factorable for $n \geq 4$.

8.26 Show, for the 4-regular graph G Figure 8.19, that for any 2-factorization of G exactly one of the 2-factors is a Hamiltonian cycle of G.

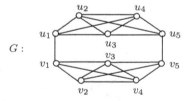

Figure 8.19: The graph G in Exercise 8.26

8.27 Figure 8.11 shows a 2-factorization of the graph $G = K_{2,2,2}$ into the 2-factors $F_1 = 2C_3$ and $F_2 = C_6$. Give an example of a 2-factorization of this graph into two 2-factors F_1^* and F_2^* where $F_i^* = C_6$ for $i = 1, 2$ and, employing the proof of Theorem 8.16, give an example of an Eulerian circuit C^* in G that produces this 2-factorization of G.

8.28 Let G be a 6-regular graph. Show that if G contains two edge-disjoint 1-factors, then G is 3-factorable.

8.29 Show for every positive even integer n that the complete graph K_n can be factored into Hamiltonian paths.

8.30 Solve the following 27-schoolgirl problem: A school mistress has 27 school-girls whom she wishes to take on a daily walk. The girls are to walk in nine rows of three girls each. Show that such walks can be made for thirteen days without two girls walking in the same row twice.

8.31 Does there exist a 2-factorization of K_7 in which no 2-factor is a Hamiltonian cycle?

8.32 Does there exist a 2-factorization of K_9 in which no two 2-factors are isomorphic?

8.3   Decompositions and Graceful Labelings

If a graph G has a factorization into subgraphs $F_1, F_2, \ldots, F_k$, then, by definition, each subgraph (factor) F_i, $1 \le i \le k$, is required to be a spanning subgraph of G. There is a related concept that we will discuss in this section.

A graph G is said to be **decomposable** into the subgraphs H_1, H_2, ..., H_k if $\{E(H_1), E(H_2), \ldots, E(H_k)\}$ is a partition of $E(G)$. Such a partition produces a **decomposition** of G. In other words, the subgraphs H_i are not required to be spanning subgraphs of G. If, on the other hand, each subgraph H_i is a spanning subgraph of G, then the decomposition is a factorization of G. If each H_i is isomorphic to some graph H, then the graph G is H-**decomposable** and the decomposition is an H-**decomposition**.

A problem concerning this concept that has attracted a great deal of attention is: Which complete graphs K_n are K_3-decomposable? In order for a complete graph K_n to be K_3-decomposable, the size of K_n must be divisible by 3, that is, 3 must divide $\binom{n}{2}$ and so $n(n-1)/6$ must be an integer. Hence either $3 \mid n$ or $3 \mid (n-1)$. Furthermore, since every vertex of K_n has degree $n-1$, each vertex must belong to $(n-1)/2$ triangles and so n must be odd. This says that either $n = 3p$ for some odd integer p or $n = 3q + 1$ for some even integer q. Therefore, either $n = 6k + 1$ or $n = 6k + 3$ for some integer k and so either $n \equiv 1 \pmod 6$ or $n \equiv 3 \pmod 6$.

A **Steiner triple system** of order n is a set S of cardinality n and a collection T of 3-element subsets, called **triples**, such that every two distinct elements of S belong to a unique triple in T. Therefore, there is a Steiner triple system of order n if and only if K_n is K_3-decomposable. Consequently, in order for a Steiner triple system of order n to exist, either $n \equiv 1 \pmod 6$ or $n \equiv 3 \pmod 6$. In 1846 Kirkman showed that the converse holds as well.

Theorem 8.21 *A Steiner triple system of order $n \ge 3$ exists if and only if $n \equiv 1 \pmod 6$ or $n \equiv 3 \pmod 6$.*

Trivially, there is a Steiner triple system of order 3 as K_3 is obviously K_3-decomposable. The first interesting and nontrivial case is K_7. One way to see that K_7 is K_3-decomposable is to let $v_1, v_2, \ldots, v_7$ be the vertices of a regular 7-gon and join each pair of vertices by a straight line segment. Consider the triangle with vertices v_1, v_2 and v_4, which we denote by H_1 (see Figure 8.20). Proceeding as we have earlier, we see that exactly one edge of H_1 is labeled with one of 1, 2 and 3. As we rotate H_1 clockwise through an angle of $2\pi/7$ radians, another triangle H_2 is produced. Continuing in this manner, we obtain a K_3-decomposition of K_7. From the K_3-decomposition of K_7, we have now produced a Steiner triple system of order 7 from the set $\{1, 2, \ldots, 7\}$, namely:

$$\{1, 2, 4\}, \{2, 3, 5\}, \{3, 4, 6\}, \{4, 5, 7\}, \{5, 6, 1\}, \{6, 7, 2\}, \{7, 1, 3\}.$$

The K_3-decomposition is a **cyclic decomposition** of K_7.

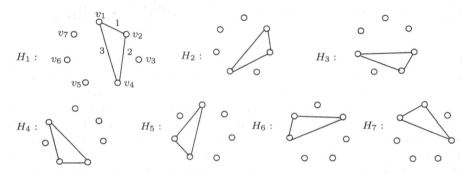

Figure 8.20: A cyclic K_3-decomposition of K_7

Steiner triple systems are named for Jakob Steiner, who was born in Utzenstorf, Switzerland on March 18, 1796. Steiner, a son of farming parents, did not learn to read or write until he was 14 years old. Within the next four years, his mathematical ability was recognized and he was permitted to start attending a prestigious school in Switzerland at age 18, even then against the wishes of his parents. In 1818 Steiner went to the University of Heidelberg, where he gave private lessons in mathematics. In 1821 he went on to the University of Berlin, where he continued to support himself by teaching. While in Berlin, he became acquainted with Niels Abel (after whom abelian groups are named), Carl Jacobi (after whom Jacobians are named) and August Crelle.

Although not as well known or as brilliant as other mathematicians of his time, Crelle nevertheless made important contributions to mathematics. Crelle was an extremely enthusiastic mathematician who had a gift for organization and who recognized mathematical ability in others. In 1826 Crelle founded a journal entirely devoted to mathematics (often referred to as *Crelle's Journal*) and titled *Journal für reine und angewandte Mathematik* (the journal still exists). Prior to 1826 other mathematics journals ordinarily reported on meetings of academies and societies where papers were read. In *Crelle's Journal*, however, the emphasis was on the mathematics. Crelle was in complete charge and was the editor-in-chief for the first 52 volumes.

Crelle recognized the importance of Abel's work and, in the first volume of his journal, published Abel's proof of the insolvability of quintic equations by radicals. Steiner was also a major contributor to the first volume of Crelle's journal. In addition to Abel and Steiner, other mathematicians had their early works made famous by publishing their first paper in the journal, including Lejeune Dirichlet and August Möbius. Almost all of Möbius' research was published in *Crelle's Journal.*

Kirkman triple systems (introduced in the preceding section) are more demanding than Steiner triple systems. Indeed, by Theorem 8.21 there is a Steiner triple system of order $n \geq 7$ if $n \equiv 1 \pmod 6$; however, none of these integers n is the order of a Kirkman triple systems as $n \equiv 3 \pmod 6$ is a necessary and sufficient condition for the existence of a Kirkman triple system of order n, as we saw in Theorem 8.19. Kirkman's first paper was published in 1846. In this paper, Kirkman solved a problem that appeared the previous year in the *Lady's and Gentleman's Diary*, namely, the existence of Steiner triple systems was verified (Theorem 8.21). The curiosity of this paper is that it was published seven years before a paper of Steiner's appeared in *Crelle's Journal* in which Steiner asked whether such systems exist. Despite the fact that Kirkman's work preceded Steiner's, these systems are named for Steiner, not Kirkman. Recall also that Kirkman thought of the concept of Hamiltonian cycles before Hamilton. Such was the fate of Kirkman.

Thomas Penyngton Kirkman was born on March 31, 1806 in Bolton, England. Although Kirkman's father did not want Kirkman to attend college, he did so and entered Trinity College Dublin in Ireland. In 1835 Kirkman returned to England and entered the Church of England. By 1839 he was vicar in the Parish of Southworth in Lancashire, a position he held for the next 52 years. The Reverend Thomas Kirkman became increasingly interested in mathematics and made contributions to combinatorics, quaternions, geometry and knot theory, although he is best remembered for his schoolgirl problem. Kirkman died on February 4, 1895 in Bowdon, England.

Another important type of decomposition problem concerns trees. In 1967 Gerhard Ringel (1919-2008) conjectured that if T is a tree of size m, then K_{2m+1} is T-decomposable. This conjecture has never been settled. However, it is related to another conjecture.

Let G be a graph of order n and size m. A one-to-one function $f : V(G) \rightarrow \{0, 1, 2, \ldots, m\}$ is called a **graceful labeling** of G if the induced edge labeling $f'\colon E(G) \rightarrow \{1, 2, \ldots, m\}$ defined by

$$f'(e) = |f(u) - f(v)| \text{ for each edge } e = uv \text{ of } G$$

is also one-to-one. If f is a graceful labeling of a graph G of order n, then so too is the **complementary labeling** $g\colon V(G) \rightarrow \{0, 1, 2, \ldots, m\}$ of f defined by $g(v) = m - f(v)$ for all $v \in V(G)$ since, for $e = uv$,

$$g'(e) = |g(u) - g(v)| = |(m - f(u)) - (m - f(v))| = |f(u) - f(v)| = f'(e).$$

A graph G possessing a graceful labeling is called a **graceful graph**.

Figure 8.21 shows five graceful graphs, including the complete graphs K_3 and K_4 and the cycle C_4, along with a graceful labeling of each of these graphs.

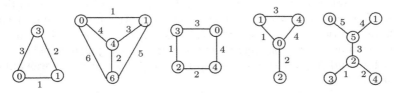

Figure 8.21: Graceful graphs

There are many graphs that are not graceful, however.

Example 8.22   *The cycle C_5 is not a graceful graph.*

Solution. Let $H = C_5$ (see Figure 8.22(a)). Assume, to the contrary, that H *is* graceful. Then there exists a graceful labeling $f : V(H) \to \{0, 1, 2, 3, 4, 5\}$. Since some edge of H is labeled 5 by the induced edge labeling, there are two adjacent vertices of H labeled 0 and 5.

The only way for an edge of H to be labeled 4 is for its incident vertices to be labeled 0 and 4 or 1 and 5. Since either f or its complementary labeling assigns adjacent vertices the labels 0 and 4, we may assume that three of the five vertices of H are labeled as in Figure 8.22(b). The vertex w cannot be labeled 1 as there is already an edge labeled 4. If x is labeled 1, then w must be labeled 2 or 3, neither of which results in a graceful labeling of H. Hence one of x and w is labeled 2 and the other is labeled 3. However, neither produces a graceful labeling of H.   ◇

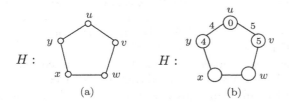

Figure 8.22: The graph H in Example 8.22

The rightmost graph shown in Figure 8.21 is, of course, a tree; in fact, it is a double star. The labeling given there shows that this tree is a graceful graph. In fact, there is a well-known conjecture due to Gerhard Ringel and Anton Kotzig.

Conjecture 8.23   *Every tree is graceful.*

If Conjecture 8.23 is true, then for each tree T of order n say, there is a graceful labeling $f : V(T) \to \{0, 1, 2, \ldots, n-1\}$, which is necessarily a bijective function. Added interest in this conjecture lies in the fact that its truth implies the truth of the earlier decomposition conjecture of Ringel. This implication was established by Alexander Rosa, who is credited with founding the subject of graph labelings. While the concept of graceful graphs is due to Rosa, who used the terminology β-valuation, the term "graceful" was introduced by Solomon Golomb.

Theorem 8.24 *If T is a graceful tree of size m, then K_{2m+1} is T-decomposable.*

Proof. Since T is a graceful tree, there exists a graceful labeling $f : V(T) \to \{0, 1, 2, \ldots, m\}$. Let $V(T) = \{v_0, v_1, \ldots, v_m\}$ where we may assume that $f(v_i) = i$ for $0 \le i \le m$. The induced edge labeling assigns the labels $1, 2, \ldots, m$ to the edges of T. Let $G = K_{2m+1}$, where $V(G) = \{v_0, v_1, \ldots, v_{2m}\}$. Let the vertices of G be the vertices of a regular $(2m+1)$-gon and draw each edge of G as a straight line segment. Consequently, T is a subgraph of G, where one edge of T is labeled i for each integer i with $1 \le i \le m$. Rotating T clockwise through an angle of $2\pi/(2m+1)$ radians a total of $2m$ times produces $2m+1$ trees isomorphic to T that form a cyclic T-decomposition of K_{2m+1}. ∎

To illustrate the proof of Theorem 8.24, consider the tree T shown in Figure 8.23, where a graceful labeling is given. A subgraph T_1 of $G = K_9$ that is isomorphic to T is also shown in Figure 8.23. A subgraph T_2 of G isomorphic to T (whose edges are indicated by dotted lines) is obtained by rotating T_1 clockwise through an angle of $2\pi/9$ radians.

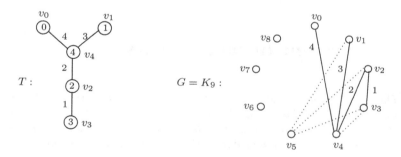

Figure 8.23: A cyclic T-decomposition of K_9

Theorem 8.24 is not only true for graceful trees, it holds for graceful graphs in general (see Exercise 8.38).

Exercises for Section 8.3

8.33 Show that the graph $K_{2,2,2}$ is not $K_{1,4}$-decomposable.

8.34 Find a P_4-decomposition of K_7.

8.35 Determine whether C_6 and C_8 are graceful.

8.36 Determine whether the graphs in Figure 8.24 are graceful.

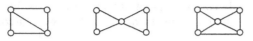

Figure 8.24: Graphs in Exercise 8.36

8.37 For the tree T in Figure 8.25, show that K_{11} is T-decomposable.

$T :$

Figure 8.25: The tree in Exercise 8.37

8.38 Let G be a graceful graph of size m.

   (a) Prove that K_{2m+1} is G-decomposable.

   (b) Prove that K_{2m+3} can be decomposed into $2m + 3$ copies of G and a Hamiltonian cycle of K_{2m+3}.

8.4 Excursion: Instant Insanity

Open the package. Notice that there are four different colors showing on each side of this stack of blocks. You may NEVER, EVER see them this way again. Now mix them up and then restack them so that there are again four colors, all different, showing on each side.

What is written above appears on an insert within packaging that contains four multi-colored cubes that make up a puzzle called **Instant Insanity**, which is manufactured by Hasbro Inc. (makers of toys and games). Each of the six faces of each cube is colored with one of the four colors red (R), blue (B), green (G) and yellow (Y). The object of the puzzle is to stack the cubes as in Figure 8.26, one on top of another, so that all four colors appear on each of the four sides.

On the reverse side of the insert is written: *Give up?* An address is supplied where a solution of the puzzle can be obtained. Reading all of this can be quite intimidating. Indeed, even before we attempt to solve the puzzle, we are being informed that it is very unlikely that we will be successful. Let's compute the number of ways in which four cubes can be stacked.

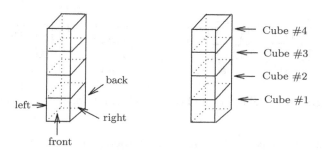

Figure 8.26: The stacking of four cubes

Select one of the cubes (which we'll call the first cube) and place it on a table, say. There are three ways this can be done, according to which pair of opposite faces will be the top and bottom of the cube. These are the "buried" faces. Select one of the other four faces as the front face. Now place the second cube on top of the first cube. Any of the six faces of the second cube can be chosen to appear directly above the front face of the first cube and each of these six faces can be positioned (rotated) in one of four ways. That is, there are $6 \cdot 4 = 24$ ways to place the second cube on top of the first cube. Consequently, the number of ways to stack all four cubes on the top of one another is $3 \cdot (24)^3 = 41,472$. Now if there is only one way to stack the cubes so that all four colors appear on all four sides, then using a trial-and-error method to solve the puzzle seems like a frustrating task and is likely to result in ... *instant insanity.*

Graph theory can help us to solve this tantalizing puzzle. Let's see how this can be done. For this purpose, it is convenient to have a way of representing a cube and the locations of the colors on its faces. See Figure 8.27.

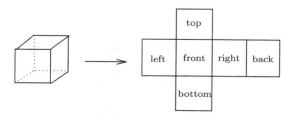

Figure 8.27: The six faces of a cube

We are now prepared to present an example.

Example 8.25 *Consider the four multi-colored cubes given in Figure 8.28.*

Solution. With each of the four cubes of Figure 8.28, we associate a pseudograph (therefore allowing both parallel edges and loops) of order 4 and size 3. The vertex set of each pseudograph is the set $\{R, B, G, Y\}$ of four colors and there is an edge joining color c_1 and color c_2 (possibly $c_1 = c_2$) whenever there is a

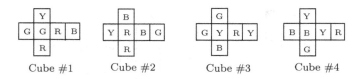

Figure 8.28: The four cubes in an Instant Insanity puzzle

pair of opposite faces colored c_1 and c_2. If there are two (or three) opposite faces colored c_1 and c_2 where $c_1 \neq c_2$, then the pseudograph has two (or three) parallel edges joining the vertices c_1 and c_2. If $c_1 = c_2$, then there are two (or three) loops at c_1. The pseudographs corresponding to the cubes of Figure 8.28 are shown in Figure 8.29.

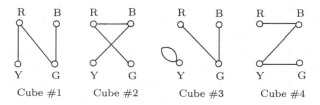

Figure 8.29: The four pseudographs in Example 8.25

A composite pseudograph M of order 4 (with vertex set {R, B, G, Y}) and size 12 and whose edge set is the union of the edge sets of these four pseudographs is shown in Figure 8.29. In order to distinguish which edges of M came from Cube #i ($i = 1, 2, 3, 4$), those three edges of M are labeled by i. The pseudograph M constructed from the pseudographs of Figure 8.29 is shown in Figure 8.30.

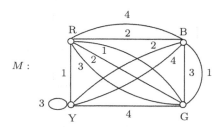

Figure 8.30: The composite pseudograph of Example 8.25

Let's pause for a moment while we review what we are seeking. Since our goal is to stack the four cubes on top of one another so that all four colors appear on all four sides, all four colors must of course appear on both the front and the back of the stack. If the front face of Cube #i ($i = 1, 2, 3, 4$) is colored c_1 and the opposite face of this cube (on the back of the stack) is colored c_2, then there must be an edge labeled i joining c_1 and c_2 in the pseudograph M.

If $c_1 = c_2$, then there must be a loop labeled i at vertex c_1. Since each color appears exactly once on the front and exactly once on the back of the stack, there must be a 2-regular spanning sub-pseudograph M' (a 2-factor) of M (where a loop is considered to have degree 2) such that there is exactly one edge labeled 1, 2, 3 and 4. Similarly, corresponding to the right and left sides of the stack, there is a 2-regular spanning sub-pseudograph M'' of M whose edge set is disjoint from that of M'. On the basis of these observations, we seek two edge-disjoint spanning 2-regular sub-pseudographs, where there is one edge labeled 1, 2, 3 and 4 in each of these two sub-pseudographs. If such a pair of pseudographs does not exist, then the puzzle can have no solution. If such a pair M', M'' of pseudographs exists, then they can be used to solve the puzzle, that is, to stack the cubes appropriately. Any 2-regular spanning sub-pseudograph must be one of the seventeen pseudographs shown in Figure 8.31.

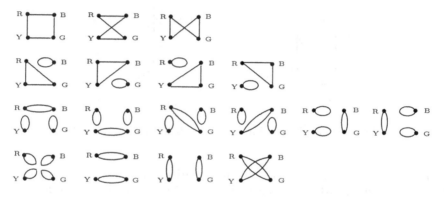

Figure 8.31: The seventeen 2-regular spanning pseudographs

Returning to our example, we see that the pseudograph M of Figure 8.30 contains the two edge-disjoint 2-regular spanning sub-pseudographs M' and M'', where the edges of these two pseudographs are labeled 1, 2, 3 and 4 (shown in Figure 8.32(a)). The pseudograph M' will correspond to the front and back of the stack to be produced and M'' will correspond to the right and left sides. (We could reverse M' and M'' if we desire.) For the purpose and convenience of stacking the cubes, we direct the edges of each component of M' and M'' so that a directed cycle results. Thus two (directed) pseudographs D' and D'' are produced, as shown in Figure 8.32(b).

With the aid of the (directed) pseudographs D' and D'' of Figure 8.32(b), we now stack the cubes. Since the arc (G, B) is labeled 1 in M', we place Cube #1 so that a green face appears in the front and a blue face on the back. Since the arc (G, R) is labeled 1 in M'', we rotate this cube (keeping a green face in the front and a blue face on the back) until we have a green face on the right and a red face on the left. We now proceed in the same way with the other three cubes and ... Voila! The puzzle has been successfully solved (see Figure 8.33). $\diamond$

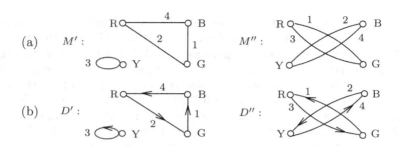

Figure 8.32: Two 2-regular spanning sub-pseudographs for Example 8.25

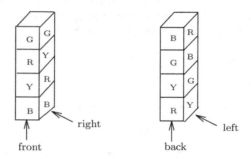

Figure 8.33: A solution for the puzzle in Example 8.25

This puzzle was conceived in 1900 by Fredrick Schossow using the four suits of playing cards (hearts, diamonds, clubs, spades) on the faces. He introduced another version during World War I where the flags of the allied nations were used to decorate the blocks. In 1900 this puzzle was called *The Great Tantalizer*. However, by far the best known and most popular version of the puzzle is based on the design by Franz (Frank) Armbruster in 1965 and consists of four plastic cubes with each face having one of four different colors. (In those days, the colors used were red, blue, green and white.)

Frank Armbruster began as an educational consultant in 1960. While working on teaching machine designs, he saw similarities between what a teaching machine does and what a game does. In particular, if the rules of the game are structured from the rules of the subject matter, then the game will teach. Each has a set of structured rules, a goal and an opportunity for strategies.

Armbruster had been interested in puzzles for much of his life and started designing games as teaching tools in 1965. He saw *Instant Insanity* as a great aid for teaching permutations and combinations at the high school level. Originally the cubes were made of wood. Thinking that the grain of wood used was giving an unintended clue to the solver, he turned to constructing the cubes from plastic. Finding a way to hold six plastic squares in place while they were cemented was the biggest challenge. Finally a method was devised and the puzzle was constructed. Armbruster was able to schedule a lunch with a representative of

Macy's in San Francisco to discuss his puzzle. This was the beginning of the puzzle as a commercial enterprise.

The first version of his famous puzzle *Instant Insanity* was licensed to the Parker Brothers Game Company, which sold over 12 million copies during 1966-1967. It was listed in the 1966 Guinness Book of Records as the best selling toy of the year, outselling the board game *Monopoly*. Although best known for his *Instant Insanity* puzzle, Armbruster is an educator, spending fifteen years at the Lockheed Corporation as an instructor and developing training methods.

Exercises for Section 8.4

8.39 Solve the Instant Insanity puzzle in Figure 8.34 by providing

(a) the pseudographs for each cube,

(b) the composite pseudograph for these four cubes,

(c) the related sub-pseudographs (Front-Back and Right-Left),

(d) a solution.

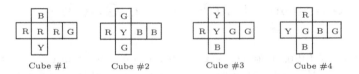

Figure 8.34: Instant Insanity puzzle for Exercise 8.39

8.40 Solve the Instant Insanity puzzle in Figure 8.35 by providing (a) - (d) as in Exercise 8.39.

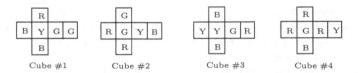

Figure 8.35: Instant Insanity puzzle for Exercise 8.40

8.41 Solve the Instant Insanity puzzle in Figure 8.36 by providing (a) - (d) as in Exercise 8.39.

8.42 Show that the cubes in Figure 8.28 can be stacked in a way different than that shown in Figure 8.33 so that all four colors appear on all four sides.

8.43 Construct a set of four multicolored cubes so that all four colors appear on each cube and such that the corresponding Instant Insanity puzzle has no solution.

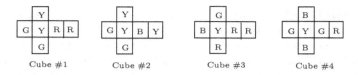

Figure 8.36: Instant Insanity puzzle for Exercise 8.41

8.44 Construct a set of four multi-colored cubes so that all four colors appear on each cube and such that the corresponding Instant Insanity puzzle has a unique solution.

8.45 For an integer k with $1 \leq k \leq 6$ and $k \neq 4$, suppose that we have a collection of k cubes where each face of each cube is assigned one of the k colors $\{1, 2, \ldots, k\}$. Investigate the puzzle of stacking these cubes on top of one another such that on each of the four sides all k colors appear.

8.5   Excursion: The Petersen Graph

Peter Christian Julius Petersen was born on June 16, 1839 in Sorø, Denmark. As a youngster, Petersen attended a private school and then the Sorø Academy, founded by King Frederick II of Denmark in 1586.

Petersen left school in 1854 because his parents couldn't afford the expense of his education. He then went to work for his uncle. When his uncle died, Petersen was left enough money to begin his studies at the Polytechnical School in Copenhagen. By 1858 Petersen had already published a book (a textbook on logarithms). Despite the fact that he passed the first part of a civil engineering examination in 1860, he decided to study mathematics at the university rather than to continue with engineering. At this time, however, the money that his uncle left him had run out and he took a position as a teacher at a private school. During the next few years Petersen had a heavy teaching load, was married and had a family but he continued to study hard.

In 1866 Petersen obtained the degree of magister of mathematics. While he was a high school teacher, Petersen learned the importance of geometric reasoning and recognized his own talent for writing textbooks. In fact, he wrote five textbooks in the 1860s, all on geometry. At the age of 30, he finally started to work seriously on his doctoral dissertation.

Petersen received his Ph.D. in 1871 from Copenhagen University. On the occasion of receiving his doctorate, Petersen wrote (translated from Danish):

> *Mathematics had, from the time I started to learn it, taken my complete interest and most of my work consisted of solving problems of*

my own and of my friends and in seeking the trisection of the angle,
a problem that has had a great influence on my whole development.

Soon afterwards, Petersen became a faculty member at Copenhagen University. He was known as an outstanding teacher. One anecdote about Petersen's teaching refers to instances when he was baffled during his lectures concerning textbooks he was using in which it was written "it is easy to see" (and he was referring to books that *he* had written). Petersen was considered a masterful writer, however. As far as his research was concerned, there were times when the elegance of his exposition took precedence over rigor. Petersen was an independent thinker and in order to be original, he rarely read the work of others with the unfortunate consequence that he occasionally obtained results that were already known. He was also quite casual about referencing the work of others.

Although Petersen worked in and made contributions to many areas of mathematics, it is only graph theory for which he is known. Indeed in his day he enjoyed an international reputation. Julius Petersen's contributions to graph theory were primarily contained within a single paper he wrote, published in 1891 and titled *Die Theorie der regulären graphs*. Prior to 1891, the important results on graph theory (including Leonhard Euler's work on Eulerian graphs and Gustav Kirchhoff's work on spanning trees) were not results expressed in terms of graphs as there really was no graph theory at that time. In the case of Petersen's paper, however, an argument could be made that for the first time a paper had been written containing fundamental results on the theory of graphs. Among the important results occurring in this paper were Theorem 8.11, referred to as Petersen's theorem (*Every 3-regular bridgeless graph contains a 1-factor.*) and Theorem 8.16 (*A graph G is 2-factorable if and only if G is r-regular for some positive even integer r*) although these theorems appeared in reverse order in Petersen's paper.

Petersen's 1891 paper was nearly co-authored with the mathematician James Joseph Sylvester (whom we will meet again in Chapter 10). The two had been working on the same problem and they corresponded extensively. Sylvester was outstanding at making conjectures, while Petersen supplied the proofs. Sylvester provided the stimulus that Petersen needed. Petersen and Sylvester were evidently making great progress towards a joint paper. In fact, Sylvester wrote to Felix Klein, editor of *Mathematische Annalen*, of their intention to submit a joint paper to the journal. When Petersen visited Sylvester in Oxford, however, it became clear that the two mathematicians were looking at the problem differently. It was decided that the two would write separate papers, even though this didn't appeal to Petersen. During Petersen's visit, he recognized that Sylvester was having health problems, both physical and mental. Sylvester never wrote his paper and would never return to graph theory again.

Sylvester was the first to use the term *graph* as it is used now. This occurred in an 1878 paper of his but Petersen's use of this term apparently caused it to be used more widely and finally it was adopted by the mathematical community.

Even though Petersen's major contribution to graph theory was his 1891 paper, that is not what he is known for. His primary fame lies not for a single paper he wrote but for a single graph that appeared in one of his papers: **the Petersen graph**. We have encountered this graph several times already and we will continue to encounter it. Petersen first mentioned this graph, not in his 28-page classic 1891 paper but in his 3-page 1898 paper "Sur le théorème de Tait" in which he presented this graph as a counterexample to Peter Guthrie Tait's "theorem": *Every 3-regular bridgeless graph is 1-factorable*. His graph did not appear in the aesthetic way in which it is commonly drawn, as shown in Figure 8.37(a), but in the less appealing way shown in Figure 8.37(b). Petersen died on August 5, 1910 in Copenhagen, Denmark.

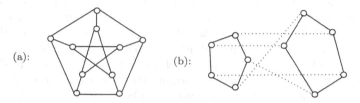

Figure 8.37: The Petersen graph

Curiously, Petersen was not the first person to use the graph that bears his name. Evidently, the first occurrence of this graph was 12 years earlier in an 1886 paper of Alfred Bray Kempe. (We will visit both Kempe and Tait again in Chapter 10.) We mentioned earlier that the Petersen graph consists of two disjoint 5-cycles joined by a particular matching of cardinality 5. If we denote the two 5-cycles by $C = (v_1, v_2, v_3, v_4, v_5, v_1)$ and $C' = (v'_1, v'_2, \ldots, v'_5, v'_1)$, then the edges of the matching are $v_1v'_1, v_2v'_3, v_3v'_5, v_4v'_2, v_5v'_4$. There are numerous interesting ways to draw the Petersen graph. Two additional ways are shown in Figure 8.38.

Figure 8.38: The Petersen graph (again)

The fact that the Petersen graph was introduced by Petersen as a counterexample would be a prelude to the way this graph is commonly encountered, as it often appears as an example or counterexample to graph theoretic statements.

Let's look at a few characteristics of this famous graph. The Petersen graph is, of course, a 3-regular graph of order 10 and therefore has size 15. We have already noted that the length of a smallest cycle in this graph is 5. The length of a smallest cycle in a graph is referred to as its **girth**. For an integer $g \geq 3$, a g-**cage** is a 3-regular graph of minimum order that has girth g. It is easy to see that K_4 is the unique 3-cage and only slightly more difficult to see that $K_{3,3}$ is the unique 4-cage. It turns out that the Petersen graph is the unique 5-cage. These three graphs are shown in Figure 8.39.

Figure 8.39: The 3-cage, 4-cage and 5-cage

Theorem 8.26 *The Petersen graph is the unique 5-cage.*

Proof. Let G be a 5-cage and let v_1 be a vertex of G. Since $\deg v_1 = 3$, the vertex v_1 has three neighbors, say v_2, v_3 and v_4. Because G contains no triangles, the set $\{v_2, v_3, v_4\}$ is independent. Since the vertices v_2, v_3 and v_4 also have degree 3, each of these vertices is adjacent to two vertices in addition to v_1, say

$$N(v_2) = \{v_1, v_5, v_6\}, \ N(v_3) = \{v_1, v_7, v_8\}, \ N(v_4) = \{v_1, v_9, v_{10}\}.$$

The vertices $v_5, v_6, \ldots, v_{10}$ are distinct since G contains no 4-cycles. Thus G contains the subgraph shown H_1 in Figure 8.40(a).

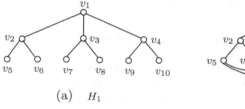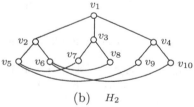

(a) H_1 (b) H_2

Figure 8.40: Subgraphs of a 5-cage G

This says that the order of G is at least 10. Since the Petersen graph PG is a 3-regular graph of order 10 having girth 5, it follows that every 5-cage has order 10 and that PG is a 5-cage. It remains to show that PG is the *only* 5-cage. Because G is a 5-cage, its order is 10 and $V(G) = \{v_1, v_2, \ldots, v_{10}\}$. Consider the vertex v_5, which must be adjacent to two vertices in the set $\{v_6, v_7, v_8, v_9, v_{10}\}$. Because G contains no triangles or 4-cycles, v_5 is not adjacent to v_6 and v_5 cannot be adjacent to both v_7 and v_8 or to both v_9 and v_{10}. Hence v_5 is adjacent to exactly one of v_7 and v_8 and adjacent to exactly one of v_9 and v_{10}. We may assume that v_5 is adjacent to v_7 and v_9. Furthermore, v_6 is adjacent to v_8 and v_{10}. Hence G contains the subgraph H_2 shown in Figure 8.40(b).

The vertex v_7 is not adjacent to v_8 or v_9 as G has no triangles, so v_7 must be adjacent to v_{10}. Consequently, v_8 and v_9 must be adjacent as well (see

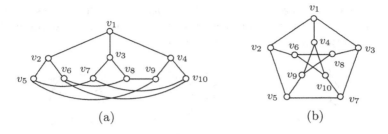

Figure 8.41: The Petersen graph: the unique 5-cage

Figure 8.41(a)). Hence there is only one possibility for the graph G, namely that G is (isomorphic to) the Petersen graph (Figure 8.41(b)).   ∎

The 6-cage, 7-cage and 8-cage are also unique and are known, respectively, as the **Heawood graph**, the **McGee graph** and the **Tutte-Coxeter graph**, shown in Figure 8.42.

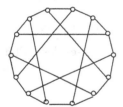 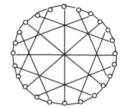 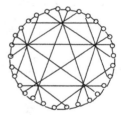

Figure 8.42: The 6-cage, 7-cage and 8-cage

We have already seen that the Petersen graph is neither 1-factorable nor Hamiltonian. Even though the Petersen graph PG is not Hamiltonian, it is close to having this property as $PG - v$ is Hamiltonian for every vertex v of PG.

8.6   Exploration: Bi-Graceful Graphs

We saw in Section 8.3 that a graph G is graceful if it has a graceful labeling. By a graceful labeling of a graph G of size m is meant a one-to-one function $f : V(G) \to \{0, 1, 2, \ldots, m\}$ having the property that the induced edge labeling $f' : E(G) \to \{1, 2, \ldots, m\}$ defined by

$$f'(e) = |f(u) - f(v)| \text{ for each edge } e = uv \text{ of } G$$

is one-to-one as well. We also saw that if G is a graceful graph of size m, then the complete graph K_{2m+1} is G-decomposable. In fact, K_{2m+1} is cyclically G-decomposable. Consider the cycle $C = (v_0, v_1, v_2, \ldots, v_{2m}, v_0)$ of length $2m + 1$

where the vertices of C are arranged cyclically in a regular $(2m + 1)$-gon. A vertex labeled i in G, where $i \in \{0, 1, 2, \ldots, 2m\}$ is placed at v_i in C. This is illustrated for the graceful graph G in Figure 8.43. By rotating G clockwise through an angle of $2\pi/9$ radians a total of eight times, a G-decomposition of K_9 is obtained.

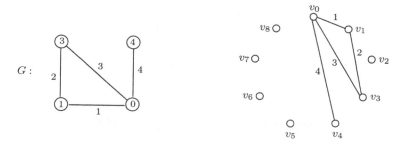

Figure 8.43: Placing the vertices of a graceful graph G
of size m on regular $(2m + 1)$-gon

Alexander Rosa determined which cycles are graceful.

Theorem 8.27 *A cycle C_n is graceful if and only if*

$$n \equiv 0 \pmod 4 \text{ or } n \equiv 3 \pmod 4.$$

We saw in Example 8.22 that C_5 is not graceful. Despite this, K_{11} is cyclically C_5-decomposable however. See Figure 8.44.

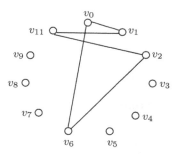

Figure 8.44: A cyclic C_5-decomposition of K_{11}

Let G be a graph of size m and let $S_i = \{i, 2m + 1 - i\}$ for $1 \le i \le m$. A **bi-graceful labeling** of G is a one-to-one function $f : V(G) \to \{0, 1, 2, \ldots, 2m\}$ that induces the edge labeling $f' : E(G) \to \{1, 2, \ldots, 2m\}$ defined by

$$f'(e) = |f(u) - f(v)| \text{ for each edge } e = uv \text{ of } G$$

such that the set $S = \{f'(e) : e \in E(G)\}$ has the property that $|S \cap S_i| = 1$ for all i $(1 \leq i \leq m)$. A bi-graceful labeling was called a ρ-valuation by Rosa. A graph admitting a bi-graceful labeling is a **bi-graceful graph**. In particular, C_5 is a bi-graceful graph (see Figure 8.45).

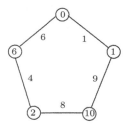

Figure 8.45: A bi-graceful labeling of C_5

A **co-graceful labeling** of a graph G is a bi-graceful labeling for which $S \cap S_i = \{2m + 1 - i\}$ for all i $(1 \leq i \leq m)$. A graph admitting a co-graceful labeling is a **co-graceful graph**. In particular, C_4 is a co-graceful graph (see Exercise 8.52).

Exercises for Section 8.6

8.46 Prove that every graceful graph has a bi-graceful labeling that is not graceful.

8.47 Prove that if G is a bi-graceful graph of size m, then K_{2m+1} is cyclically G-decomposable.

8.49 Is C_6 bi-graceful?

8.50 Is C_7 bi-graceful?

8.51 Is C_9 bi-graceful?

8.52 Show that C_4 is a co-graceful graph.

Chapter 9

Planarity

9.1 Planar Graphs

The directors of an amusement center have decided to open a new theme park in the center. The initial plan for the theme park is to build six attractions, which are temporarily denoted by A1, A2, ..., A6. Figure 9.1(a) shows the initial location of the attractions.

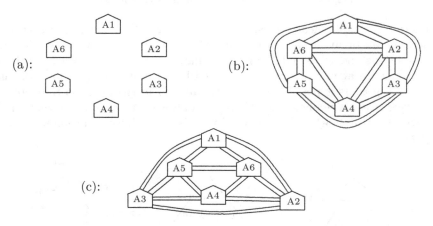

Figure 9.1: Six attractions in a theme park

In the summer, the amusement center often becomes very hot and walking between attractions can be uncomfortable. Preliminary studies indicate that the least amount of traffic is likely to occur between attractions (1) A1 and A4, (2) A2 and A5 and (3) A3 and A6. The designers feel that, despite the expense, it would be good for business to build an air-conditioned tube enclosing moving walkways in both directions between all pairs of attractions except those in (1)–

(3). One possible concern is whether this can be done without any two tubes interfering with each other. Figure 9.1(b) shows that the tubes can indeed be built without any pair intersecting. Figure 9.1(c) shows that if the attractions are relocated, then an even better design for the location of the tubes can be given.

After time passes, it is decided that the attractions A1, A2, ..., A6 need to be modified and they are now called B1, B2, ..., B6. Furthermore, it is decided to add a seventh attraction B7. (See Figure 9.2.) In addition, it is decided that moving walkway tubes should be built between every pair of attractions, except the pairs {B1, B4}, {B1, B5}, {B2, B5}, {B2, B6}, {B3, B6}, {B3, B7} and {B4, B7}. How should this be done?

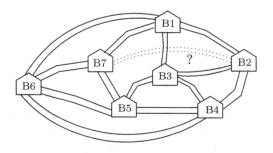

Figure 9.2: Seven attractions in a theme park

There is a graph theory question (or two) here. The theme park with six attractions can be modeled by the graph G_6 of Figure 9.3(a) whose vertices are the attractions and whose edges are the walkways. In a similar way, the theme park with seven attractions can be modeled by a graph G_7. The question that we are asking is: Is it possible to draw G_6 and G_7 in the plane so that none of their edges cross. The answer for G_6 is certainly yes and such a drawing for G_6 is shown in Figure 9.3(b). The answer for G_7 is ... well, we don't know (at least not yet).

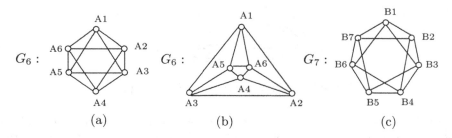

Figure 9.3: The graphs G_6 and G_7

A graph G is called a **planar graph** if G can be drawn in the plane so that no two of its edges cross each other. Therefore, the graph G_6 of Figure 9.3 is planar. We have yet to determine whether the graph G_7 is planar. A graph that is not planar is called **nonplanar**. A graph G is called a **plane graph** if it is drawn in the plane so that no two edges of G cross. Thus, while a graph may be planar, as drawn it may not be a plane graph, such as the graph G_6 in Figure 9.3(a).

There is a well-known puzzle that has appeared in a number of books, magazines and comic books. There are three utilities (gas, water and electricity) that need to be connected to three houses by gas lines, water mains and electrical lines. Can this be done without any of the lines or mains crossing each other? This situation is shown in Figure 9.4(a). This problem is referred to as the **Three Houses and Three Utilities Problem**. There are reports that this problem may have been introduced by the American puzzle-maker Sam Loyd, Sr. in 1900. The situation described in this problem can be modeled by the graph of Figure 9.4(b), which, in fact, is the graph $K_{3,3}$. In graph theory terms then, the Three Houses and Three Utilities Problem asks whether the graph $K_{3,3}$ is planar.

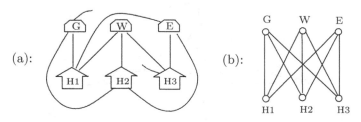

Figure 9.4: The Three Houses and Three Utilities Problem

Before providing a solution to this problem, it is useful to make some observations about planar graphs (actually connected planar graphs). First there are some well-known classes of planar graphs. Every cycle is planar. Every path and every star are planar. Indeed, every tree is planar. Of course, every graph that can be drawn in the plane without any two of its edges crossing is planar, as this is the definition of a planar graph. It may appear that $K_{3,3}$ is not planar but if it is not planar, then how do we show this? After all, just because we don't see how to draw a graph without its edges crossing doesn't mean that the graph is nonplanar. We are about to deal with questions such as this.

Consider the graph H shown in Figure 9.5(a). Of course, H is connected. But H is also planar as we can see from Figure 9.5(b), where H is drawn as a plane graph.

A plane graph divides the plane into connected pieces called **regions**. For example, in the case of the plane graph H of Figure 9.5(b), there are six regions. This graph H is redrawn in Figure 9.6, where the six regions are denoted by $R_1, R_2, \ldots, R_6$. In every plane graph, there is always one region that is unbounded. This is the **exterior region**. For the graph H of Figure 9.6, R_6 is the exterior region. The subgraph of a plane graph whose vertices and edges are

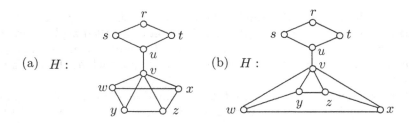

(a) H :   (b) H :

Figure 9.5: A planar graph and a plane graph

incident with a given region R is the **boundary** of R. The boundaries of the six regions of the graph H of Figure 9.6 are also shown in that figure.

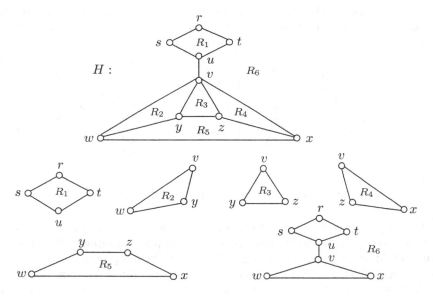

Figure 9.6: A plane graph and its regions

Notice that uv is a bridge in the graph H of Figure 9.6 and is on the boundary of one region only, namely the exterior region. In fact, a bridge is always on the boundary of exactly one region (though not necessarily the exterior region). An edge that is not a bridge lies on the boundary of two regions. For example, vy lies on the the boundary of both R_2 and R_3. If we were to remove the edge vy, then the resulting graph is a plane graph as well but has one less region as R_2 and R_3 become part of a single region. On the other hand, the graph $H - uv$ is disconnected but there is no change in the number of regions. Another observation is useful.

If G is a connected plane graph with at least three edges, then the boundary of every region of G has at least three edges.

Looking at the plane graph H of Figure 9.6 yet again, we see that H is a connected graph of order 9 and size 13 having six regions. Furthermore, the bridge is on the boundary of a single region (the exterior region) and all other edges are on the boundaries of two regions. Letting n, m and r denote the order, size and number of regions, respectively, we then have $n = 9$, $m = 13$ and $r = 6$. So in this case, $n - m + r = 2$. In fact, $n - m + r = 2$ is true for all connected plane graphs. This is a consequence of an observation made by Leonhard Euler who reported this to the German mathematician Christian Goldbach in a letter dated November 14 , 1750. This result first appeared in print in 1752. This is referred to as the **Euler Identity**.

Theorem 9.1 (**The Euler Identity**) *If G is a connected plane graph of order n, size m and having r regions, then $n - m + r = 2$.*

Proof. First, if G is a tree of order n, then $m = n - 1$ (by Theorem 4.4) and $r = 1$; so $n - m + r = 2$. Therefore, we need only be concerned with connected graphs that are not trees. Assume, to the contrary, that the theorem does not hold. Then there exists a connected plane graph G of smallest size for which the Euler Identity does not hold. Suppose that G has order n, size m and r regions. So $n - m + r \neq 2$. Since G is not a tree, there is an edge e that is not a bridge. Thus $G - e$ is a connected plane graph of order n and size $m - 1$ having $r - 1$ regions. Because the size of $G - e$ is less than m, the Euler Identity holds for $G - e$. So $n - (m - 1) + (r - 1) = 2$ but then $n - m + r = 2$, which is a contradiction. $\blacksquare$

Figure 9.7 shows a planar graph G and several ways of drawing G as a plane graph. However, since G has a fixed order $n = 7$ and fixed size $m = 9$ and the Euler Identity holds ($n - m + r = 7 - 9 + r = 2$), each drawing of G as a plane graph always produces the same number of regions, namely $r = 4$.

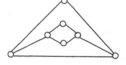

Figure 9.7: Different drawings of a planar graph

The Euler Identity has many useful and interesting consequences. One of these (which will allow us to *prove* that some graphs are *not* planar) tells us that planar graphs cannot have too many edges.

Theorem 9.2 *If G is a planar graph of order $n \geq 3$ and size m, then*

$$m \leq 3n - 6.$$

Proof. First, suppose that G is connected. If $G = P_3$, then the inequality holds. So we can assume that G has at least three edges. Draw G as a plane graph, where G has r regions denoted by $R_1, R_2, \ldots, R_r$. The boundary of each region contains at least three edges. So if m_i is the number of edges on the boundary of R_i $(1 \leq i \leq r)$, then $m_i \geq 3$. Let

$$M = \sum_{i=1}^{r} m_i \geq 3r.$$

The number M counts an edge once if the edge is a bridge and counts it twice if the edge is not a bridge. So $M \leq 2m$. Therefore, $3r \leq M \leq 2m$ and so $3r \leq 2m$.

Applying the Euler Identity to G, we have

$$6 = 3n - 3m + 3r \leq 3n - 3m + 2m = 3n - m. \tag{9.1}$$

Solving the inequality (9.1) for m, we get $m \leq 3n - 6$.

If G is disconnected, then edges can be added to G to produce a connected plane graph of order n and size m', where $m' > m$. From what we have just shown, $m' \leq 3n - 6$ and so $m < 3n - 6$. ∎

Theorem 9.2 provides a necessary condition for a graph to be planar and so provides a sufficient condition for a graph to be nonplanar. In particular, the contrapositive of Theorem 9.2 gives us the following:

If G is a graph of order $n \geq 3$ and size m such that $m > 3n - 6$, then G is nonplanar.

There are now some immediate but interesting consequences of Theorem 9.2.

Corollary 9.3 *Every planar graph contains a vertex of degree 5 or less.*

Proof. Suppose that G is a graph every vertex of which has degree 6 or more. Let G have order n and size m. Certainly, $n \geq 7$. Then

$$2m = \sum_{v \in V(G)} \deg v \geq 6n.$$

Thus $m \geq 3n > 3n - 6$. By Theorem 9.2, G is nonplanar. ∎

We can now give an example of a nonplanar graph.

Corollary 9.4 *The complete graph K_5 is nonplanar.*

Proof. The graph K_5 has order $n = 5$ and size $m = 10$. Since $m = 10 > 9 = 3n - 6$, it follows that K_5 is nonplanar by Theorem 9.2. ∎

Let's revisit the proof of Theorem 9.2, where we were discussing the r regions $R_1, R_2, \ldots, R_r$ of a plane graph G. We mentioned that the number m_i of edges on the boundary of R_i $(1 \leq i \leq r)$ is at least 3. Of course, if there are *exactly* three edges on the boundary of each region, then $m_i = 3$ for each i $(1 \leq i \leq r)$ and so $M = 3r$. We also mentioned that M counts an edge once if the edge is a bridge and counts it twice if the edge is not a bridge. The only way for the inequality in (9.1) to be an equality is for $3r = M = 2m$, which means that the boundary of every region must contain exactly three edges *and* that there are no bridges in G (which eliminates the possibility that $G = K_{1,3}$). Under these conditions, $m = 3n - 6$. Therefore, the only way for the equality $m = 3n - 6$ to occur in a connected plane graph G of order $n \geq 3$ and size m is that the boundary of every region of G (including the exterior region) is a triangle. Three examples of this are shown in Figure 9.8.

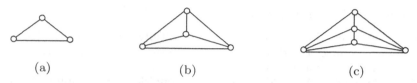

(a)   (b)   (c)

Figure 9.8: Maximal planar graphs

Notice that the graphs in Figures 9.8(a) and 9.8(b) are complete graphs but the graph in Figure 9.8(c) is $K_5 - e$ for some edge e. Indeed, if we have a connected planar graph G of order n and size m, where $m = 3n - 6$ and $n \geq 5$, then G is not complete. If we were to add an edge e between any two nonadjacent vertices of G, then the graph $G+e$ cannot be planar because its size $m+1$ exceeds $3n - 6$. This confirms what we learned in Corollary 9.4, namely, that K_5 is *not* planar.

A graph G is **maximal planar** if G is planar but the addition of an edge between any two nonadjacent vertices of G results in a nonplanar graph. Another way to say this is that a graph G is maximal planar if G is planar but G is not a proper spanning subgraph of any other planar graph. For $1 \leq n \leq 4$, the only maximal planar graph of order n is K_n. Thus all of the graphs in Figure 9.8 are maximal planar. Necessarily then, if a maximal planar graph G of order $n \geq 3$ and size m is drawn as a plane graph, then the boundary of every region of G is a triangle and $m = 3n - 6$.

Now that we have a sufficient condition for a graph to be nonplanar, let's return to the Three Houses and Three Utilities Problem, which, if you recall, is equivalent to determining whether $K_{3,3}$ is planar. Attempts to draw $K_{3,3}$ in the plane without edges crossing were unsuccessful, thereby leading us to believe that $K_{3,3}$ is nonplanar. The graph $K_{3,3}$ has order $n = 6$ and size $m = 9$. Thus $3n - 6 = 12$. Therefore, $m \leq 3n - 6$, which says ... nothing. We only know that a graph is nonplanar if $m > 3n - 6$. Therefore, we can make no conclusion about the planarity of $K_{3,3}$ from Theorem 9.2. On the other hand, if we recall that $K_{3,3}$ is bipartite (and therefore contains no odd cycles) and look at the proof of Theorem 9.2 in more detail, then we can finally prove what we believed to be true all along.

Theorem 9.5 *The graph $K_{3,3}$ is nonplanar.*

Proof. Assume, to the contrary, that $K_{3,3}$ is planar and draw $K_{3,3}$ as a plane graph. Since $n = 6$ and $m = 9$, it follows by the Euler Identity that $n - m + r = 6 - 9 + r = 2$ and so $r = 5$. Let $R_1, R_2, \ldots, R_5$ be the five regions and let m_i be the number of edges on the boundary of R_i ($1 \leq i \leq 5$). Since $K_{3,3}$ has no triangles, $m_i \geq 4$ for $1 \leq i \leq 5$ and because $K_{3,3}$ contains no bridges, it follows that

$$2m = \sum_{i=1}^{5} m_i \geq 20$$

and so $m \geq 10$. This is a contradiction. ∎

Therefore, the Three Houses and Three Utilities Problem is solved! It is *impossible* to connect the three utilities to the three houses without any gas or electrical lines or water mains crossing.

While Theorem 9.2 provides a necessary condition for a graph to be planar, it is only just that: a necessary condition. This theorem can help us to prove that a graph is nonplanar but there are nonplanar graphs that cannot be proved to be nonplanar by this theorem (such as $K_{3,3}$). However, there is a remarkable theorem that gives a condition that is both necessary *and* sufficient for a graph to be planar. The theorem that we are about to present is due to Kazimierz Kuratowski.

Kazimierz Kuratowski was born in Warsaw, Poland on February 2, 1896. Because of the political situation in 1913, Kuratowski left Poland then to study engineering at the University of Glasgow in Scotland. He completed his first year at the university but while back home preparing to begin his second year of studies, World War I broke out, making it impossible for him to return to Scotland.

The University of Warsaw in Poland had been under Russian control for decades and even became an underground university. However, during 1915, the University of Warsaw was reformulated as a Polish university. One of its first students was Kuratowski who took up mathematics. He was greatly influenced by the faculty there, several of whom were interested in the evolving mathematical field of topology. Kuratowski wrote his first research paper in 1917 and was awarded his Ph.D. in 1921. In 1927, Kuratowski became a mathematics professor at the Technical University of Lvov. The mathematicians there did a great deal of research, often working at tea shops and cafés. It is while Kuratowski was in Lvov that he discovered and proved an important theorem in graph theory, which we are about to state. In 1934 Kuratowski returned to Warsaw and became a professor at the University of Warsaw. Life was disrupted greatly in 1939 during the German invasion of Poland. An underground university was developed during World War II and Kuratowski taught there. After World War II the entire Polish educational system had to be rebuilt and Kuratowski took a leadership role in doing this. Although Kuratowski did a great deal of research throughout his life, primarily in set theory and topology, it was his work in directing schools of mathematical research and education that are perhaps most notable. He died where he was born, in Warsaw, on June 18, 1980.

So what exactly is Kuratowski's contribution to graph theory? It turns out that to determine which graphs are planar and which are nonplanar, it is Corollary 9.4 and Theorem 9.5 that provide the keys, where it is shown that K_5 and $K_{3,3}$ are nonplanar. Certainly, if H is a nonplanar subgraph of a graph G, then G is nonplanar as well. In particular, if G contains either K_5 or $K_{3,3}$ as a subgraph, then G is nonplanar.

We now know that a graph G of order $n \geq 3$ and size m is nonplanar if any of the following occurs: (1) $m > 3n - 6$, (2) G contains K_5 as a subgraph, (3) G contains $K_{3,3}$ as a subgraph. We have already seen that $K_{3,3}$ is a nonplanar graph of order n and size m for which the inequality $m > 3n - 6$ does not hold. Furthermore, even if $m > 3n - 6$, then there is no guarantee that G contains either K_5 or $K_{3,3}$ as a subgraph, as we now show.

Example 9.6 *There exists a graph of order $n \geq 3$ and size $m > 3n - 6$ that contains neither K_5 nor $K_{3,3}$ as a subgraph.*

Solution. Consider the graph G of order $n = 7$ and size $m = 16$ shown in Figure 9.9. Since $m = 16 > 15 = 3n - 6$, it follows that G is nonplanar. In fact, $G - uv$ is a maximal planar graph.

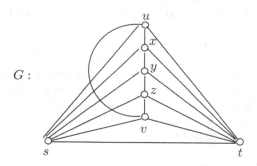

Figure 9.9: A nonplanar graph

Assume, to the contrary, that G contains a subgraph F such that $F = K_5$ or $F = K_{3,3}$. Necessarily, F must contain the edge uv, for otherwise, F is a subgraph of $G - uv$, which is impossible. If $F = K_5$, then $u, v \in V(F)$ and the remaining three vertices of F must be adjacent to both u and v. However, only s and t are adjacent to both u and v. Thus $F \neq K_5$ and so we must have $F = K_{3,3}$. Let U and W be the two partite sets of F. Since $uv \in E(F)$, one of u and v belongs to U and the other belongs to W, say $u \in U$ and $v \in W$. Since the remaining two vertices of W are adjacent to every vertex in U, either $W = \{v, s, t\}$ or $W = \{v, s, x\}$. If $W = \{v, s, t\}$, then only u and z are adjacent to all three vertices of W, which is impossible. If $W = \{v, s, x\}$, then only u and t are adjacent to all three vertices of W, again which is impossible. Thus $F \neq K_{3,3}$, which produces a contradiction.   ◇

Because the graph G of Figure 9.9 is nonplanar and contains neither K_5 nor $K_{3,3}$ as a subgraph, it follows that the nonplanarity of a graph doesn't depend on the graph containing K_5 or $K_{3,3}$ as a subgraph. As we are about to see, it does depend on something very close to this however.

Let's now turn our attention to the graph G of Figure 9.10. If we replace the edge uv by a vertex s of degree 2 and join s to u and v, then we obtain the graph G_1. The graph G_1 is referred to as a subdivision of G. We might also think of producing G_1 by inserting a vertex of degree 2 into the edge uv of G.

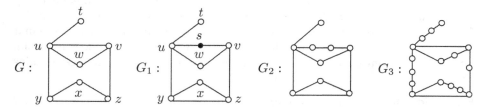

Figure 9.10: Subdivisions of a graph

More formally, a graph G' is called a **subdivision** of a graph G if $G' = G$ or one or more vertices of degree 2 are inserted into one or more edges of G. Consequently, all of G, G_1, G_2 and G_3 are subdivisions of G. In fact, G_2 is a subdivision of G_1 as well.

Perhaps it is clear that every subdivision of a planar graph is planar and that every subdivision of a nonplanar graph is nonplanar. This says that if G' if a subdivision of a graph G, then G' is planar if and only if G is planar. Therefore, if H is a graph that contains a subdivision of K_5 or a subdivision of $K_{3,3}$ as a subgraph, then H is nonplanar. Kuratowski's amazing theorem states that the converse of this statement is true as well. Proving Kuratowski's theorem is a complex task and, consequently, no proof of it is presented here.

Theorem 9.7 (**Kuratowski's Theorem**) *A graph G is planar if and only if G does not contain a subdivision of K_5 or $K_{3,3}$ as a subgraph.*

Let's summarize what we've learned at this point. Suppose that we are given a graph G of order $n \geq 3$ and size m and we wish to determine whether G is planar. To show that G is planar, certainly one option is to draw G as a plane graph. One way to verify that G is nonplanar is to show that $m > 3n - 6$. However, if $m \leq 3n - 6$, it may still be the case that G is nonplanar. A surefire way to verify that G is nonplanar is to show that a subdivision of K_5 or $K_{3,3}$ is a subgraph of G. To show that G contains a subdivision of K_5 as a subgraph, we need to find a subgraph H containing five vertices of degree 4, every two of which are connected by a path, all of whose interior vertices have degree 2 in the subgraph H (see Figure 9.11(a)). To show that G contains a subdivision of $K_{3,3}$ as a subgraph, we need to find a subgraph F containing six vertices of degree 3, partitioned into two sets V_1 and V_2 of three vertices each, such that every vertex in V_1 is connected to every vertex in V_2 by a path, all of whose interior vertices have degree 2 in the subgraph F (see Figure 9.11(b)). What this also says is that if G is a graph that contains (1) at most four vertices of degree 4 or more *and* (2) at most five vertices of degree 3 or more, then G must be planar.

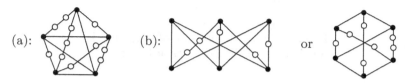

Figure 9.11: Subdivisions of K_5 and $K_{3,3}$

We now present an application of Theorem 9.7.

Example 9.8 *Determine whether the graph G of Figure 9.12 is planar.*

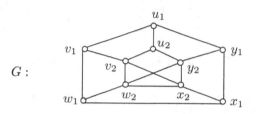

Figure 9.12: The graph G in Example 9.8

Solution. Certainly as drawn, G is not a plane graph. Of course, this neither proves nor disproves that G is nonplanar (although one may be suspicious that this is the case). The graph G has order $n = 10$ and size $m = 17$. Since $m = 17 \le 3n - 6 = 24$, we cannot use Theorem 9.2 to show that G is nonplanar. On the other hand, just because $m \le 3n - 6$, this certainly does not imply that G is planar either.

Next, let's see if we can find a subgraph of G that is either a subdivision of K_5 or a subdivision of $K_{3,3}$. Actually, G resembles K_5 as it is drawn. However, only four vertices of G have degree 4 or more. Therefore, it is impossible that G contains a subdivision of K_5 as a subgraph. On the other hand, the graph F shown in Figure 9.13 *is* a subgraph of G. Since F is a subdivision of $K_{3,3}$, it follows by Kuratowski's theorem that G is indeed nonplanar.   ◇

We will see another characterization of planar graphs (Theorem 9.15), due to Klaus Wagner, in Section 9.3.

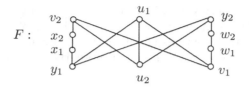

F :

Figure 9.13: A subdivision of $K_{3,3}$ that is
a subgraph of the graph G of Figure 9.12

Exercises for Section 9.1

9.1 Show that each of the graphs in Figure 9.14 is planar by drawing it as a
plane graph. Verify that the Euler Identity holds for each graph.

 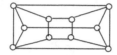 

Figure 9.14: Graphs for Exercise 9.1

9.2 A connected k-regular graph of order 12 is embedded in the plane, resulting
in eight regions. What is k?

9.3 (a) The vertices of a certain graph G have degrees 3, 4, 4, 4, 5, 6, 6. Prove
that G is nonplanar.

(b) The vertices of a certain graph G have degrees 4, 4, 4, 5, 5, 5, 6, 6, 6,
7, 7, 7. Prove that G is nonplanar.

9.4 (a) Find all integers n such that K_n is planar.

(b) Find all pairs r, s of integers for which $K_{r,s}$ is planar.

9.5 Show that there exists

(a) a 4-regular planar graph and a 4-regular nonplanar graph.

(b) a 5-regular planar graph and a 5-regular nonplanar graph.

(c) no r-regular planar graph for $r \geq 6$.

9.6 Prove or disprove the following.

(a) Every subgraph of a planar graph is planar.

(b) Every subgraph of a nonplanar graph is nonplanar.

(c) If G is a nonplanar graph, then G contains a proper nonplanar subgraph.

(d) If G does not contain K_5 or $K_{3,3}$ as a subgraph, then G is planar.

(e) If G is a graph of order n and size m with $m \leq 3n - 6$, then G is planar.

(f) If G is a graph with one or more triangles and contains no subdivision of K_5 as a subgraph, then G is planar.

9.7 Give an example of each of the following or explain why no such example exists.

(a) a planar graph of order 4.

(b) a nonplanar graph of order 4.

(c) a nonplanar graph of order 6 that contains neither K_5 nor $K_{3,3}$ as a subgraph.

(d) a plane graph having 5 vertices, 10 edges and 7 regions.

(e) a planar graph of order $n \geq 3$ and size m with $m = 3n - 6$.

(f) a nonplanar graph of order $n \geq 3$ and size m with $m = 3n - 6$.

9.8 Determine, with explanation, whether the graph $K_4 \times K_2$ is planar.

9.9 Determine, with explanation, whether the graph of Figure 9.15 is planar.

Figure 9.15: The graph in Exercise 9.9

9.10 Determine, with explanation, whether the graph of Figure 9.16 is planar.

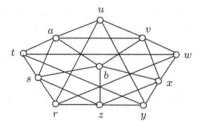

Figure 9.16: The graph in Exercise 9.10

9.11 Determine, with explanation, whether the graph G of Figure 9.17 is planar.

9.12 Determine, with explanation, whether the graph G of Figure 9.18 is planar. (See the graph G_7 of Figure 9.3(c).)

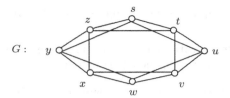

Figure 9.17: The graph in Exercise 9.11

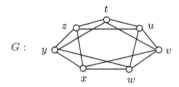

Figure 9.18: The graph in Exercise 9.12

9.13 (a) Prove that if G is a connected planar graph of order $n \geq 3$ and size m without triangles, then $m \leq 2n - 4$.

   (b) Use (a) to show that $K_{3,3}$ is nonplanar.

   (c) Prove or disprove: If G is a planar bipartite graph, then G has a vertex of degree 3 or less.

9.14 Let G be a connected plane graph of order $n \geq 5$ and size m.

   (a) Prove that if the length of a smallest cycle in G is 5, then $m \leq \frac{5}{3}(n-2)$.

   (b) Use (a) to show that the Petersen graph is nonplanar.

   (c) Use Kuratowski's theorem to show that the Petersen graph is nonplanar.

   (d) Prove or disprove: If $n < 20$ and the length of a smallest cycle in G is 5, then G has a vertex of degree 2 or less.

9.15 Prove that if G is a planar graph of order $n \leq 11$, then G has a vertex of degree 4 or less.

9.16 Do there exist two non-isomorphic maximal planar graphs of the same order?

9.17 It is not difficult to show that $\overline{C}_n$ is planar if $3 \leq n \leq 5$ and that $\overline{C}_n$ is nonplanar if $n \geq 9$. This leaves only $\overline{C}_6$, $\overline{C}_7$ and $\overline{C}_8$ in question. For each of these three graphs, determine, with justification, whether it is planar or nonplanar.

9.18 Prove that there exists no maximal planar graph G of order $n \geq 3$ whose complement $\overline{G}$ is also maximal planar.

9.19 If a maximal planar graph of order 100 is embedded in the plane, how many regions result?

9.20 Determine all maximal planar graphs, if any, where one-third of their vertices have degree 3, one-third have degree 4 and one-third have degree 5.

9.21 Prove that if G is a maximal planar graph of order at least 4, then $\delta(G) \geq 3$.

9.22 Prove that there exists only one 4-regular maximal planar graph.

9.2 Embedding Graphs on Surfaces

If G is a planar graph, then we know that G can be drawn in the plane in such a way that no two edges cross. Such a "drawing" is also called an **embedding** of G in the plane. In addition, we say that G can be **embedded** in the plane. On the other hand, if G is nonplanar, then G cannot be embedded in the plane, that is, it is impossible to draw G in the plane without some of its edges crossing.

Perhaps it is clear that if a graph G is planar, then G can be embedded on the sphere as well as the plane. Furthermore, if a graph G can be embedded on a sphere, then it must be planar. Although these observations may not seem particularly enlightening, this brings up the question of considering surfaces other than the sphere on which a graph might be embedded. But what other surfaces are there? A common surface is the **torus**, a doughnut-shaped surface (see Figure 9.19(a)). In Figure 9.19(b), we see that the graph K_4 can be embedded on the torus. In fact, there is more than one way to embed K_4 on the torus (see Figure 9.19(c)).

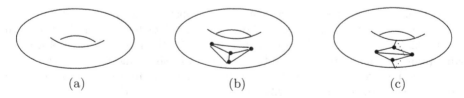

(a)   (b)   (c)

Figure 9.19: Embedding K_4 on the torus

Not only can K_4 be embedded on the torus, so can K_5. Figure 9.20(a) shows an embedding of K_5 on the torus; Figure 9.20(b) shows an embedding of $K_{3,3}$ on the torus.

Embedding graphs on a torus, as we did in Figure 9.20, can be difficult to visualize. However, there are alternative ways to represent these embeddings as we will now explain. How is a torus constructed? One way is to begin with a rectangular piece of material (the more flexible the better) as in Figure 9.21 and

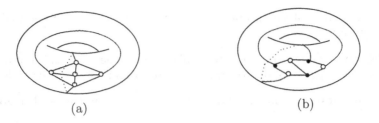

Figure 9.20: Embedding K_5 and $K_{3,3}$ on the torus

first make a cylinder from it by identifying sides a and c, which are the same after the identification occurs. The sides b and d then become circles. These circles are then identified to produce a torus.

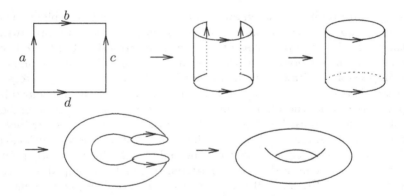

Figure 9.21: Constructing a torus

Now that we have seen how a torus can be constructed, we see that the torus can be represented by a rectangle whose opposite sides have been identified. In Figure 9.22(a), the rectangle represents a torus. So the points labeled A are the same point on the torus, the points labeled B are the same point on the torus and the points labeled C are the same point. Embeddings of K_5 and $K_{3,3}$ on the torus are shown in Figures 9.22(b) and 9.22(c), respectively.

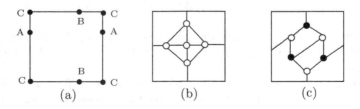

Figure 9.22: Embedding K_5 and $K_{3,3}$ on the torus

There is another way to represent a torus and embedding a graph on a torus. We begin with a sphere and drill two holes in its surface, as in Figure 9.23(a). Then we attach a handle on the sphere, where the ends of handle are placed over the two holes as in Figure 9.23(b). The surface that we have just constructed, namely a sphere with one handle, is, in actuality, a torus, although it looks different than the previous way we constructed a torus. In Figure 9.23(c), an embedding of K_5 on the torus is shown, where one of the edges of K_5 passes over the handle of the torus.

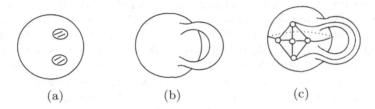

(a) (b) (c)

Figure 9.23: Embedding K_5 on the torus

Just as a torus is a sphere with a handle, we can consider a sphere on which a number of handles have been attached. We denote a sphere with k handles by S_k. The surface S_k is also called a **surface of genus** k. Thus, S_1 is the torus, while S_0 is the sphere itself.

We are now prepared to make an important observation. Let G be a graph (any graph) and draw G on the sphere. Of course, if G is planar, then we can draw G on the sphere in such a way that no two of its edges cross; while if G is nonplanar, then we cannot do this. On the other hand, if G is nonplanar, then we can draw G on the sphere so that only two edges cross at any point of intersection. Whenever such a crossing of two edges occurs, we can attach a handle at an appropriate position and pass one of the two edges over the handle so that these two edges no longer cross in the surface we have just constructed. What we have just observed then is that every graph can be **embedded** on some surface. The smallest nonnegative integer k such that a graph G can be embedded on S_k is called the **genus** of G and is denoted by $\gamma(G)$. Therefore, $\gamma(G) = 0$ if and only if G is planar; while $\gamma(G) = 1$ if and only if G is nonplanar but G can be embedded on the torus. In particular, $\gamma(K_5) = 1$ and $\gamma(K_{3,3}) = 1$.

Throughout this chapter, we have only discussed connected graphs. Of course, disconnected graphs are either planar or nonplanar, just as connected graphs are. In fact, perhaps it is clear that a disconnected graph G is planar if and only if every component of G is planar. For this reason, as far as studying planar graphs is concerned, we need only be concerned with studying connected graphs. However, as we are about to see, this is not the only reason why we restrict ourselves to connected graphs when studying planar graphs. Consider the disconnected plane graph $G = 2K_3$ shown in Figure 9.24. Certainly G has order $n = 6$, size $m = 6$ and $r = 3$, whose three regions are denoted by R_1, R_2 and R_3. Therefore, in this case, $n - m + r = 6 - 6 + 3 = 3$ and so the

Euler Identity does not hold. This may not be surprising since, after all, in the hypothesis of Theorem 9.1, the plane graph G is required to be connected.

$2K_3$:
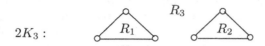

Figure 9.24: A disconnected plane graph

Let's return to the two embeddings of K_4 on the torus in Figures 9.19(b) and 9.19(c). These are shown again in Figures 9.25(a) and 9.25(b), respectively, where, in addition, each embedding is given when the torus is represented as a rectangle with opposite sides identified. We also add a third embedding of K_4 on the torus in Figure 9.25(c). Just as regions of the plane (or sphere) are created when a planar graph is embedded in the plane (or sphere), so too are regions of the torus (or any other surface) created when a graph is embedded on the torus (or other surface). Since these surfaces are more complex, it may be more difficult to determine what the regions are in this case. A fact that may be helpful is the following: Let G be a graph that is embedded on some surface. If a point A lies in region R and a point A' lies in region R' and A can be connected to A' by a curve on the surface that never intersects a vertex or an edge of G, then R and R' are the same region. If no such curve connecting A and A' exists, then R and R' are different regions. Therefore, the embedding of K_4 on the torus shown in Figure 9.25(a) produces four regions (denoted by R_1, R_2, R_3, R_4). The embedding of K_4 on the torus shown in Figure 9.25(b) produces three regions (denoted by R_1, R_2, R_3), while the embedding of K_4 in Figure 9.25(c) produces only two regions (denoted by R_1, R_2). So the number of regions is $r = 4$ for the embedding in Figure 9.25(a), $r = 3$ for the embedding in Figure 9.25(b) and $r = 2$ for the embedding shown in Figure 9.25(c). That is, even though K_4 is connected, we don't always obtain the same value of $n - m + r$ when the embedding takes place on the torus. Although this may seem disappointing, there is a property that the embedding of K_4 on the torus has in Figure 9.25(c) that the other two embeddings in Figure 9.25 do not have and, as it turns out, this is a critical difference.

Let G be a graph that is embedded on some surface S_k, where $k \geq 0$. Then, of course, regions on S_k are produced. A region is called a **2-cell** if any closed curve that is drawn in that region can be continuously contracted (or shrunk) in that region to a single point. An embedding, every region of which is a 2-cell, is called a **2-cell embedding**. Let's return to the embedding of $G = 2K_3$ in the plane (or on the sphere) that we considered in Figure 9.24. In that embedding, the two regions each of whose boundary is a triangle are 2-cells, while the remaining region is not a 2-cell. This embedding is shown on the sphere in Figure 9.26. Although the closed curve C in the exterior region can be continuously contracted to a single point in that region, the closed curve C' cannot.

In general, no embedding of a disconnected graph in the plane is a 2-cell embedding, while every embedding of a connected graph in the plane is a 2-cell

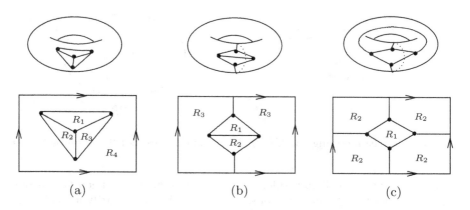

Figure 9.25: Embedding K_4 on the torus

Figure 9.26: An embedding of $2K_3$
on the sphere that is not a 2-cell embedding

embedding. However, if G is a connected graph that is embedded on a surface of positive genus, then the embedding may or may not be a 2-cell embedding. Let's recall the embeddings of K_4 on the torus given in Figures 9.25(a) and 9.25(b). Neither of these embeddings is a 2-cell embedding. Indeed, the curves C and C' shown in Figures 9.27(a) and 9.27(b), respectively, cannot be continuously contracted in that region to a single point.

The embedding of K_4 on the torus given in Figure 9.25(c) *is* a 2-cell embedding, however. In that embedding, there are two regions and so $n - m - r = 4 - 6 + 2 = 0$. As we are about to see, whenever a connected graph of order n and size m is 2-cell embedded on a torus resulting in r regions, then $n - m - r = 0$. We present an even more general result.

Theorem 9.9 *Let G be a connected graph that is 2-cell embedded on a surface of genus $k \geq 0$. If G has order n, size m and r regions, then*

$$n - m + r = 2 - 2k.$$

Proof. We proceed by induction on k. Let G be a connected graph of order n

Figure 9.27: Embeddings of K_4 on the torus that are not 2-cell embeddings

and size m that is 2-cell embedded on a surface of genus 0. Then G is a plane graph. Suppose that G has r regions, each of which is necessarily a 2-cell. Then $n - m + r = 2 = 2 - 2 \cdot 0$ by the Euler Identity. Hence the theorem holds when $k = 0$.

Assume, for every connected graph G' of order n' and size m' that is 2-cell embedded on a surface S_k, where $k \geq 0$, resulting in r' regions, that $n' - m' + r' = 2 - 2k$. Let G be a connected graph of order n and size m that is 2-cell embedded on S_{k+1}, resulting in r regions.

Let H be one of the $k+1$ handles of S_{k+1}. We may assume that no vertices of G lie on H. However, since the embedding of G on S_{k+1} is a 2-cell embedding, there are edges of G on H. Draw a closed curve C around H, which must intersect some edges of G. Suppose that there are $t \geq 1$ points of intersection of C and edges on H. Let the points of intersection be vertices, where then each of the t edges becomes two edges. Furthermore, the segments of C between vertices become edges. We add two vertices of degree 2 along C to produce two additional edges. (This guarantees that there are no parallel edges and that a graph results.) Denote the resulting graph by G_1, which has order n_1, size m_1 and r_1 regions. Observe that $n_1 = n + t + 2$ and $m_1 = m + 2t + 2$. Since each portion of C that became an edge of G_1 is in a region of G, the addition of such an edge divides that region into two regions, each of which is a 2-cell. Since there are t such edges, $r_1 = r + t$.

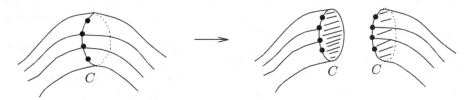

Figure 9.28: A step in the proof of Theorem 9.9

Next we cut the handle H along C and "patch" the two resulting holes, producing two duplicate copies of the vertices and edges along C. Denote the resulting graph by G_2, which is now 2-cell embedded on S_k. Let G_2 have order n_2, size m_2 and r_2 regions, all of which are 2-cells. Then $n_2 = n_1 + t + 2$,

$m_2 = m_1 + t + 2$ and $r_2 = r_1 + 2$. Therefore,

$$n_2 = n + 2t + 4, \ m_2 = m + 3t + 4 \text{ and } r_2 = r + t + 2.$$

By the induction hypothesis, $n_2 - m_2 + r_2 = 2 - 2k$ and so

$$(n + 2t + 4) - (m + 3t + 4) + (r + t + 2) = 2 - 2k.$$

Therefore, $n - m + r + 2 = 2 - 2k$ and $n - m + r = 2 - 2(k + 1)$. ∎

It turns out that if G is a connected graph that is embedded on a surface of genus $\gamma(G)$, then this embedding is necessarily a 2-cell embedding. Hence we have the following corollary.

Corollary 9.10 *If G is a connected graph of order n and size m that is embedded on a surface of genus $\gamma(G)$, resulting in r regions, then*

$$n - m + r = 2 - 2\gamma(G).$$

We now have a corollary of Corollary 9.10.

Corollary 9.11 *If G is a connected graph of order $n \geq 3$ and size m, then*

$$\gamma(G) \geq \frac{m}{6} - \frac{n}{2} + 1.$$

Proof. First, let G be embedded on a surface of genus $\gamma(G)$ resulting in r regions. By Corollary 9.10, $n - m + r = 2 - 2\gamma(G)$. Let $R_1, R_2, \ldots, R_r$ be the regions of G and let m_i be the number of edges on the boundary of R_i $(1 \leq i \leq r)$. So $m_i \geq 3$. Since every edge is on the boundary of one or two regions, it follows that

$$3r \leq \sum_{i=1}^{r} m_i \leq 2m$$

and so $3r \leq 2m$. Therefore,

$$6 - 6\gamma(G) = 3n - 3m + 3r \leq 3n - 3m + 2m = 3n - m. \tag{9.2}$$

Solving (9.2) for $\gamma(G)$, we have

$$\gamma(G) \geq \frac{m}{6} - \frac{n}{2} + 1,$$

as desired. ∎

If the graph G in the statement of Corollary 9.11 is planar, then $\gamma(G) = 0$ and the conclusion of this corollary states that $0 \geq \frac{m}{6} - \frac{n}{2} + 1$ or, equivalently, $m \leq 3n - 6$, which returns us to Theorem 9.2.

The graph K_5 has order $n = 5$ and size $m = 10$. By Corollary 9.11, $\gamma(K_5) \geq 1/6$ and so once again we see that K_5 is nonplanar. Of course, we have already seen that $\gamma(K_5) = 1$. By Corollary 9.11, $\gamma(K_6) \geq 1/2$ and $\gamma(K_7) \geq 1$. In fact, it can be shown that $\gamma(K_6) = \gamma(K_7) = 1$. Indeed, the following formula was obtained by Gerhard Ringel and J. W. T. (Ted) Youngs (1910-1970). (This formula will be revisited in Chapter 10.)

Theorem 9.12 *For* $n \geq 3$,

$$\gamma(K_n) = \left\lceil \frac{(n-3)(n-4)}{12} \right\rceil .$$

Exercises for Section 9.2

9.23 Embed each of the following graphs in Figure 9.29 on the torus (represented as a rectangle with opposite sides identified).

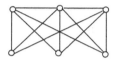 

Figure 9.29: The graphs in Exercise 9.23

9.24 Determine, with explanation, the genus of K_6.

9.25 Determine, with explanation, the genus of $K_{4,4}$.

9.26 Determine, with explanation, the genus of the Petersen graph.

9.27 Prove or disprove:

   (a) There exists a planar graph that cannot be embedded on the torus.

   (b) There exists a nonplanar graph that cannot be embedded on the torus.

   (c) If G is a graph of order n and size m, then G can be embedded on a sphere with m handles.

   (d) If a graph G can be embedded on the torus, then $\gamma(G) = 1$.

9.28 By Theorem 9.12, $\gamma(K_7) = 1$.

   (a) Show that the boundary of every region in an embedding of K_7 on the torus is a triangle.

   (b) Let there be an embedding of K_7 on the torus and let R_1 and R_2 be two neighboring regions. Let G be the graph obtained by adding a new vertex y in R_1 and joining y to the vertices on the boundaries of both R_1 and R_2. Prove that $\gamma(G) = 2$.

9.3   Excursion: Graph Minors

As we have seen, Kuratowski's theorem (Theorem 9.7) provides a characterization of planar graphs: *A graph G is planar if and only if G does not contain a subdivision of K_5 or $K_{3,3}$ as a subgraph.* This is not the only characterization of planar graphs, however.

For a graph G and an edge $e = uv$ of G, a graph G' is said to be obtained from G by **contracting the edge** e (or **identifying the vertices** u and v) if G' is (isomorphic to) the graph obtained by joining u in the graph $G - v$ to any neighbor of v not already adjacent to u. We also say that G' is obtained from G by an **edge contraction**. (By symmetry, G' is also the graph obtained by joining v in $G - u$ to any neighbor of u in G not already adjacent to v.) This is illustrated for the graph G of Figure 9.30 and the edge $e = uv$. The graph G'' is obtained by contracting the edge xy in G'

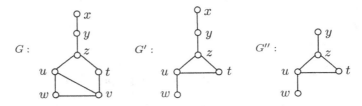

Figure 9.30: Contracting an edge

If we were to begin with a graph G, contract an edge in the graph G to produce the graph G', contract an edge in G' to produce the graph G'' and so on, until finally obtaining a graph H, then it is possible to describe such a graph H more simply. Let G be a graph where $\{V_1, V_2, \ldots, V_k\}$ is a partition of $V(G)$ such that $G[V_i]$ is connected for every integer i $(1 \le i \le k)$. Let H be the graph with vertex set $\{V_1, V_2, \ldots, V_k\}$ where V_i is adjacent to V_j $(i \ne j)$ if some vertex in V_i is adjacent to some vertex in V_j in G. For example, in the graph G of Figure 9.30, if we let $V_1 = \{x, y\}$, $V_2 = \{z\}$, $V_3 = \{u, v\}$, $V_4 = \{t\}$ and $V_5 = \{w\}$, then the resulting graph H in Figure 9.31 is isomorphic to the graph G'' of Figure 9.30.

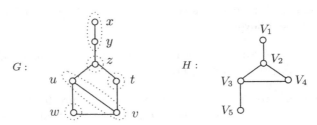

Figure 9.31: Edge contractions

A graph H is called a **minor** of a graph G if (a graph isomorphic to) H can be obtained from G by a succession of edge contractions, edge deletions or vertex deletions (in any order). Consequently, the graph H of Figure 9.31 is a minor of the graph G of that figure. Let's consider another example, namely the graph G_1 of Figure 9.32, where we let $V_1 = \{u_1\}$, $V_2 = \{u_2\}$, $V_3 = \{u_3\}$, $V_4 = \{v_1, w_1, x_1\}$, $V_5 = \{w_2, x_2\}$, $V_6 = \{v_3, w_3, x_3\}$ and $V_7 = \{x_4\}$. Then the graph H_1 can be obtained from G_1 by successive edge contractions. The graph H_1 is consequently a minor of G_1. By deleting the edge V_1V_2 and the vertex V_7, we see that $K_{3,3}$ is also a minor of G_1.

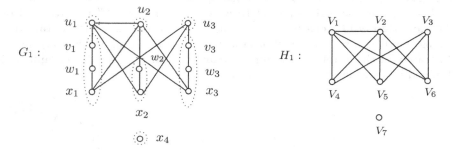

Figure 9.32: Minors of graphs

Minors of graphs have some interesting implications. As the example in Figure 9.32 may suggest, we have the following.

Theorem 9.13 *If a graph G is a subdivision of a graph H, then H is a minor of G.*

Also, if G is a graph that is embeddable on a surface S_k, where $k \geq 0$, then any graph obtained from G by an edge contraction, edge deletion or vertex deletion can also be embedded on S_k. This leads to the following observation.

Theorem 9.14 *If a graph H is a minor of a graph G, then $\gamma(H) \leq \gamma(G)$.*

With the aid of these two observations, a characterization of planar graphs, due to Klaus Wagner (1910-2000), can now be stated. In 1937, only a year after obtaining his Ph.D. from Universität zu Köln (the University of Cologne), Wagner proved the following.

Theorem 9.15 (**Wagner's Theorem**) *A graph G is planar if and only if neither K_5 nor $K_{3,3}$ is a minor of G.*

In Example 9.8 we showed that the graph of Figure 9.12 (redrawn as the graph G of Figure 9.33) is nonplanar. Despite the fact that the appearance of G might remind one of K_5, this graph does not contain a subdivision of K_5 as a subgraph. In fact, we verified that G is nonplanar by showing that G contains a subdivision of $K_{3,3}$ as a subgraph. On the other hand, if we were to consider the parition $\{V_1, V_2, V_3, V_4, V_5\}$ of $V(G)$ for which $V_1 = \{u_1, v_1, w_1, x_1, y_1\}$, $V_2 = \{v_2\}$, $V_3 = \{u_2, y_2\}$, $V_4 = \{w_2\}$ and $V_5 = \{x_2\}$, then we see that K_5 is a minor of G and so G is nonplanar by Wagner's theorem.

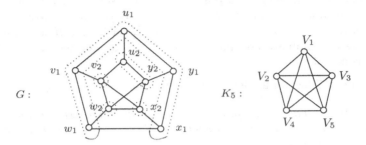

Figure 9.33: The graph G having K_5 as a minor

Undoubtedly, the major theorem concerning minors is one obtained by Neil Robertson and Paul Seymour in 1990.

Theorem 9.16 (**The Graph Minor Theorem**) *For any infinite set S of graphs, there exist two distinct graphs in S such that one of these graphs is a minor of the other.*

Paul Seymour received his doctoral degree from Oxford University in 1975. He became a professor at Princeton University after having spent several years working at Bellcore (Bell Communications Research). Neil Robertson received his Ph. D. in 1969 from the University of Waterloo under the direction of William Tutte and became a professor at Ohio State University. Much of Robertson's motivation comes from furthering the work of Tutte. His work on graph minors is aimed at highlighting the structural features of graphs that are obtained by excluding a fixed graph as a minor. Robertson, with interests in the fine arts, poetry and drama, considered it a privilege to work in an area that is creativity driven.

We now describe a remarkable consequence of Theorem 9.16. Consider the surface S_k of genus $k \geq 0$. Certainly, if G is a sufficiently small graph (in terms of order and/or size), then G can be embedded on S_k. Hence if we begin with a graph F that cannot be embedded on S_k and perform successive edge contractions, edge deletions and vertex deletions, then eventually we arrive at a graph F' that also cannot be embedded on S_k but such that any additional edge contraction, edge deletion or vertex deletion of F' produces a graph that *can* be embedded on S_k. We say that such a graph F' is **minimally nonembeddable on S_k**. Consequently, every minor of F' that is distinct from F' *can* be embedded on S_k. In particular, the graphs K_5 and $K_{3,3}$ are minimally nonembeddable on S_0 (or more simply, **minimally nonplanar**). Indeed, these are the only minimally nonplanar graphs. This leads us to the following remarkable consequence of the Graph Minor Theorem and another discovery of Robertson and Seymour.

Theorem 9.17 *For each integer $k \geq 0$, the set of minimally nonembeddable graphs on S_k is finite.*

Proof. By Wagner's theorem, the statement is true for $k = 0$. Assume, to the contrary, that there exists a positive integer k such that the set S of minimally nonembeddable graphs on S_k is infinite. By the Graph Minor Theorem, S contains two non-isomorphic graphs G and H such that H is a minor of G. However, G is minimally nonembeddable and H cannot be embedded on S_k. This is a contradiction.   ∎

Theorem 9.17 has an immediate consequence.

Corollary 9.18 *For every nonnegative integer k, there exists a finite set S of graphs such that a graph G is embeddable on S_k if and only if H is not a minor of G for every graph H in S.*

Although the number of minimally nonembeddable graphs on the torus is finite, it is known that this number exceeds 800.

Exercises for Section 9.3

9.29 For the graphs G and G' in Figure 9.34, show that G' is a minor of G.

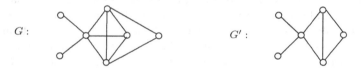

Figure 9.34: The graph G in Exercise 9.29

9.30 Show that the graph K_5 is a minor of the Petersen graph, thereby showing that the Petersen graph is nonplanar.

9.31 Prove that every nonplanar graph has K_5 or $K_{3,3}$ as a minor.

9.32 What graph results from

    (a) a single edge contraction in K_5?

    (b) a single edge contraction in $K_{3,3}$?

    (c) two edge contractions in $K_{3,3}$?

9.33 Theorem 9.13 states that: *If a graph G is a subdivision of a graph H, then H is a minor of G.* Show that its converse is false.

9.4 Exploration: Embedding Graphs in Graphs

We saw in Theorem 2.7 that for every graph G and every integer $r \geq \Delta(G)$, there exists an r-regular graph H containing G as an induced subgraph. We say that G is **embedded** as an induced subgraph in H. This result was presented by Dénes König in his 1936 book. Let's recall how this result was proved when $r = \Delta(G)$.

If G is r-regular, then $H = G$ has the desired properties. Otherwise, let G' be another copy of G and join corresponding vertices in G and G' whose degrees are less than r, producing the graph G_1. If G_1 is r-regular, then $H = G_1$ has desired properties. If G_1 is not r-regular, then we continue this procedure until arriving at an r-regular graph G_k, where $k = \Delta(G) - \delta(G)$. This is illustrated for the graph G of Figure 9.35, where $\Delta(G) = 3$ and $\delta(G) = 1$.

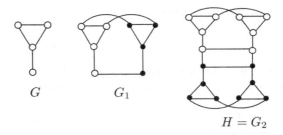

Figure 9.35: A graph H containing G as an induced subgraph

The construction presented by König to embed a graph G with maximum degree r as an induced subgraph in an r-regular graph H doesn't produce an r-regular graph of smallest order in general. In fact, while the graph H of Figure 9.35 has order 16, the minimum order of a 3-regular graph containing G as an induced subgraph is 6.

In 1963 Paul Erdős and Paul J. Kelly developed a method for determining the minimum order of an r-regular graph H in which a given G with $r = \Delta(G)$ can be embedded as an induced subgraph. In order to state their theorem, we need to present a few definitions.

Let G be a graph with maximum degree r whose vertex set is $V(G) = \{v_1, v_2, \ldots, v_n\}$. Let d_i denote the degree of v_i and let $e_i = r - d_i$ denote the **deficiency** of v_i. In addition, let $e = \max\{e_i : 1 \leq i \leq n\}$ be the **maximum deficiency** and $s = \sum_{i=1}^{n} e_i$ the **total deficiency**. We can now state the theorem of Erdős and Kelly.

Theorem 9.19 *Let G be a graph of order n, where $r = \Delta(G)$. Then $k + n$ is the minimum order of an r-regular graph H in which G can be embedded as an induced subgraph where k is the least integer satisfying (1) $kr \geq s$, (2) $k^2 - (r+1)k + s \geq 0$, (3) $k \geq e$ and (4) $(k+n)r$ is even.*

Figure 9.36 shows four non-regular graphs G_i $(1 \le i \le 4)$ and graphs H_i that are $\Delta(G_i)$-regular graphs of smallest order containing G_i as an induced subgraph. The first three pairs G_i, H_i of graphs appear as illustrations in the article by Erdös and Kelly.

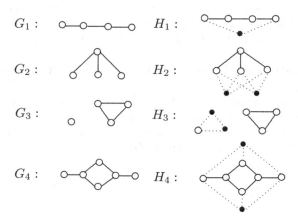

Figure 9.36: Four $\Delta(G_i)$-regular graphs H_i $(1 \le i \le 4)$
of minimum order containing G_i as an induced subgraph

For each vertex v of each graph H_i $(1 \le i \le 4)$ of Figure 9.36, there is an induced subgraph of H_i containing v that is isomorphic to G_i. Of course, if v is a vertex of the original graph G_i, then this is obvious. However, this is also true for each vertex added to G_i to produce H_i. Indeed, it is often the case that for every vertex v of a $\Delta(G)$-regular graph H of minimum order containing a given graph G as an induced subgraph, there is an induced subgraph of H containing v that is isomorphic to G. This does not always happen, however.

Figure 9.37 shows a graph G of order 10 obtained from the complete bipartite graph $K_{3,3}$ by subdividing two of its edges. Figure 9.37 also shows a 3-regular graph H of minimum order 12 containing G as an induced subgraph. However, there is no induced subgraph of H containing u that is isomorphic to G.

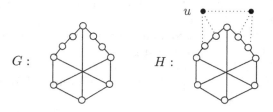

Figure 9.37: A graph H having no induced subgraph
containing u that is isomorphic to G

These observations lead to the following concept. A graph G is said to be

uniformly embedded in a graph H if for every vertex v of H, there is an induced subgraph of H containing v that is isomorphic to G. Therefore, each of the graphs G_i ($1 \leq i \leq 4$) of Figure 9.36 is uniformly embedded in the graph H_i.

For a graph G with maximum degree r, the r-regular graph H constructed by König has the property that G is uniformly embedded in H. However, the graph G of Figure 9.37 is not uniformly embedded in H.

Embeddings can be even more demanding. A graph G is **homogeneously embedded** in a graph H if for each vertex x of G and each vertex y of H, there exists an embedding of G in H as an induced subgraph with x at y. Equivalently, a graph G is **homogeneously embedded** in a graph H if for each vertex x of G and each vertex y of H there exists an induced subgraph H' of H and an isomorphism ϕ from G to H' such that $\phi(x) = y$.

A graph F of minimum order in which G can be homogeneously embedded is called a **frame** of (or for) G and the order of F is called the **framing number** $\mathrm{fr}(G)$ of G. Therefore, if G is a graph of order n, then $\mathrm{fr}(G) \geq n$.

For example, the framing number of the path P_3 is 4 since P_3 can be homogeneously embedded in C_4 (see Figure 9.38) and P_3 cannot be homogeneously embedded any graph of order 3. The cycle C_4 also has framing number 4 since C_4 can be homogeneously embedded in itself.

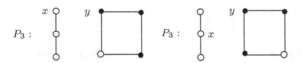

Figure 9.38: Homogeneously embedding P_3 in C_4

The graph $G = K_1 \cup K_2$ can be homogeneously embedded in C_5 (see Figure 9.39); however, $\mathrm{fr}(G) = 4$ since G can also be homogeneously embedded in the graph $2K_2$ of order 4 and G cannot be homogeneously embedded in any graph of order 3.

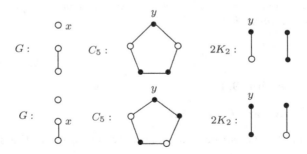

Figure 9.39: Homogeneously embedding $K_1 \cup K_2$ in C_5 and $2K_2$

If a graph G can be homogeneously embedded in a graph H, then $\delta(H) \geq \Delta(G)$. As another illustration, consider the graph G of Figure 9.40.

Example 9.20 *The graph G of Figure 9.40 has framing number 6.*

Solution. Since $\Delta(G) = 3$, it follows that if H is a graph in which G can be homogeneously embedded, then $\delta(H) \geq 3$. Certainly, $\mathrm{fr}(G) \geq 5$. If $\mathrm{fr}(G) = 5$, then a frame H of G can be constructed by adding a new vertex y_1 that is adjacent to at least u and v. However, $\deg_H y_1 \geq 3$; so y_1 must be also adjacent to at least one of x and w, say y_1 is adjacent to x. Let H_1 be the graph constructed thus far (see Figure 9.40). However, H_1 fails to contain an induced subgraph isomorphic to G with u at x_1 and so G cannot be homogeneously embedded in H_1. Therefore, $\mathrm{fr}(G) \geq 6$. Since G can be homogeneously embedded in the graph $H_2 = K_{2,2,2}$ of order 6 shown in Figure 9.40, it follows that $\mathrm{fr}(G) = 6$. $\Diamond$

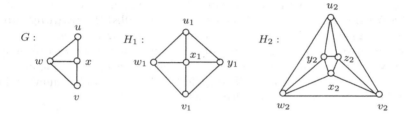

Figure 9.40: Homogeneously embedding a graph

There is an important question that might have occurred to you: For a given graph G, how do we know that the framing number of G exists? This is answered in the following theorem.

Theorem 9.21 *For every graph G, there exists a graph in which G can be homogeneously embedded.*

In fact, even more can be said.

Theorem 9.22 *For every graph G, there exists a regular graph in which G can be homogeneously embedded.*

Exercises for Section 9.4

9.34 Consider the graph G of Figure 9.41.

    (a) Show that G can be homogeneously embedded in the 3-cube Q_3.

    (b) Determine the framing number of G.

9.35 Determine the framing number of $2K_1 \cup K_2$.

9.36 (a) Determine a frame for the graph $K_{1,3}$.

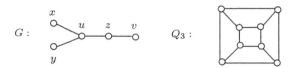

Figure 9.41: The graphs in Exercise 9.34

    (b) Does there exist a planar graph H in which $K_{1,3}$ can be homogeneously embedded?

9.37 Determine the framing number of $K_{1,t}$, where $t \geq 4$.

9.38 Prove that if G is a graph of order n with $\mathrm{fr}(G) = n$, then G has a unique frame.

9.39 Use Theorem 9.21 to prove Theorem 9.22.

9.40 Let $G = \overline{K}_4$. Give an example of an r-regular graph H_r of minimum order such that G can be homogeneously embedded in H_r when

    (a) $r = 0$,   (b) $r = 1$,   (c) $r = 2$,   (d) $r = 3$.

9.41 Let $G = P_4$. Does there exist a graph H such that (1) for every two adjacent vertices x_1 and x_2 of G and every two adjacent vertices y_1 and y_2 of H and (2) for every two nonadjacent vertices x_1 and x_2 of G and every two nonadjacent vertices y_1 and y_2 of H, there exists an embedding of G in H as an induced subgraph with x_1 at y_1 and x_2 at y_2?

9.42 Ask and answer a question of your own dealing with homogeneous embeddings of graphs.

Chapter 10

Coloring

10.1   The Four Color Problem

Wolfgang Haken
Smote the Kraken
One! Two! Three! Four!
Quoth he: 'The monster is no more'.

What is *this* all about? We are about to explain. In the article "The mathematics of map coloring," which was published in a 1969 issue of the *Journal of Recreational Mathematics*, its author, the distinguished mathematician H. S. M. (Donald) Coxeter, mentioned that in nearly every instance when a map of the United States is colored to distinguish neighboring states, at most five or six colors are used. What is the minimum number of colors that can be used to color the states in the United States if every two states that share a common border are required to be colored differently? Two states that share only a common point, however, such as Utah and New Mexico, are permitted to be colored the same (see Figure 10.1). Since Nevada and Utah are neighboring states, that is, they share a common boundary, they must be assigned different colors. In fact, Nevada has a ring of five neighboring states, namely, Utah, Idaho, Oregon, California and Arizona. Therefore, each of these five states must be assigned a color different from that used for Nevada. On the other hand, three colors are needed to color the five states bordering Nevada. So four colors are needed in all to color these six states. Indeed, all states in the United States can be colored with four colors.

Coloring regions (whether these are states, countries or counties) in a map with a minimum number of colors such that neighboring regions (those sharing a common boundary) are colored differently does not appear to be a question with which map-makers of the past were concerned. Indeed, the mathematical

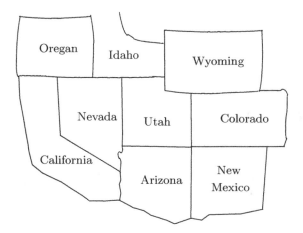

Figure 10.1: Western United States

historian Kenneth May found no evidence of this when he studied books on map-making.

So, if a map is divided into regions in some manner, what *is* the minimum number of colors required if neighboring regions are to be colored differently? And why is this a question that should even concern us? You might think that the answer to the first question depends on the map and you'd be right – although we have already mentioned that coloring the states in the United States requires four colors and that four colors suffice. One might expect, however, that this question would have a very different answer if the map consisted of many regions (say billions) and the map was designed so that many of the regions had a large number of neighboring regions.

Evidently, this question originated not with map-makers but with a mathematician. In 1852 Francis Guthrie (1831–1899), a recent graduate of University College London, observed that the counties of England could be colored with four colors so that neighboring counties were colored differently. Francis Guthrie found maps where three colors weren't enough but he felt that four colors were enough for all maps and he attempted to prove this. He showed his "proof" to his younger brother Frederick, who was taking a class at the time from the well-known Augustus De Morgan. Francis was not completely happy with the proof he had given, however. With Francis' permission, Frederick showed what Francis had written to De Morgan on October 23, 1852. De Morgan was pleased with this and felt it was new. The very same day, De Morgan wrote the following letter to the celebrated mathematician William Rowan Hamilton:

> *A student of mine asked me to day to give him a reason for a fact which I did not know was a fact – and do not yet. He says that if a figure be anyhow divided and the compartments differently coloured so that figures with any portion of common boundary <u>line</u> are differently*

*coloured – four colours may be wanted but not more – the following
is his case in which four <u>are</u> wanted.*

A B C D are
names of
colours

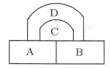

Query cannot a necessity for five or more be invented.

Hamilton replied on October 26, 1852:

I am not likely to attempt your "quaternion" of colours very soon.

Although this problem (which was to become known as the **Four Color
Problem**) apparently did not excite Hamilton, De Morgan continued to be
interested in it. Overall interest in this problem subsided during the next several
years however.

During this period, in 1865, the London Mathematical Society was founded
at University College London and Augustus De Morgan served as its first president. The second president of the Society was the mathematician James Joseph
Sylvester and Arthur Cayley became the third president of the Society. Much of
Cayley's early work on algebra had in fact been done with Sylvester.

James Joseph Sylvester was born in London on September 3, 1814. In 1833
Sylvester attended St. John's College, Cambridge where he excelled in mathematics. This was a time when a student was required to sign a religious oath to
the Church of England before he could graduate. Being Jewish, Sylvester refused
to take the oath and was not permitted to graduate. For the same reason, he
was ineligible for a fellowship. During 1838-1841 Sylvester taught physics at the
University of London where his religion did not work against him. One of his
colleagues there was De Morgan, who had earlier been Sylvester's teacher.

After a short stay in the United States in 1841, Sylvester returned to England
where he worked as a lawyer and actuary. He did tutoring in mathematics,
however and, curiously, one of his students was Florence Nightingale, well known
for her pioneering work in nursing and hospital reform. Arthur Cayley was also
a lawyer at that time and he and Sylvester often discussed mathematics. Despite
having very different personalities, the two became life-long friends.

Sylvester tried hard to obtain a mathematics position and only secured one,
at the Royal Military Academy at Woolwich, when the successful applicant died.
Sylvester went on to do important work in matrix theory and the theory of equations and he introduced the terms *matrix* and *discriminant* into mathematics.
Being at a military academy, however, Sylvester was required to retire at half-pay
at age 55.

Sylvester was about to give up mathematics and concentrate on writing poetry when his life took a major turn in 1876, with the founding of a new university
in the United States: the Johns Hopkins University in Baltimore, Maryland.

He was offered the position of the University's first professor of mathematics. Sylvester became the most senior of the original faculty of this university, both in terms of age and prior accomplishments. Employing Sylvester turned out to be a wise decision for the University as he was instrumental in hiring many faculty members who became prominent mathematicians themselves. In 1878 he founded the *American Journal of Mathematics*, the first mathematics journal in the United States. Cayley was among the first to publish his research in this journal.

Meanwhile, interest in the Four Color Problem was revived during an 1878 meeting of the London Mathematical Society presided over by Henry Smith of Exeter College, Oxford. One of Smith's students there was Percy John Heawood, whom we will soon encounter. On June 13, 1878 Cayley asked if this problem had been solved. One of the people attending this meeting was a bright but amateur mathematician by the name of Alfred Bray Kempe (pronounced KEMP).

The next year on July 17, 1879 Kempe announced in the magazine *Nature* that he had solved the problem and that every map could indeed be colored with four or fewer colors. Kempe (1849–1922) had graduated from Cambridge University in 1872 and studied mathematics under Cayley. He later became a barrister (a lawyer) with a specialty in ecclestical law. He continued his interest in mathematics, however. Cayley suggested that Kempe should publish his discovery, which he did in 1879 in the second volume of the *American Journal of Mathematics*.

In his approach, Kempe used a technique which involved a concept that was later to be called **Kempe chains**. The idea goes something like the following: Suppose that we have a map in which all regions except one, say region X, have been colored with four colors (red, blue, green and yellow) and we would like to color region X with one of these colors as well. Of course, if not all four colors have been used to color the regions that surround X, then there is a color available for X. Hence we may assume that all four colors have been used for the ring of regions that surround X. It may occur that there are exactly four regions in such a ring, say A, B, C, D (in clockwise order), which are colored red, blue, green and yellow, respectively. (See Figure 10.2.)

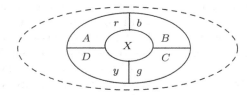

Figure 10.2: A region surrounded by a ring of four regions

There are two possibilities: (1) There is no chain of neighboring regions from A to C that are alternately colored red and green. (2) There *is* a chain of neighboring regions from A to C that are alternately colored red and green. Should (1) occur, then interchange the colors red and green in all chains of neighboring regions beginning at A and which are alternately colored red and green. Since C does not appear in any such chain, once the colors are interchanged, both A and C are colored green and the color red is available for X. If (2) should occur, then there cannot be a chain of neighboring regions from B to D whose colors are alternately blue and yellow. The technique used in (1) can now be applied to chains of neighboring regions colored blue and yellow to produce a new coloring of all regions of the map (except X) in which both B and D are colored yellow, leaving blue available for X.

Of course, this only takes care of the situation where exactly four regions occur in the ring about X (and where all four colors are used to color these surrounding regions). What if there are more than four regions in this ring? We will see later that we can assume that the number of regions in a ring about X does not exceed 5. But this still leaves us with one other situation: There are five regions in a ring about X and all four colors are used to color these surrounding regions. Of course, some color would have to be used twice in this case (though not for neighboring regions). Kempe believed that an argument similar to the one used for when four regions surround X could be used for the situation when five regions surround X and this was the key to (and the downfall of) his proof.

Kempe was honored for his accomplishment. He was elected Fellow of the Royal Society in 1881 and served as its treasurer for a number of years. He published two refinements of his proof, one of which sparked the interest of the mathematician Peter Guthrie Tait (1831–1901), who provided his own solutions of the Four Color Problem. Tait had many interests and many friends (including Hamilton). Despite being involved with so many projects, he always seemed to find time to get things done. One of Tait's four sons, Frederick Guthrie Tait, excelled in golf and was known to the golfing world as Freddie Tait. Frederick was the amateur champion of the British Open Golf Tournaments in 1896 and 1898. There is an international tournament held in his honor in Kimberley, South Africa as Frederick was killed in the Anglo-Boer War of 1899-1902. Peter Tait, in fact, wrote a research paper on the trajectory of golf balls. He was also interested in knots and how they crossed. One of Tait's personal problems, however, was that he was often involved in arguments with his colleagues.

The next major event in the history of the Four Color Problem involved the British mathematician Percy John Heawood (pronounced HAY-wood). Heawood (1861–1955) studied at Oxford University and was a faculty member at Durham College (later Durham University) for more than fifty years. Heawood was known for his large moustache and for permitting his dog to attend his lectures. He did not retire until he was 78 and lived for another 16 years. In 1890 Heawood published a paper "Map colouring theorem" in which he pointed out a "defect" in Kempe's celebrated solution of the Four Color Problem by producing a counterexample to Kempe's proof in the case where region X is surrounded by five regions. (Additional discussion of Heawood's paper occurs in Section 10.4.)

Kempe agreed that Heawood had discovered an error in his paper. In what must have been a very difficult thing for him to do, Kempe reported Heawood's work to the London Mathematical Society himself and said that he was unable to correct the error he had made. Heawood couldn't correct Kempe's error either. However, Heawood became so intrigued with the Four Color Problem that he continued working on it for the rest of his life – another six decades.

It turned out that Tait's "solutions" were also incorrect. In addition to the "proofs" by Kempe and Tait, there was yet another published incorrect proof – by Frederick Temple, who was Bishop of London at the time and who would later become the Archbishop of Canterbury.

The conjecture that every map can be colored with four or fewer colors became known as the **Four Color Conjecture**. Many believed the conjecture to be true. Heawood's example was only a counterexample to Kempe's technique, not a counterexample to the Four Color Conjecture. Indeed, it was not that difficult to show that the map constructed by Heawood could be colored with four colors. Even though Kempe's method was unsuccessful, Heawood was able to use this method to prove that every map could be colored with *five* or fewer colors. (This topic will be discussed in more detail in Section 10.4.) On the other hand, whether any map actually required five colors was not known and would not be known for another 86 years.

After Heawood's paper, attempts to settle the Four Color Conjecture slowed considerably. In his attempted proof, Kempe used the fact that some region was surrounded by a ring of five or fewer regions. Only the case where there was a region surrounded by exactly five regions caused any difficulties. However, the obstacles in that case were severe. As time went on, mathematicians who attempted to prove the conjecture divided the proof into a larger number of more detailed cases dealing with configurations of regions that might occur in the map. The idea was to find a set of configurations of regions (rather than a single region) surrounded by a ring of regions such that every map contains at least one of these configurations and such that if the regions on and outside the ring can be colored with four colors, then the entire map can be colored with four colors. Such a set was later referred to as an **unavoidable set of reducible configurations**.

The concept of reducibility was introduced in a paper of one of the well-known mathematicians of the 20th century: George David Birkhoff (1884–1944). His son, Garrett Birkhoff, would also become a prominent mathematician. Garrett became interested in algebra and obtained a Ph.D. from Cambridge University under the supervision of Philip Hall. Garrett Birkhoff went on to join Saunders Mac Lane to co-author the popular textbook *Survey of Modern Algebra*, which introduced abstract algebra to many undergraduate mathematics students.

George David Birkhoff received his Ph.D. from the University of Chicago in 1907, with his thesis in the areas of ordinary differential equations and boundary value problems. He went on to become a faculty member at the University of Wisconsin at Madison, Princeton and Harvard, where he became Dean of the Faculty of Arts and Science. Although Birkhoff worked in a wide range of areas, he was fascinated with the work of Jules Henri Poincaré on dynamical systems, an area that Birkhoff studied extensively. Poincaré died in 1912 and in his last paper, he showed that the existence of periodic solutions of the restricted problem of three bodies could be deduced from a certain geometric theorem. He was unable to prove this theorem, however, except in a special case. Within a few months, Birkhoff had given a simple but insightful proof of "Poincaré's Last Geometric Theorem," which was published in the January 1913 issue of the *Transactions of the American Mathematical Society*, an accomplishment that would bring Birkhoff lasting fame. In that same year, Birkhoff published his paper on the reducibility of graphs.

As we have mentioned, in Kempe's attempted proof of the Four Color Theorem, he used the fact that every map contains a region surrounded by five or fewer regions. That is, such configurations were "unavoidable." If a proof by minimum counterexample of the Four Color Theorem was being attempted and we were assured that a map with a minimum number of regions could not be colored with four or fewer colors and that map contained a region surrounded by four or fewer regions, then we could show that the entire map could be colored with four colors, producing a contradiction and arriving at a proof of the Four Color Theorem. Therefore, in a minimum counterexample, all regions must be surrounded by five or more regions.

A **reducible configuration** is any arrangement of regions that cannot occur in a minimum counterexample. For example, a single region surrounded by exactly four regions is reducible. This means that if some map under consideration contains a certain reducible configuration, then any coloring of the regions of the map that lie outside this configuration with four or fewer colors can be extended to a coloring of the entire map with four or fewer colors. Consequently, if an unavoidable set of reducible configurations could be found, the Four Color Theorem would be proved. This was the approach that eventually would prove to be successful.

Philip Franklin used Birkhoff's idea to show that every map with 25 or fewer regions was 4-colorable. This was increased to 39 by Oystein Ore and Joel Stemple in 1970 and to 95 by Jean Mayer in 1976. By the 1970s, configurations with ring sizes ranging from 13 to 15 were being studied.

The German mathematician Heinrich Heesch (1906–1995) developed a more systematic way to search for an unavoidable set of reducible configurations. He became increasingly convinced that this was a method that would lead to a solution of the Four Color Problem and presented this view at seminars at the Universities of Hamburg and Kiel shortly after World War II. One of the students at the University of Kiel who attended these seminars was Wolfgang Haken.

Heesch estimated that an unavoidable set of reducible configurations could possibly contain as many as 10,000 elements. Furthermore, manually verifying that such a large number of configurations were reducible did not appear to be practical. One of Heesch's techniques, called *D*-**reducibility**, to show that a configuration is reducible was sufficiently algorithmic to lend itself to a computerized approach. This technique required consideration of every 4-coloring of the regions in the ring and showing that each of these could be extended to a 4-coloring of the entire map. For example, if the ring contained 14 regions, then there would be a total of 199,291 different colorings of the regions of the ring. Computers were used to verify D-reduction in the 1960s. Even for a single configuration, the computer took many hours. To make matters worse, even if the D-reduction program failed on a configuration, this did not mean that the configuration was not reducible. Heesch discovered another method for establishing reducibility, which he called C-**reducibility**. This was a major positive step.

By the late 1960s, a major effort was underway to solve the Four Color Problem with the aid of computers. The people involved in this approach included Heesch, Haken, Karl Dürre and Yoshio Shimamoto, Chair of the Applied Mathematics Department of the United States Atomic Energy Commission at their Brookhaven Laboratory. The Commission had access to a Control Data 6600 (developed by Stephen Cray), the fastest computer at that time. Shimamoto constructed a configuration (that eventually became known as the *Shimamoto horseshoe*) which if D-reducible would establish the truth of the Four Color Conjecture. For some time it seemed that this configuration was D-reducible. In fact, Heesch had encountered this configuration earlier and had been convinced by testing it on a computer that it was D-reducible. However, William Tutte and Hassler Whitney had believed for some time that if Shimamoto's approach was correct, then there must be a considerably simpler proof. Since they could find no error in Shimamoto's logic, they became convinced that the problem was with the computer. Since the proof of the Four Color Conjecture was now resting on the D-reducibility of this single configuration, it became essential that its D-reducibility be confirmed. Haken had studied this and also found nothing wrong with Shimamoto's reasoning. Even as this configuration was being tested for D-reducibility, a rumor surfaced that the Four Color Problem had been solved. At this moment the final step in a proof of the Four Color Conjecture was relying on a computer. Finally the Cray computer came up with the sad news that the Shimamoto horseshoe was *not* D-reducible, a fact that Tutte and Whitney would also establish.

During the early 1970s, those attempting a reducibility proof of the Four Color Conjecture included Heesch, Frank Allaire and Edward Swart, Frank Bernhart and Haken. After the failure of Shimamoto's attempt to find an easier way to solve the Four Color Problem, Haken's interest in using the computer for a solution diminished. However, Haken had a doctoral student at the time and one of the members of the thesis committee was Kenneth Appel. Unlike Haken, Appel was a very knowledgeable computer programmer. Appel was more optimistic and suggested to Haken that the two of them should return to a computer approach. Appel and Haken developed an algorithm in which they tested for "reduction obstacles." This gave them an approach that saved a great deal of computer time. They were also assisted by John Koch, a computer science graduate student, who wrote efficient programs to test for reducibility. While Appel and Haken were working feverishly on the problem, Appel was able to gain access to the IBM 370-168, an extremely powerful computer used by the administration at the University of Illinois. While all of this was going on, others continued working on a solution and it wasn't clear who would obtain a proof first. In June 1976,

Appel and Haken finally succeeded in constructing an unavoidable set of 1936 reducible configurations, which was verified using 1200 hours of computer time on three computers. Appel (now retired from the University of New Hampshire) and Haken (now retired from the University of Illinois) announced their success to the world at the 1976 Summer Meeting of the American Mathematical Society and the Mathematical Association of America at the University of Toronto.

To be sure, not all mathematicians were happy with this proof. In fact, many were highly skeptical and uncomfortable with it. This initiated numerous discussions of what a mathematical proof is. A major step in the acceptance of their proof, however, was its acceptance by the distinguished mathematician William Tutte, who, as Blanche Descartes, wrote in the third issue of the first volume of the *Journal of Graph Theory* the poem that we stated earlier:

> *Wolfgang Haken*
> *Smote the Kraken*
> *One! Two! Three! Four!*
> *Quoth he: 'The monster is no more'.*

It is perhaps comical and an understatement of major proportions that Tutte titled his short poem *"Some Recent Progress in Combinatorics."*

Thus the Four Color Problem had been solved. In fact, a simpler solution (still computer-assisted), employing an unavoidable set of 633 reducible configurations, was given in 1993 by Neil Robertson, Daniel P. Sanders, Paul Seymour and Robin Thomas.

10.2   Vertex Coloring

It is not particularly difficult to show that the map drawn in Figure 10.3 can be colored with four colors, that is, each region of the map can be assigned one of four given colors such that neighboring regions are colored differently. Indeed, one such coloring is shown in the figure, where r, b, g and y denote red, blue, green and yellow, respectively. What does coloring the regions of a map have to do with graphs? Actually, there is a close connection. With each map, there is associated a graph G, called the **dual** of the map, whose vertices are the regions of the map and such that two vertices of G are adjacent if the corresponding regions are neighboring regions. The dual of the map in Figure 10.3 is also shown in the figure. Observe that the graph G of Figure 10.3 is a connected planar graph. In fact, the dual of every map is a connected planar graph. Conversely, every connected planar graph is the dual of some map. Indeed, representing the regions of a map and adjacency of regions by a graph actually occurred in the 1879 paper of Kempe. The term "graph" was evidently used for the first time only a year earlier by James Joseph Sylvester.

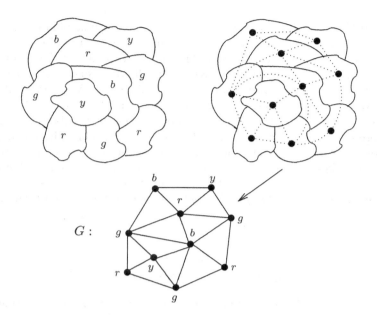

Figure 10.3: A map and its dual

Coloring the regions of a map suggests coloring the vertices of its dual. Indeed, it suggests coloring the vertices of any graph. By a **proper coloring** (or, more simply, a **coloring**) of a graph G, we mean an assignment of colors (elements of some set) to the vertices of G, one color to each vertex, such that adjacent vertices are colored differently. The smallest number of colors in any coloring of a graph G is called the **chromatic number** of G and is denoted by $\chi(G)$. (The symbol χ is the Greek letter "chi.") If it is possible to color (the vertices of) G from a set of k colors, then G is said to be k-**colorable**. A coloring that uses k colors is called a k-**coloring**. If $\chi(G) = k$, then G is said to be k-**chromatic** and every k-coloring of G is a **minimum coloring** of G.

Figure 10.3 shows a coloring of a graph G, namely, a coloring of the dual of the map in Figure 10.3. Necessarily then, G is 4-colorable; indeed, G is 4-chromatic. In fact, the coloring of G in Figure 10.3 is suggested by the coloring of the map. Hence the Four Color Theorem gives us the following result, which is then a restatement of this famous theorem.

Theorem 10.1 (The Four Color Theorem) *The chromatic number of every planar graph is at most 4.*

We have already mentioned that the origin of the Four Color Conjecture goes back to 1852. At one time, however, there were some who believed the origin went back even farther. This is not so, but the confusion may be understandable. In 1840 August Möbius (1790-1868) posed a problem that is sometimes referred to as:

The Problem of the Five Princes

There once was a king who had five sons. In his will he stated that on his death his kingdom should be divided into five regions for his sons in such a way that each region should have a common boundary with the other four. How can this be done?

If a solution to this problem is not evident, then it might be interesting to look at an extension of this problem that was posed by Heinrich Tietze (1880-1964), a prominent topologist:

The Problem of the Five Palaces

The king additionally required each of his five sons to build a palace in his region and the sons should link each pair of palaces by roads so that no two roads cross. How can this be done?

The Problem of the Five Palaces can be modeled by a graph G whose vertices are the palaces and whose edges are the roads. Then $G = K_5$. A solution to this problem requires G to be planar, which it is not. The graph G is the dual graph of any solution to the Problem of the Five Princes. This says that neither problem has a solution. To divide the kingdom into five regions in this manner would require five colors to color the regions. However, there is no such configuration of five regions. What this statement says is that no map can contain five regions, every two of which are neighboring. This does not solve the Four Color Problem, however, because what the statement does not say is that there can't be some other configuration of regions that might require five colors. This is where the difficulty lies.

Let's make a few observations about coloring graphs and the chromatic numbers of some familiar graphs. First, if a graph G contains even one edge, then at least two colors are required to color G. That is, $\chi(G) = 1$ if and only if $G = \overline{K}_n$ for some positive integer n.

In any coloring of a graph G, no two vertices that are colored the same can be adjacent. Sets of vertices, no two of which are adjacent, are necessarily of interest to us when discussing coloring. Recall that a set S of vertices in a graph G is **independent** if no two vertices of S are adjacent. Ordinarily, a graph has many independent sets of vertices. Recall also that a **maximum independent set** is an independent set of maximum cardinality. The number of vertices in a maximum independent set of G is denoted by $\alpha(G)$ and is called the **vertex independence number** (or, more simply, the **independence number**) of G. For the graph $G = C_6$ of Figure 10.4, $S_1 = \{v_1, v_4\}$ and $S_2 = \{v_2, v_4, v_6\}$ are both independent sets. Since no independent set of G contains more than three vertices, $\alpha(G) = 3$.

If G is a k-chromatic graph, then it is possible to partition $V(G)$ into k independent sets $V_1, V_2, \ldots, V_k$, called **color classes**, but it is *not* possible to partition $V(G)$ into $k - 1$ independent sets. Typically, we think of the vertices in the color class V_i $(1 \leq i \leq k)$ as being assigned color i. Conversely, if G is a

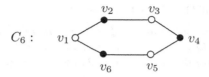

Figure 10.4: Independent sets

graph whose vertex set can be partitioned into k independent sets, but no fewer, then $\chi(G) = k$. Therefore, in order for a graph G to have chromatic number 2, the graph G must be nonempty and it must be possible to partition $V(G)$ into two independent sets V_1 and V_2. Consequently, every edge of G must join a vertex of V_1 and a vertex of V_2. But this means that G is a bipartite graph with partite sets V_1 and V_2.

Theorem 10.2 *A graph G has chromatic number 2 if and only if G is a nonempty bipartite graph.*

By Theorem 1.12, a graph G is bipartite if and only if G has no odd cycles. So if a graph G contains an odd cycle, then $\chi(G) \geq 3$. Of course, if $G = C_n$ for some even integer $n \geq 4$, then $\chi(G) = 2$. On the other hand, if $n \geq 3$ is an odd integer, then $\chi(C_n) = 3$. We already know that $\chi(C_n) \geq 3$ when $n \geq 3$ is an odd integer. To show that $\chi(C_n) = 3$, we need only show that there exists a 3-coloring of C_n. Actually, we can color some vertex of C_n with the color 3 and alternate the colors 1 and 2 for the remaining vertices (see Figure 10.5).

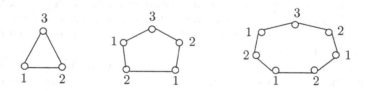

Figure 10.5: The chromatic numbers of odd cycles

The way we established the chromatic number of C_n for each odd integer $n \geq 3$ illustrates how to establish the chromatic number of any graph. To show that $\chi(G) = k$ for some graph G and some integer $k \geq 3$, we must show that

(1) at least k colors are needed to color G (or, equivalently, show that it is impossible to color G with $k - 1$ colors) and

(2) there is a k-coloring of G.

Example 10.3 *The graph G shown in Figure 10.6 is 3-chromatic.*

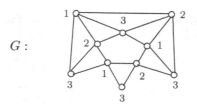

G :

Figure 10.6: A 3-chromatic graph G

Solution. Since G contains a triangle, it follows that $\chi(G) \geq 3$. On the other hand, a 3-coloring of G is shown in Figure 10.6, implying that $\chi(G) \leq 3$. Therefore, $\chi(G) = 3$. ◇

Some comments concerning the solution of Example 10.3 may be useful. As we said, G contains a triangle, which implies that $\chi(G) \geq 3$. Therefore, to show that $\chi(G) = 3$, it suffices to present a 3-coloring of G. Of course, a 3-coloring of G is shown in Figure 10.6. However, we didn't initially know that such a coloring was possible. This graph G also contains a 5-cycle, which requires three colors. Perhaps we thought of 3-coloring this 5-cycle first. There is a number of ways to 3-color this 5-cycle. One of these is shown in Figure 10.7(a). If we now attempt to color the remaining vertices of G so that only three colors are used, then the resulting coloring produces two adjacent vertices that are colored 3 as shown in Figure 10.7(b), which is impossible. This might have led us to conclude (incorrectly) that $\chi(G) = 4$. This indicates that care must be taken when coloring a graph.

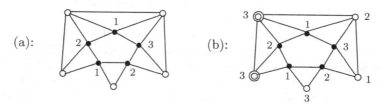

(a):   (b):

Figure 10.7: Coloring a graph

A useful application of coloring occurs in certain kinds of scheduling problems.

Example 10.4 *The mathematics department of a certain college plans to schedule the classes Graph Theory (GT), Statistics (S), Linear Algebra (LA), Advanced Calculus (AC), Geometry (G) and Modern Algebra (MA) this summer. Ten students (see below) have indicated the courses they plan to take. With this information, use graph theory to determine the minimum number of time periods needed to offer these courses so that every two classes having a student in com-*

mon are taught at different time periods during the day. Of course, two classes
having no students in common can be taught during the same period.

Anden: LA, S	Brynn: MA, LA, G
Chase: MA, G, LA	Denise: G, LA, AC
Everett: AC, LA, S	François: G, AC
Greg: GT, MA, LA	Harper: LA, GT, S
Irene: AC, S, LA	Jennie: GT, S

Solution. First, we construct a graph H whose vertices are the six subjects.
Two vertices (subjects) are joined by an edge if some student is taking both
classes (see Figure 10.8). The minimum number of time periods is $\chi(H)$. Since
H contains the odd cycle (GT, S, AC, G, MA, GT), it follows that three colors
are needed to color the vertices on this cycle. Since LA is adjacent to all vertices
of this cycle, a fourth color is needed for LA. Thus $\chi(H) \geq 4$. However, there
is a 4-coloring of H shown in Figure 10.8 and so $\chi(H) = 4$. This also tells us
one way to schedule these six classes during four time periods, namely, Period 1:
Graph Theory, Advanced Calculus; Period 2: Geometry; Period 3: Statistics,
Modern Algebra; Period 4: Linear Algebra.   ◇

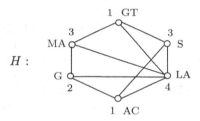

Figure 10.8: The graph of Example 10.4

Certainly, every graph G of order n is n-colorable. If $G = K_n$, then every two
vertices must be assigned different colors and so $\chi(K_n) = n$. If G has order n
and $G \neq K_n$, then G contains two nonadjacent vertices, say u and v. Assigning
u and v the color 1 and the remaining $n - 2$ vertices the colors $2, 3, \ldots, n - 1$
produces an $(n - 1)$-coloring of G and so $\chi(G) \leq n - 1$. That is, a graph G of
order n has chromatic number n if and only if $G = K_n$.

Another observation is useful. If H is a subgraph of a graph G, then any
coloring of G produces a coloring of H as well. Since it may be possible to color
H with even fewer colors, it follows that

$$\chi(H) \leq \chi(G).$$

Opposite to an independent set of vertices in a graph is a clique. A **clique** in a
graph G is a complete subgraph of G. The order of the largest clique in a graph
G is its **clique number**, which is denoted by $\omega(G)$. (The symbol ω is the Greek
letter "omega.") In fact,

$$\alpha(G) = k \text{ if and only if } \omega(\overline{G}) = k.$$

In general, there is no formula for the chromatic number of a graph. In fact, determining the chromatic number of even a relatively small graph is often an extremely challenging problem. However, lower bounds for the chromatic number of a graph G can be given in terms of the clique number and the independence number and order of G.

Theorem 10.5 *For every graph G of order n,*

$$\chi(G) \geq \omega(G) \quad and \quad \chi(G) \geq \frac{n}{\alpha(G)}.$$

Proof. Let H be a clique of G having order $\omega(G)$. Then $\chi(H) = \omega(G)$. Since H is a subgraph of G, it follows that $\chi(H) \leq \chi(G)$, that is, $\omega(G) \leq \chi(G)$.

Suppose that $\chi(G) = k$. Then $V(G)$ can be partitioned into k independent sets $V_1, V_2, \ldots, V_k$. Hence

$$n = |V(G)| = |V_1 \cup V_2 \cup \ldots \cup V_k| = \sum_{i=1}^{k} |V_i| \leq k\alpha(G).$$

Therefore, $\chi(G) = k \geq n/\alpha(G)$. ∎

To establish sharpness for the bounds given in Theorem 10.5, consider the graph $G = K_{3,3,3,3}$ with partite sets V_1, V_2, V_3, V_4, as illustrated in Figure 10.9. The order of G is $n = 12$, its independence number is $\alpha(G) = 3$ and its clique number is $\omega(G) = 4$. In fact, $\chi(G) = 4 = \omega(G) = n/\alpha(G)$. A 4-coloring of G can be given by coloring the vertices in V_i with the color i for $1 \leq i \leq 4$.

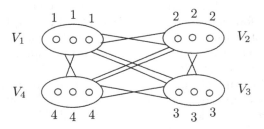

Figure 10.9: Coloring the graph $K_{3,3,3,3}$

In Example 1.5 (in Chapter 1) we showed how a graph could model the traffic lanes at the intersection of two streets. We now revisit this example and consider a question that is relevant to this situation.

Example 10.6 *Figure 10.10 shows the traffic lanes L1, L2, ..., L9 at the intersection of two busy streets. A traffic light is located at this intersection. During a certain phase of the traffic light, those cars in lanes for which the light is green may proceed safely through the intersection. What is the minimum number of phases needed for the traffic light so that (eventually) all cars may proceed through the intersection?*

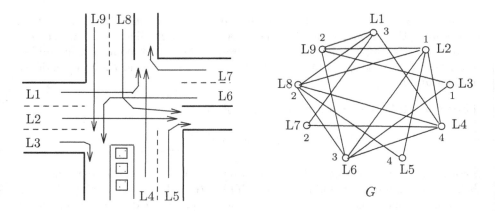

Figure 10.10: Traffic lanes at street intersections

Solution. Construct a graph G to model this situation (see Figure 10.10), where $V(G) = \{L1, L2, \ldots, L9\}$ and two vertices (lanes) are joined by an edge if vehicles in these two lanes cannot safely enter the intersection at the same time, as there is a possibility of an accident.

Answering this question requires determining the chromatic number of the graph G in Figure 10.10. First notice that if $S = \{L2, L4, L6, L8\}$, then $G[S] = K_4$ and so $\chi(G) \geq 4$. Since there exists a 4-coloring of G, as indicated in the graph of Figure 10.10, $\chi(G) = 4$.   ◇

Since the vertex set of the graph G of Figure 10.10 can be partitioned quite obviously into the four independent sets $\{L1, L2, L3\}$, $\{L4, L5\}$, $\{L6, L7\}$ and $\{L8, L9\}$, one might be led to say that the answer to the question in Example 10.6 is clearly 4. However, this observation (while also providing another 4-coloring of G) again only shows that $\chi(G) \leq 4$.

A **coloring** of a graph G can also be thought of as a function c from $V(G)$ to the set $\mathbf{N}$ of positive integers (or natural numbers) such that adjacent vertices have distinct functional values, that is, a **coloring** of G is a function $c : V(G) \to \mathbf{N}$ such that $uv \in E(G)$ implies that $c(u) \neq c(v)$.

We now present an upper bound for the chromatic number of a graph in terms of its maximum degree. The technique used in the proof of the following result is algorithmic and Greedy in nature, in the sense that colors are assigned to the vertices, one vertex at a time, in what appears to be an optimal manner.

Theorem 10.7 *For every graph G,*

$$\chi(G) \leq 1 + \Delta(G).$$

Proof. Let $V(G) = \{v_1, v_2, \ldots, v_n\}$. Define a coloring $c : V(G) \to \mathbf{N}$ recursively as follows: $c(v_1) = 1$. Once $c(v_i)$ has been defined, $1 \leq i < n$, define $c(v_{i+1})$ as the smallest positive integer not already used to color any of the neighbors of v_{i+1}. Since v_{i+1} has $\deg v_{i+1}$ neighbors, at least one of the integers $1, 2, \ldots, 1 + \deg v_{i+1}$ is available for $c(v_{i+1})$. Therefore, $c(v_{i+1}) \leq 1 + \deg v_{i+1}$. If the maximum color assigned to the vertices of G is $c(v_j)$, say, then

$$\chi(G) \leq c(v_j) \leq 1 + \deg v_j \leq 1 + \Delta(G),$$

as desired. ∎

We have already noted that if $n \geq 3$ is odd, then $\chi(C_n) = 3$. Also, $\chi(K_n) = n$. Hence $\chi(C_n) = 1 + \Delta(C_n)$ if $n \geq 3$ is odd and $\chi(K_n) = 1 + \Delta(K_n)$. Therefore, the bound presented in Theorem 10.7 is attained for odd cycles and complete graphs. There is a theorem due to Rowland Leonard Brooks, which tells us that these are the only connected graphs for which the bound is attained. Since the proof is a bit involved, we do not include it.

Theorem 10.8 (**Brooks' Theorem**) *For every connected graph G that is not an odd cycle or a complete graph,*

$$\chi(G) \leq \Delta(G).$$

There is an upper bound for the chromatic number of a graph that is ordinarily superior to that given in Theorems 10.7 and 10.8 and which can be proved using an approach similar to that given for the proof of Theorem 10.7.

Theorem 10.9 *For every graph G,*

$$\chi(G) \leq 1 + \max\{\delta(H)\},$$

where the maximum is taken over all induced subgraphs H of G.

Proof. Among all induced subgraphs of G, let k denote the maximum of their minimum degrees. Suppose that G has order n and let v_n be a vertex of $G_n = G$ such that $\deg_G v_n = \delta(G)$. Thus $\deg_G v_n \leq k$. Therefore, $G_{n-1} = G - v_n$ contains a vertex v_{n-1} such that $\deg_{G_{n-1}} v_{n-1} \leq k$. Continuing in this manner, we construct a sequence $v_1, v_2, \ldots, v_n$ of all vertices of G and a sequence $G_1, G_2, \ldots, G_n$ of induced subgraphs of G such that $v_i \in V(G_i)$ for $1 \leq i \leq n$ and $\deg_{G_i} v_i \leq k$.

Define a coloring $c : V(G) \to \mathbf{N}$ recursively as follows: $c(v_1) = 1$. Once $c(v_i)$ has been defined, $1 \leq i < n$, define $c(v_{i+1})$ as the smallest positive integer not already used to color any of the neighbors of v_{i+1}. Since v_{i+1} has $\deg_{G_{i+1}} v_{i+1}$ neighbors among the vertices $v_1, v_2, \ldots, v_i$ and $\deg_{G_{i+1}} v_{i+1} \leq k$, at least one of the integers $1, 2, \ldots, k + 1$ is available for $c(v_{i+1})$. Hence every vertex of G is assigned one of the colors $1, 2, \ldots, 1 + k$ and so $\chi(G) \leq 1 + k$, as desired. ∎

Let's return to the inequality $\chi(G) \geq \omega(G)$ in Theorem 10.5. Although the lower bound $\omega(G)$ for $\chi(G)$ is attained when $G = K_{3,3,3,3}$ and, in fact, is attained for every complete multipartite graph, there are numerous examples of graphs G for which $\chi(G) \neq \omega(G)$. For example, for $G = C_5$, we have seen that $\chi(G) = 3$;

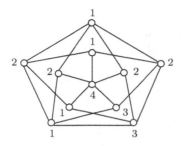

Figure 10.11: The Grötzsch graph: A triangle-free
graph with chromatic number 4

yet $\omega(G) = 2$. The graph G of Figure 10.11 is known as the **Grötzsch graph**. This graph is **triangle-free** (it has no triangles) but has chromatic number 4. So $\chi(G) = 4$ and $\omega(G) = 2$.

It may seem that graphs with a large chromatic number must have large cliques, but this is not so. In fact, triangle-free graphs can have large chromatic numbers. This fact has been observed by a number of mathematicians. The following proof is due to Jan Mycielski, a mathematician who spent many years at the University of Colorado and who is known for his work in sets and logic. In the proof, we make use of a graph, sometimes called the shadow graph of a graph. The **shadow graph** $S(G)$ of a graph G is obtained from G by adding, for each vertex v of G, a new vertex v', called the **shadow vertex** of v, and joining v' to the neighbors of v in G. Observe that (1) a vertex of G and its shadow vertex are not adjacent in $S(G)$ and (2) no two shadow vertices are adjacent in $S(G)$. The shadow graph $S(C_5)$ of C_5 is shown in Figure 10.12. The Grötzsch graph of Figure 10.11 is then obtained by adding a new vertex z to $S(C_5)$ and joining z to the shadow vertices in $S(C_5)$.

$S(C_5)$:

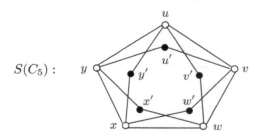

Figure 10.12: The shadow graph of C_5

Theorem 10.10 *For every integer $k \geq 3$, there exists a triangle-free graph with chromatic number k.*

Proof. We proceed by induction on k. We have already seen that the result is true for $k = 3$ and $k = 4$. Assume that there is a triangle-free $(k-1)$-chromatic graph F, where $k \geq 5$ is an integer. Let G be the graph obtained by adding a new vertex z to the shadow graph $S(F)$ of F and joining z to the shadow vertices in $S(F)$. We show that G is a triangle-free graph with $\chi(G) = k$.

First, we verify that G is triangle-free. Assume, to the contrary, that there is a set U of three vertices of G such that $G[U] = K_3$. Since no two shadow vertices are adjacent in G, it follows that U contains at most one shadow vertex. Because z is adjacent only to shadow vertices and U contains at least one vertex that is not a shadow vertex, $z \notin U$. On the other hand, F is triangle-free and so at least one vertex of U is not in F. Therefore, $U = \{u, v, w'\}$, where u and v are adjacent vertices of F and w' is a shadow vertex that is adjacent to u and v. Thus $w \neq u, v$. However then, w is adjacent to u and v, producing a triangle in F, which is impossible since F is triangle-free.

It remains to show that $\chi(G) = k$. Let c^* be a $(k-1)$-coloring of F. We extend c^* to a k-coloring of G by defining $c^*(x') = c^*(x)$ for each $x \in V(F)$ and defining $c^*(z) = k$. Thus $\chi(G) \leq k$. Next we show that $\chi(G) \geq k$. Since F is a subgraph of G, it follows that $k - 1 = \chi(F) \leq \chi(G)$. Assume, to the contrary, that $\chi(G) = k - 1$. Let there be given a $(k-1)$-coloring c of G, say with colors $1, 2, \ldots, k-1$. We may assume that $c(z) = k - 1$. Since z is adjacent to every shadow vertex in G, it follows that the shadow vertices are colored with the colors $1, 2, \ldots, k-2$. For every shadow vertex x' of G, the color $c(x')$ is different from the colors assigned to the neighbors of x. Therefore, if for each vertex y of G belonging to F, the color $c(y)$ is replaced by $c(y')$, we have a $(k-2)$-coloring of F. This is impossible, however, since $\chi(F) = k - 1$. ∎

Theorem 10.10 therefore shows the existence of graphs G for which $\chi(G)$ is considerably larger than $\omega(G)$. While there has been much interest in graphs G for which $\chi(G) > \omega(G)$, there has even been more interest in graphs G for which not only $\chi(G) = \omega(G)$ but $\chi(H) = \omega(H)$ for every induced subgraph H of G. A graph G is called **perfect** if $\chi(H) = \omega(H)$ for every induced subgraph H of G. This concept was introduced in 1963 by the French mathematician Claude Berge. From our earlier remarks, every complete multipartite graph is perfect and so every complete bipartite graph is perfect. In fact, every bipartite graph is perfect (see Exercise 10.16).

Claude Berge made two conjectures concerning perfect graphs. The first of these was verified in 1972 by László Lovász and the second was verified in 2002 by Maria Chudnovsky, Neil Robertson, Paul Seymour and Robin Thomas.

The Perfect Graph Theorem A graph is perfect if and only if its complement is perfect.

The Strong Perfect Graph Theorem A graph G is perfect if and only if neither G nor $\overline{G}$ contains an induced odd cycle of length 5 or more.

Exercises for Section 10.2

10.1 Determine the chromatic number of each of the graphs in Figure 10.13.

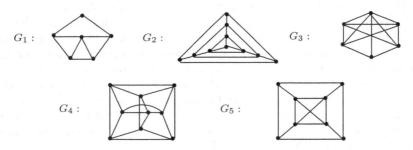

Figure 10.13: Graphs in Exercise 10.1

10.2 Determine the chromatic number of each of the following:

(a) the Petersen graph, (b) the n-cube Q_n, (c) the wheel $W_n = C_n + K_1$.

10.3 What is the chromatic number of a tree?

10.4 Prove or disprove:

(a) If a planar graph contains a triangle, then its chromatic number is 3.

(b) If there is a 4-coloring of a graph G, then $\chi(G) = 4$.

(c) If it can be shown that there is no a 3-coloring of a graph G, then $\chi(G) = 4$.

(d) If G is a graph with $\chi(G) \leq 4$, then G is planar.

10.5 Prove that every graph of order 6 with chromatic number 3 has at most 12 edges.

10.6 Prove or disprove:

(a) If a graph G contains a subgraph isomorphic to the complete graph K_r, then $\chi(G) \geq r$.

(b) If G is a graph with $\chi(G) \geq r$, then G contains a subgraph isomorphic to the complete graph K_r.

10.7 Show that there exists no graph G with $\chi(G) = 6$ whose vertices have degrees $3, 3, 3, 3, 3, 3, 4, 4, 5, 5, 5, 5$.

10.8 Give an example of the following or explain why no such example exists:

(a) a planar graph with chromatic number 5,

(b) a nonplanar graph with chromatic number 3,

(c) a graph G with $\Delta(G) = 2\chi(G)$,

(d) a graph G with $\chi(G) = 2\Delta(G)$,

(e) a noncomplete graph of order n with chromatic number n.

10.9 Prove or disprove:

(a) There exists a nonplanar graph G such that $G - v$ is planar and $\chi(G) = \chi(G - v) + 1$ for every vertex v of G.

(b) There exists a nonplanar graph G such that $G - v$ is planar and $\chi(G) = \chi(G - v)$ for every vertex v of G.

10.10 Eight mathematics majors at a small college are permitted to attend a meeting dealing with undergraduate research during final exam week provided they make up all the exams missed on the Monday after they return. The possible time periods for these exams on Monday are

(1) 8:00 - 10:00   (2) 10:15 - 12:15   (3) 12:30 - 2:30
(4) 2:45 - 4:45   (5) 5:00 - 7:00   (6) 7:15 - 9:15.

Use graph theory to determine the earliest time on Monday that all eight students can finish their exams if two exams cannot be given during the same time period if some student must take both exams. The eight students and the courses [Advanced Calculus (AC), Differential Equations (DE), Geometry (G), Graph Theory (GT), Linear Programming (LP), Modern Algebra (MA), Statistics (S), Topology (T)] each student is taking are listed below:

Alicia: AC, DE, LP	Brian: AC, G, LP
Carla: G, LP, MA	Diane: GT, LP, MA
Edward: DE, GT, LP	Faith: DE, GT, T
Grace: DE, S, T	Henry: AC, DE, S

10.11 Eight chemicals are to be shipped across country by air express. The cost of doing this depends on the number of containers shipped. The cost of shipping one container is \$125. For each additional container the cost increases by \$85. Some chemicals interact with one another and it is too risky to ship them in the same container. The chemicals are labeled by $c_1, c_2, \ldots, c_8$ and chemicals that interact with a given chemical are given below:

$c_1 : c_2, c_5, c_6$ $c_2 : c_1, c_3, c_5, c_7$ $c_3 : c_2, c_4, c_7$
$c_4 : c_3, c_6, c_7, c_8$ $c_5 : c_1, c_2, c_6, c_7, c_8$ $c_6 : c_1, c_4, c_5, c_8$
$c_7 : c_2, c_3, c_4, c_5, c_8$ $c_8 : c_4, c_5, c_6, c_7$.

What is the minimum cost of shipping the chemicals and how should the chemicals be packed into containers?

10.12 The road intersection shown in Figure 10.14 needs a traffic signal to handle the traffic flow. If cars from two different lanes could collide, then cars from these two lanes will not be permitted to enter the intersection at the same time. What is the minimum number of signal phases that are needed to ensure safe traffic?

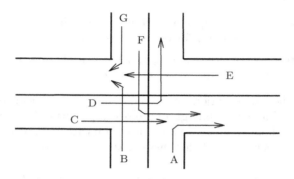

Figure 10.14: A road intersection in Exercise 10.12

10.13 A graph G of order n has $\chi(G) = \alpha(G) = k$, where $\alpha(G)$ is the independence number of G. Furthermore, for every k-coloring of G, there is a unique partition of $V(G)$ into color classes such that every two distinct color classes have different cardinality. Show that $\Delta(G) = n - 1$, where $\Delta(G)$ is the maximum degree of G.

10.14 For a class $\mathcal{F}$ of graphs, define $\chi(\mathcal{F})$ as the maximum value of $\chi(G)$ among all graphs G in $\mathcal{F}$ (if this maximum exists). A class $\mathcal{G}$ of graphs consists of all graphs G of order n and size m for which $m \leq 4n - 4$ and having the property that if $G \in \mathcal{G}$ and H is an induced subgraph of G, then $H \in \mathcal{G}$. Determine $\chi(\mathcal{G})$. [Hint: First determine an upper bound for $\chi(G)$ from the fact that $\chi(G) \leq 1 + \max\{\delta(H)\}$, where the maximum is taken over all induced subgraphs H of G.]

10.15 Let A be a set of cardinality $n \geq 1$, say $A = \{a_1, a_2, \ldots, a_n\}$ and let S be a collection of pairs of elements of A. Consider those functions $f : A \to \mathbf{N}$ having the property that the two elements in each pair in S have distinct images. Among these functions, let g be one whose range has minimum cardinality. How is this related to a problem in graph theory?

10.16 (a) Prove that every bipartite graph is perfect.

(b) Determine whether the complement of the Petersen graph is perfect.

10.3 Edge Coloring

In addition to coloring the regions of a map and coloring the vertices of a graph, it is also of interest to color the edges of a graph. An **edge coloring** of a nonempty graph G is an assignment of colors to the edges of G, one color to each edge, such that adjacent edges are assigned different colors. The minimum number of colors that can be used to color the edges of G is called the **chromatic index** (or sometimes the **edge chromatic number**) and is denoted by $\chi'(G)$. An edge coloring that uses k colors is a k-**edge coloring**. In Figure 10.15, a 4-edge coloring of a graph G is given.

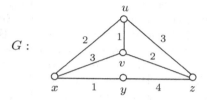

Figure 10.15: A 4-edge coloring of a graph

Let G be a graph containing a vertex v with $\deg v = k \geq 1$. Then there are k edges incident with v. Any edge coloring must assign k distinct colors to the edges incident with v and so $\chi'(G) \geq \deg v = k$. In particular,

$$\chi'(G) \geq \Delta(G)$$

for every nonempty graph G.

Example 10.11 *The graph G of Figure* 10.15 *has chromatic index* 4.

Solution. Since $\Delta(G) = 3$ for the graph G of Figure 10.15, it follows that $\chi'(G) \geq 3$. Since there is a 4-edge coloring of G shown in Figure 10.15, it follows that $\chi'(G) \leq 4$. Hence $\chi'(G) = 3$ or $\chi'(G) = 4$. But which is it? We show that $\chi'(G) = 4$.

Assume, to the contrary, that there exists a 3-edge coloring of G, using the colors 1, 2 and 3, say. Then every vertex of degree 3 is incident with edges colored 1, 2 and 3. In particular, the three edges incident with u are colored 1, 2 and 3. We may assume, without loss of generality, that uv is colored 1, ux is colored 2 and uz is colored 3. Since vx is adjacent to both uv and ux, the edge vx must be colored 3. Similarly, vz must be colored 2. Since xy is adjacent to both ux and vx, the edge xy must be colored 1. By the same reasoning, the edge yz must be colored 1. However, xy and yz are adjacent edges colored 1. This is impossible. Therefore, $\chi'(G) = 4$, as we claimed. ◇

For the graph G of Figure 10.15, we now know that $\chi'(G) = 1 + \Delta(G)$. Of course, we knew that $\chi'(G) \geq \Delta(G)$, as this inequality holds for all nonempty graphs. In what must be considered *the* fundamental theorem on edge colorings, Vadim G. Vizing showed that if one knows the maximum degree of a graph (which, of course, is very easy to determine), then we are very close to knowing the chromatic index of the graph. Since the proof of Vizing's theorem is quite involved, we omit it.

Theorem 10.12 (**Vizing's Theorem**) *For every nonempty graph G, either*

$$\Delta(G) \leq \chi'(G) \leq 1 + \Delta(G).$$

Vadim Vizing was born on March 25, 1937 in Kiev in Ukraine. After World War II his family was forced to move to the Novosibirsk region of Siberia because his mother was half-German. He began his study of mathematics at the University of Tomsk in 1954 and graduated in 1959. He was then sent to the famous Steklov Institute in Moscow to study for a Ph.D. His area of research was function approximation, which he did not like. Because his request to change his area of research was not granted, he did not complete his degree and returned to Novosibirsk in 1962. He then studied at the Mathemtical Institute of the Academy of Sciences in Academgorodoc, where he obtained a Ph.D. in 1966 without a formal supervisor but with the assistance of Alexander Zykov. (Like many mathematicians, Vizing's interests included music, books and chess.)

While in Novosibirsk, Vizing started working on a problem that involved coloring the wires of a network. As he studied the problem, he became interested in more theoretical questions. He then proved the famous theorem which bears his name (Theorem 10.12). This paper was submitted to the prestigious journal *Doklady*, only to have it rejected because of the referee found it to be uninteresting. The paper was finally published in 1964 in the journal *Metody Diskretnogo Analiza*, a local journal in Novosibirsk. By the time it appeared, the result had already become quite well known, primarily because it had been mentioned to others by Zykov. When Vizing was asked what makes a mathematical result outstanding, he replied:

A mathematician should do research and find new results, and then time will decide what is important and what is not!

By Vizing's theorem then, the chromatic index of every nonempty graph G is one of two numbers, namely $\Delta(G)$ or $\Delta(G) + 1$. Let's look at some well-known graphs, beginning with cycles. For every cycle C_n, $n \geq 3$, we have $\Delta(C_n) = 2$, so $\chi'(C_n) = 2$ or $\chi'(C_n) = 3$. If n is even, then we may simply alternate the colors 1 and 2 about the edges of C_n, arriving at an edge coloring. On the other hand, if n is odd and we attempt to alternate the colors 1 and 2 about the edges of C_n, then the final edge of C_n will be colored the same as the first edge. Since these two edges are adjacent, a contradiction is produced. Thus, $\chi'(C_n) = 3$ if $n \geq 3$ is odd. Edge colorings of some cycles are illustrated in Figure 10.16.

A simple observation concerning edge colorings will be useful. Let there be given an edge coloring of a graph G of order n. Any two edges of G that are

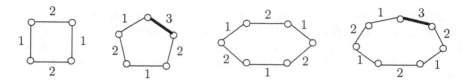

Figure 10.16: Coloring the edges of cycles

colored the same cannot be adjacent, of course. This says that we can never have more than $n/2$ edges of G that are colored the same. If n is odd, then the maximum number of edges that can be colored the same is therefore $(n-1)/2$. In particular, for the graph C_n, where $n \geq 3$ is odd, no more than $(n-1)/2$ edges can be colored the same and no more $(n-1)/2 + (n-1)/2 = n - 1$ edges can be colored with two colors. This observation says that $\chi'(C_n) \geq 3$ if $n \geq 3$ is odd.

The observation above provides us with the following more general result.

Theorem 10.13 *Let G be a graph of odd order n and size m. If*

$$m > \frac{(n-1)\Delta(G)}{2},$$

then $\chi'(G) = 1 + \Delta(G)$.

Proof. In any edge coloring of G, no more than $(n-1)/2$ edges can be colored the same. Therefore, no more than $(n-1)\Delta(G)/2$ edges can be colored with $\Delta(G)$ colors. Since $m > (n-1)\Delta(G)/2$, it follows that $\chi'(G) > \Delta(G)$. By Vizing's theorem, $\chi'(G) \leq 1 + \Delta(G)$ and so $\chi'(G) = 1 + \Delta(G)$. ∎

Perhaps you have already observed that determining the chromatic index of a graph G is the same as attempting to partition the edge set of G into a minimum number of independent sets of edges (matchings) or to decompose G into a minimum number of 1-regular subgraphs. If each 1-regular subgraph in the decomposition is a spanning subgraph of G, then we are now asking is whether G is 1-factorable. Consequently, there is a very close connection between the chromatic index of a graph and several topics we discussed in Chapter 8. As with 1-factorizations of graphs, there are applications of graphs involving scheduling that are related to edge colorings.

Example 10.14 *Alvin (A) has invited three married couples to his summer house for a week: Bob (B) and Carrie (C) Hanson, David (D) and Edith (E) Irwin and Frank (F) and Gena (G) Jackson. Since all six guests enjoy playing tennis, he decides to set up some tennis matches. Each of his six guests will play a tennis match against every other guest except his/her spouse. In addition, Alvin will play a match against each of David, Edith, Frank and Gena. If no one is to play two matches on the same day, what is a schedule of matches over the smallest number of days?*

Solution. First, we construct a graph H whose vertices are the people at Alvin's summer house, so $V(H) = \{A, B, C, D, E, F, G\}$, where two vertices of H are adjacent if the two vertices (people) are to play a tennis match. (The graph H is shown in Figure 10.17.) To answer the question, we determine the chromatic index of H.

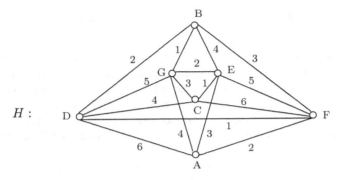

Figure 10.17: The graph H in Example 10.14

First, observe that $\Delta(H) = 5$. By Theorem 10.12, $\chi'(H) = 5$ or $\chi'(H) = 6$. Also, the order of H is $n = 7$ and its size is $m = 16$. Since

$$m = 16 > 15 = \frac{(7-1) \cdot 5}{2} = \frac{(n-1)\Delta(H)}{2},$$

it follows by Theorem 10.13 that $\chi'(H) = 6$. Figure 10.17 gives a 6-edge coloring of H, which provides a schedule of matches

Day 1 :	Bob-Gena, Carrie-Edith, David-Frank
Day 2 :	Alvin-Frank, Bob-David, Edith-Gena
Day 3 :	Alvin-Edith, Bob-Frank, Carrie-Gena
Day 4 :	Alvin-Gena, Edith-Frank Carrie-David
Day 5 :	David-Gena, Edith-Frank
Day 6 :	Alvin-David, Carrie-Frank

that takes place over the smallest number of days (namely six).   ◇

We now make an observation about the chromatic index of regular graphs. By Vizing's theorem, if G is an r-regular graph, $r \geq 1$, then $\chi'(G) = r$ or $\chi'(G) = r + 1$. Furthermore, $\chi'(G) = r$ if and only if G is 1-factorable. Specifically,

An r-regular graph G, $r \geq 1$, has chromatic index r if and only if G is 1-factorable.

Let's now turn to edge colorings of complete graphs. Certainly, $\chi'(K_2) = 1 = \Delta(K_2)$. Since $K_3 = C_3$, it follows that $\chi'(K_3) = 3 = 1 + \Delta(K_3)$. However, $\chi'(K_4) = 3$ as the edge coloring in Figure 10.18 shows.

Since K_n is $(n-1)$-regular, it follows that $\chi'(K_n) = n - 1$ or $\chi'(K_n) = n$. Necessarily, $\chi'(K_n) = n - 1$ if and only if K_n is 1-factorable. By Theorem 8.18, this only occurs when n is even.

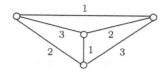

Figure 10.18: An edge coloring of K_4

Theorem 10.15 *For every integer $n \geq 2$,*

$$\chi'(K_n) \;=\; \begin{cases} n & \text{if } n \text{ is odd} \\ n-1 & \text{if } n \text{ is even.} \end{cases}$$

We now consider another scheduling problem.

Example 10.16 *Five individuals have been invited to a bridge tournament (bridge is a game of cards): Allen (A), Brian (B), Charles (C), Doug (D), Ed (E). A game of bridge is played between two 2-person teams. Every 2-person team $\{X, Y\}$ is to play against all other 2-person teams $\{W, Z\}$, where, of course, neither W nor Z is X or Y. If the same team cannot play bridge more than once on the same day, then what is the fewest number of days needed for all possible games of bridge to be played. Set up a schedule for doing this in the smallest number of days. What graph models this situation?*

Solution. We construct a graph G whose vertices consist of all 2-person teams, where we denote a vertex by XY rather than $\{X, Y\}$. Two vertices (2-person teams) XY and WZ are adjacent in G if they will be playing a game of bridge. The graph G is shown in Figure 10.19. Observe that G is isomorphic to the Petersen graph.

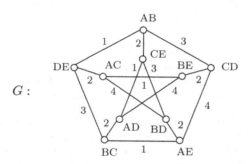

Figure 10.19: The graph in Example 10.16

We now determine the chromatic index of G. Since G is 3-regular, $\chi'(G) = 3$ if and only if G is 1-factorable. However, as Petersen himself pointed out, G is not 1-factorable. Therefore, $\chi'(G) = 4$. A 4-edge coloring of G is shown in Figure 10.19. This creates a schedule of games

Day 1 :   AB-DE, AE-BC, AC-BE, AD-CE
Day 2 :   AB-CE, AC-DE, AE-BD, AD-BC, BE-CD
Day 3 :   AB-CD, BC-DE, BD-CE
Day 4 :   AC-BD, AD-BE, AE-CD

that takes place over the smallest number of days.   ◇

We have already observed that $\chi'(C_n) = \Delta(C_n)$ if $n \geq 4$ is even. Of course, C_n is a bipartite graph if $n \geq 4$ is even. Denés König observed that $\chi'(G) = \Delta(G)$ for *every* bipartite graph. There are several different proofs of this result. In the proof given below, we use the fact that every r-regular bipartite graph, $r \geq 1$, is 1-factorable (Theorem 8.15). This proof uses an argument that is reminiscent of the proof of Theorem 2.7.

Theorem 10.17 (**König's Theorem**) *If G is a nonempty bipartite graph, then*

$$\chi'(G) = \Delta(G).$$

Proof. Suppose that $\Delta(G) = r \geq 1$. First, we show that there exists an r-regular bipartite graph H containing G as a subgraph. This is certainly true if G is r-regular, in which case, we let $H = G$. So we can assume that $\delta(G) < r$. Suppose that the partite sets of G are U_0 and W_0. Let G' be another copy of G where the partite set U_0 is denoted by U_0' and W_0 is denoted by W_0' in G'. Join each vertex in G whose degree is less than r to the corresponding vertex in G', producing a bipartite graph G_1 with partite sets $U_1 = U_0 \cup W_0'$ and $W_1 = W_0 \cup U_0'$ such that $\delta(G_1) = \delta(G) + 1$. If G_1 is r-regular, then $H = G_1$ has the desired property. If $\delta(G_1) < r$, then let G_1' be another copy of G_1 where the partite set U_1 is denoted by U_1' and W_1 is denoted by W_1' in G_1'. Join each vertex in G_1 whose degree is less than r to the corresponding vertex in G_1', producing a bipartite graph G_2 with partite sets $U_2 = U_1 \cup W_1'$ and $W_2 = W_1 \cup U_1'$ such that $\delta(G_2) = 2 + \delta(G)$. We continue this until we arrive at an r-regular bipartite graph G_k, where $k = r - \delta(G)$, containing G as a subgraph, in which case we let $H = G_k$.

By Theorem 8.15, H is 1-factorable and so contains r 1-factors $F_1', F_2', \ldots, F_r'$. For $1 \leq i \leq r$, let F_i be the 1-regular subgraph of G where $E(F_i) = E(F_i') \cap E(G)$. Then $E(F_1), E(F_2), \ldots, E(F_r)$ are edge color classes of G and so $\chi'(G) \leq r$. Since $\Delta(G) = r$, it follows that $\chi'(G) \geq r$ and so $\chi'(G) = r$.   ∎

Interest in edge colorings of graphs was undoubtedly inspired by the Four Color Problem. We have seen that coloring the regions of a map is equivalent to coloring the vertices of a certain connected planar graph (the dual of the map). To be sure, research in coloring graphs was motivated by the Four Color Problem. We have seen that Alfred Bray Kempe's "proof" of the Four Color Theorem, although incorrect, contained some important ideas, which were expanded upon and eventually led to a correct proof of the theorem. We mentioned earlier that another mathematician who gave an incorrect "proof" of the Four Color Theorem was Peter Guthrie Tait. However, he too developed several interesting ideas that gave rise to new areas of study.

By a **cubic map** is meant a connected 3-regular, bridgeless, plane graph. Tait observed that if the regions of all cubic maps can be colored with four or fewer colors, then the regions of all plane graphs can be colored with four or fewer colors. Certainly, there is no reason to consider maps (plane graphs) containing vertices of degree 1 or 2. If a plane graph H contains vertices of degree 4 or more, then a cubic map G can be constructed from H by drawing a sufficiently small circle C about each such vertex v of H, identifying a new vertex at each point of intersection of C with the edges incident with v and deleting v and its incident edges (see Figure 10.20(a)). Now if the regions of the resulting cubic map G can be colored with four or fewer colors, then such a coloring can be used to produce a coloring of the regions of H using four or fewer colors (see Figure 10.20(b)).

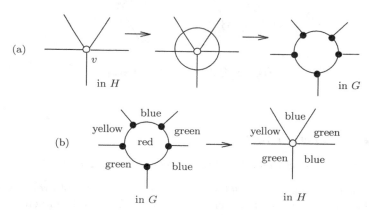

Figure 10.20: Coloring the regions of cubic maps

Of course, concentrating on cubic maps (rather than arbitrary maps) would only be useful if it could be proved that the regions of every cubic map could be colored with four or fewer colors. What Tait proved is the following.

Theorem 10.18 *The regions of a cubic map G can be colored with four or fewer colors if and only if $\chi'(G) = 3$.*

Indeed, 3-edge colorings of 3-regular graphs became known as **Tait colorings**. Certainly, every 3-regular graph G has even order and if G is Hamiltonian, then the edges of a Hamiltonian cycle C can be colored alternately red and blue, say. Removing these edges yields a 1-factor, whose edges can be colored green, say. Tait was able to show that if all 3-connected, 3-regular planar graphs are 3-edge colorable, then all cubic maps are 3-edge colorable. By Theorem 10.18, to prove the Four Color Theorem, it suffices to show therefore that every 3-connected, 3-regular planar graph has chromatic index 3. Of course, if every 3-connected, 3-regular planar graph is Hamiltonian, then the proof is complete. Since Tait believed these graphs were Hamiltonian, he had convinced himself that he had proved the Four Color Theorem. Although the defect in Tait's "proof" was eventually recognized, it wasn't until 1946 when William Tutte presented an example of a 3-connected, 3-regular planar graph that is not Hamiltonian. See Figure 10.21.

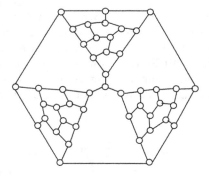

Figure 10.21: The Tutte graph

Exercises for Section 10.3

10.17 Determine the chromatic index of each graph in Exercise 10.1.

10.18 For a positive integer k, let H be a $2k$-regular graph of order $4k + 1$. Let G be obtained from H by removing a set of $k - 1$ independent edges from H. Prove that $\chi'(G) = \Delta(G) + 1$.

10.19 Seven softball teams from Atlanta, Boston, Chicago, Denver, Louisville, Miami and Nashville have been invited to participate in tournaments, where each team is scheduled to play a certain number of the other teams (given below). No team is to play more than one game each day. Set up a schedule of games over the smallest number of days.

Atlanta :	Boston, Chicago, Miami, Nashville
Boston :	Atlanta, Chicago, Nashville
Chicago :	Atlanta, Boston, Denver, Louisville
Denver :	Chicago, Louisville, Miami, Nashville
Louisville :	Chicago, Denver, Miami
Miami :	Atlanta, Denver, Louisville, Nashville
Nashville :	Atlanta, Boston, Denver, Miami

10.20 Let G be a bipartite graph with partite sets U and W where $\Delta(G) = r \geq 1$ and $\delta(G) < r$.

    (a) Use Theorem 8.15 to show that if there is an r-regular bipartite graph H containing G as a subgraph such that at least one of the partite sets of H is U or W, then $\chi'(G) = \Delta(G)$ (thereby giving an alternative proof of König's theorem 10.17 for such graphs G).

(b) Show that there need not be an r-regular bipartite graph H containing G as a subgraph such that at least one of the partite sets of H is U or W.

10.4 Excursion: The Heawood Map Coloring Theorem

We mentioned that during an 11-year period in the 19th century (1879-1890), the Four Color Theorem was considered to have been verified by Alfred Bray Kempe. However, all this changed in 1890 when Percy John Heawood wrote that he had discovered an error Kempe had made in the way he interchanged colors in what were to be called Kempe chains. It was not accidental that Heawood had read Kempe's paper. When Arthur Cayley asked, at a meeting of the London Mathematical Society in 1878, for the status of the Four Color Conjecture, Henry Smith was presiding over the meeting. Smith was a Professor of Geometry at Oxford University who would mention this conjecture during his lectures. Soon afterwards, Heawood became a student of Smith and Heawood became interested in this problem after hearing about it from Smith.

In his paper, Heawood produced a counterexample (see Figure 10.22), not to the statement Kempe was trying to prove (the Four Color Theorem) but to the proof Kempe had given. Indeed, Kempe's proof was quite ingenious and Heawood was able to use Kempe's technique to show that every map could be colored with five or fewer colors. We've seen that this is equivalent to showing that every planar graph can be colored with five or fewer colors.

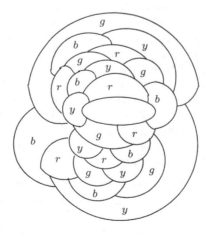

Figure 10.22: A counterexample to Kempe's proof

Theorem 10.19 (**The Five Color Theorem**) *Every planar graph is 5-colorable.*

Proof. Assume, to the contrary, that this statement is false. Then among all planar graphs that are not 5-colorable, let G be the one of smallest order. Since G is not 5-colorable, the order of G is necessarily 6 or more.

By Corollary 9.3, the minimum degree of every planar graph never exceeds 5. Now let v be a vertex of G such that $\deg v = \delta(G)$. Therefore, $\deg v \le 5$. The graph $G - v$ is clearly planar and since the order of $G - v$ is less than the order of G, the graph $G - v$ is 5-colorable. Let a 5-coloring of $G - v$ be given. If either $\deg v \le 4$ or $\deg v = 5$ and the number of colors used to color the neighbors of v is less than 5, then one of these five colors is available for v. Assigning v this color produces a 5-coloring of G, which is a contradiction. Hence we may assume that $\deg v = 5$ and all five colors have been used for the neighbors of v. Consequently, we have the situation pictured in Figure 10.23.

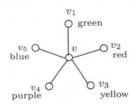

Figure 10.23: A step in the proof of Theorem 10.19

Suppose that there is no $v_2 - v_5$ path in $G - v$, all of whose vertices are colored red or blue (so there is no red-blue Kempe chain in $G - v$ containing both v_2 to v_5). In this case, let S be the set of all red and blue vertices of $G - v$ connected to v_5 by a red-blue path. Certainly, $v_5 \in S$ but, by assumption, $v_2 \notin S$. Now interchange the colors of the vertices belonging to S. Therefore, v_5 is now colored red but v_2 remains red. Hence the color blue is now available for v. Coloring v blue produces a 5-coloring of G. However, this is impossible; so there must be a $v_2 - v_5$ red-blue path in $G - v$.

Since $G - v$ contains a red-blue path from v_2 to v_5, there can be no green-yellow path from v_1 to v_3. Let S' be the set of vertices in $G - v$ connected v_1 by a green-yellow path. Then $v_1 \in S'$ but $v_3 \notin S'$. Interchanging the colors of the vertices in S' results in v_1 being colored yellow but does not change the color of v_3. However, the color green in now available for v. Coloring v green produces a 5-coloring of G, which is a contradiction. ∎

Heawood's paper, which pointed out Kempe's error and which contained a proof of the Five Color Theorem, did not stop with these however. Heawood went on to consider other ideas in his paper. Of course, a major consequence of Heawood's paper is that doubt had returned to the value of the largest chromatic number of a graph that could be embedded on the sphere. In his paper Heawood turned his attention to determining these values for other surfaces.

For a nonnegative integer k, let

$$\chi(S_k) = \max\{\chi(G)\},$$

where the maximum is taken over all graphs G that can be embedded on S_k. After Kempe's 1879 paper, it was believed that $\chi(S_0) = 4$. Following Heawood's 1890 paper, it was only known that $\chi(S_0) = 4$ or $\chi(S_0) = 5$. After Appel and Haken's announcement in 1976, it was known once and for all that $\chi(S_0) = 4$ (the Four Color Theorem). In his 1890 paper, Heawood attempted to obtain a formula for $\chi(S_k)$ when k is a positive integer; in fact, he thought he had done this. What he did do, however, was to obtain an upper bound for $\chi(S_k)$.

Theorem 10.20 *For every positive integer k,*

$$\chi(S_k) \le \frac{7 + \sqrt{1 + 48k}}{2}.$$

Proof. Let G be a graph that can be embedded on S_k and let

$$h = \frac{7 + \sqrt{1 + 48k}}{2}.$$

From this definition of h, one can show that

$$6 + \frac{12(k-1)}{h} = h - 1.$$

(This observation will be useful later.) We show that $\chi(G) \le h$.

Among the induced subgraphs of G, let H be one having the largest minimum degree. By Theorem 10.9, $\chi(G) \le 1 + \delta(H)$. Suppose that H has order n and size m. If $n \le h$, then $\delta(H) \le n - 1$ and $\chi(G) \le n \le h$. So we may assume that $n > h$.

Since G is embeddable on S_k, so too is H. Hence by Corollary 9.11,

$$k \ge \gamma(H) \ge \frac{m}{6} - \frac{n}{2} + 1.$$

Thus $m \le 3n + 6(k-1)$. Therefore,

$$n\delta(H) \le \sum_{v \in V(H)} \deg v = 2m \le 6n + 12(k-1)$$

and so

$$\delta(H) \le 6 + \frac{12(k-1)}{n} \le 6 + \frac{12(k-1)}{h} = h - 1.$$

Consequently,

$$\chi(G) \le 1 + \delta(H) = h = \frac{7 + \sqrt{1 + 48k}}{2},$$

giving the desired result. ∎

According to Theorem 10.20, $\chi(S_1) \leq 7$. To show that $\chi(S_1) = 7$, it is necessary to show that there is a graph having chromatic number 7 that can be embedded on the torus. As it turns out, there is such a graph, namely K_7 (see Figure 10.24). Therefore, $\chi(S_k) = \left\lfloor \frac{7+\sqrt{1+48k}}{2} \right\rfloor$ for $k = 1$, as Heawood showed. To show that this formula holds for every positive integer k would take another 78 years. The proof was accomplished by considering a large number of cases involving many individuals. The two mathematicians who were the most instrumental in completing the proof, however, were Gerhard Ringel and Ted Youngs.

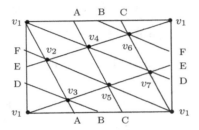

Figure 10.24: Embedding K_7 on S_1

Theorem 10.21 (**The Heawood Map Coloring Theorem**) *For every positive integer k,*

$$\chi(S_k) = \left\lfloor \frac{7 + \sqrt{1 + 48k}}{2} \right\rfloor.$$

Proof. By Theorem 10.20, $\chi(S_k) \leq \frac{7+\sqrt{1+48k}}{2}$. Since $\chi(S_k)$ is an integer, it follows that

$$\chi(S_k) \leq \left\lfloor \frac{7 + \sqrt{1 + 48k}}{2} \right\rfloor.$$

To verify that $\chi(S_k) = \left\lfloor \frac{7+\sqrt{1+48k}}{2} \right\rfloor$, we need only show that there exists a graph that can be embedded on S_k and has chromatic number $\left\lfloor \frac{7+\sqrt{1+48k}}{2} \right\rfloor$. There is a natural candidate. Let

$$n = \left\lfloor \frac{7 + \sqrt{1 + 48k}}{2} \right\rfloor.$$

Certainly,

$$\chi(K_n) = n = \left\lfloor \frac{7 + \sqrt{1 + 48k}}{2} \right\rfloor. \tag{10.1}$$

Next, we show that K_n can be embedded on S_k. By Theorem 9.12,

$$\gamma(K_n) = \left\lceil \frac{(n-3)(n-4)}{12} \right\rceil.$$

By (10.1),

$$n \le \frac{7 + \sqrt{1 + 48k}}{2}$$

and so $2n - 7 \le \sqrt{1 + 48k}$. Solving this inequality for k, we obtain

$$k \ge \frac{(n - 3)(n - 4)}{12}.$$

Since k is an integer,

$$k \ge \left\lceil \frac{(n - 3)(n - 4)}{12} \right\rceil = \gamma(K_n),$$

that is, the genus of K_n is at most k and K_n can be can be embedded on S_k. ∎

Therefore, we see that the proof we gave of Theorem 10.21 relies on knowing a formula for $\gamma(K_n)$. This is precisely what required the efforts of so many for so long. Combining Theorem 10.21 and the Four Color Theorem allows us to state the following.

Corollary 10.22 *For every nonnegative integer k,*

$$\chi(S_k) = \left\lfloor \frac{1 + \sqrt{1 + 48k}}{2} \right\rfloor.$$

Exercises for Section 10.4

10.21 Show that the proof of Theorem 10.20 fails when $k = 0$.

10.22 It is known that the Petersen graph P is not planar. Thus P cannot be embedded on the sphere.

(a) Show that P can be embedded on the torus however.

(b) How many regions does P have when it's embedded on the torus?

(c) What is the minimum number of colors that can be assigned to these regions so that every two adjacent regions are colored differently?

10.23 Prove or disprove: If G is a graph such that $\chi(G) \le \chi(S_k)$ for some positive integer k, then G can be embedded on S_k.

10.5 Exploration: Modular Coloring

The fundamental problem concerning coloring the vertices of a graph G deals with finding the smallest positive integer k such that each vertex of G can be assigned one of the "colors" $1, 2, \ldots, k$ in such a way that every two adjacent vertices of G are assigned different colors. This smallest integer k is then the chromatic number $\chi(G)$ of G. The idea here is to distinguish every pair of adjacent vertices of G in some way. In this manner, this has been accomplished by means of a proper coloring the vertices of G. That is, every two adjacent vertices are distinguished because they are colored differently. There are many ways, however, of distinguishing every two adjacent vertices of a graph by means of coloring and many ways of generalizing the idea of coloring. Some of these ways involve different choices for the elements that are used for colors.

For an integer $k \geq 2$, the set $\mathbf{Z}_k$ of integers modulo k consists of the elements $0, 1, 2, \ldots, k - 1$, where addition in $\mathbf{Z}_k$ is defined for $a, b \in \mathbf{Z}_k$ by $a + b = c$ if $0 \leq c \leq k - 1$ and $c \equiv a + b \pmod{k}$. For example, $1 + 1 = 0$ in $\mathbf{Z}_2$, $2 + 3 = 1$ in $\mathbf{Z}_4$ and $3 + 4 = 2$ in $\mathbf{Z}_5$.

For a nontrivial connected graph G, let $c : V(G) \to \mathbf{Z}_k$ be a vertex coloring where adjacent vertices of G can be assigned the same color. By the **color sum** $\sigma(v)$ of a vertex v in G is meant the sum in $\mathbf{Z}_k$ of the colors of the vertices adjacent to v, that is,

$$\sigma(v) = \sum_{u \in N(v)} c(u).$$

The coloring c is called a **modular k-coloring** if $\sigma(x) \neq \sigma(y)$ for each pair x, y of adjacent vertices of G. Thus every two adjacent vertices of G are distinguished by the fact that they have different color sums. The **modular chromatic number** $mc(G)$ of a graph G is the minimum k for which G has a modular k-coloring.

Theorem 10.23 *Every nontrivial connected graph has a modular k-coloring for some integer $k \geq 2$.*

Proof. Let G be a nontrivial connected graph with $V(G) = \{v_1, v_2, \ldots, v_n\}$. Define a coloring c of G by $c(v_i) = 2^{i-1}$ for $1 \leq i \leq n$. Let

$$k = \sum_{i=1}^{n-1} 2^i = 2\left(2^{n-1} - 1\right).$$

Considering $c : V(G) \to \mathbf{Z}_k$, it follows that $1 \leq \sigma(v_i) \leq k$ for all i $(1 \leq i \leq n)$ and $\sigma(v_i) \neq \sigma(v_j)$ when v_i and v_j are adjacent. Hence c is a modular k-coloring of G. ∎

As a consequence of Theorem 10.23, we have the following.

Corollary 10.24 *For every nontrivial connected graph G, the modular chromatic number of G exists.*

If $mc(G) = k$ for a nontrivial connected graph G, then every modular k-coloring c of G results in a vertex coloring using the elements of $\mathbf{Z}_k$ as colors and such that every two adjacent vertices are colored differently. Therefore, this is a proper coloring of G using k colors. From this, it follows that

$$\chi(G) \leq mc(G)$$

for every nontrivial connected graph G.

For example, the modular chromatic number of the graph $G = P_4 \times P_2$ in Figure 10.25 is 2. The color assigned to a vertex v is placed within the circle representing v and $\sigma(v)$ is placed next to the vertex.

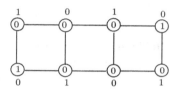

Figure 10.25: A modular coloring of $P_4 \times P_2$

That $mc(P_4 \times P_2) = 2$ may not be surprising since $P_4 \times P_2$ is bipartite. However, the graph G of Figure 10.25 is also bipartite but $mc(G) \neq 2$.

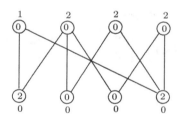

Figure 10.26: A modular coloring of a bipartite graph

While each nontrivial path is a tree with modular chromatic number 2, not every tree has modular chromatic number 2. We show for the tree T in Figure 10.27(a) that $mc(T) = 3$. Assume, to the contrary, that $mc(T) = 2$. Then there exists a modular 2-coloring c of T. Because of the symmetry of the structure of T, the color sums of the vertices of T are those shown in Figure 10.27(b). Since $\sigma(w_2) = \sigma(v_8) = 1$, it follows that $c(v_5) = c(v_7) = 1$. This, however, contradicts the fact that $\sigma(v_6) = 1$. Hence $mc(T) \neq 2$ and so $mc(T) = 3$. A modular 3-coloring of T is shown in Figure 10.27(c). On the other hand, every nontrivial tree has modular chromatic number 2 or 3.

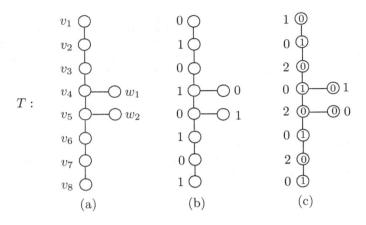

Figure 10.27: A tree T with $mc(T) = 3$

Theorem 10.25 *If T is a nontrivial tree, then $mc(T) = 2$ or $mc(T) = 3$.*

The modular chromatic number of each cycle is also either 2 or 3. However, it may be a bit surprising which cycles have modular chromatic number 3 (see Exercise 10.25).

Theorem 10.26 *For each integer $n \geq 3$,*

$$mc(C_n) = \begin{cases} 2 & \text{if } n \equiv 0 \ (\text{mod } 4) \\ 3 & \text{otherwise.} \end{cases}$$

Exercises for Section 10.5

10.24 Prove that each nontrivial path has modular chromatic number 2.

10.25 For each $n \geq 3$, prove that $mc(C_n) = 2$ if $n \equiv 0$ (mod 4) and $mc(C_n) = 3$ if $n \not\equiv 0$ (mod 4).

10.26 Prove that if G is a complete multipartite graph, then $mc(G) = \chi(G)$.

Chapter 11

Ramsey Numbers

11.1 The Ramsey Number of Graphs

Except for a 3-year period during World War II, the William Lowell Putnam mathematical competition for undergraduates has taken place every year since 1938. This exam, administered by the Mathematical Association of America, consists of (since 1962) twelve challenging mathematical problems. This competition was designed to stimulate a healthy rivalry in colleges and universities throughout the United States and Canada. The 1953 exam contained the following problem.

Problem A2 *The complete graph with 6 points (vertices) and 15 edges has each edge colored red or blue. Show that we can find 3 points such that the 3 edges joining them are the same color.*

This problem concerns the topic of Ramsey numbers in graph theory, which is named for Frank Plumpton Ramsey, who was born on February 22, 1903 in Cambridge, Cambridgeshire, England. Ramsey entered Winchester College in 1915. After completing his education there in 1920, he went to Trinity College, Cambridge on a scholarship to study mathematics. In 1924 he was elected as a Fellow of King's College, Cambridge, an especially notable honor since he never attended King's College.

In 1925 Ramsey published his first major work: "The Foundations of Mathematics." It was Ramsey's intent to improve on *Principia Mathematica* by Bertrand Russell and Alfred North Whitehead. His second paper "On a problem of formal logic" was presented to the London Mathematical Society. It was this paper that was to lead to concepts and a theory that bear his name.

In addition to mathematics, Ramsey was deeply interested in economics and philosophy. His work on economics included probability, the theory of taxation and optimal saving. Philosophy was his main interest, however. He worked so intensely that he only studied four hours per day, which left him time to do other things he enjoyed: tennis, walking, listening to music. His promising future ended abruptly at age 26 when he died on January 19, 1930.

Although Frank Ramsey proved an even more general theorem, we state a restricted version that is more closely connected to our investigation of graphs.

Theorem 11.1 (Ramsey's Theorem) *For any $k + 1 \geq 3$ positive integers t, n_1, n_2, ..., n_k, there exists a positive integer n such that if each of the t-element subsets of the set $\{1, 2, ..., n\}$ is colored with one of the k colors $1, 2, ..., k$, then for some integer i with $1 \leq i \leq k$, there is a subset S of $\{1, 2, ..., n\}$ containing n_i elements such that every t-element subset of S is colored i.*

To understand how Ramsey's Theorem ties in with graph theory, suppose that $\{1, 2, ..., n\}$ is the vertex set of the complete graph K_n. Let's see what Ramsey's Theorem says when $t = 1$. According to Ramsey's Theorem, there is a positive integer n such that if each of the 1-element subsets of the set $\{1, 2, ..., n\}$, that is, if each of the vertices of K_n, is colored with one of the k colors $1, 2, ..., k$, then there are n_i vertices colored i for some integer i ($1 \leq i \leq k$). This is simply a variation of the Pigeonhole Principle (see Appendix 1). In fact, the integer

$$n = 1 + \sum_{i=1}^{k}(n_i - 1)$$

satisfies the condition.

What Ramsey's Theorem says when $t = 2$ is considerably more intriguing. In this case, each 2-element subset of the set $\{1, 2, ..., n\}$ is assigned one of the colors $1, 2, ..., k$, which can be interpreted as coloring the edges of the complete graph K_n. The statement of Ramsey's Theorem in this case is now given.

Theorem 11.2 (Ramsey's Theorem) *For any $k \geq 2$ positive integers n_1, n_2, ..., n_k, there exists a positive integer n such that if each edge of K_n is colored with one of the colors $1, 2, ..., k$, then for some integer i with $1 \leq i \leq k$, there exists a complete subgraph K_{n_i} such that every edge of K_{n_i} is colored i.*

A nontechnical interpretation of Ramsey's Theorem might go something like this: Every sufficiently large structure, regardless of how disorderly it may appear to be, contains an orderly substructure of any prescribed size.

The special case of Ramsey's Theorem when there are two colors ($k = 2$) will be of particular interest to us. By a **red-blue coloring** of a graph G is meant an assignment of the colors red and blue to the edges of G, one color to each edge. Now let F be a graph. A subgraph of G that is isomorphic to F all of whose edges are colored red is called a **red** F. If all of the edges of a subgraph of G that is isomorphic to F are colored blue, then the subgraph is called a **blue** F. Let F_1 and F_2 be two nonempty graphs and let K_n be the complete graph of order n for some integer n. Now let there be given a red-blue coloring of K_n.

It's possible that a red F_1 has been produced in K_n or a blue F_2 or both – or neither!

By Ramsey's theorem then, if we begin with two complete graphs K_s and K_t and color the edges of a sufficiently large complete graph K_n red or blue in any manner whatsoever, then we must have either a red K_s or a blue K_t. Therefore, if F_1 is a graph of order s and F_2 is a graph of order t, then in K_n we must have either a red F_1 or a blue F_2.

For two nonempty graphs F_1 and F_2, the **Ramsey number** $r(F_1, F_2)$ is defined as the smallest positive integer n such that if every edge of K_n is colored red or blue in any manner whatsoever, then either a red F_1 or a blue F_2 is produced. From what was mentioned above, the Ramsey number $r(F_1, F_2)$ is defined.

In order to show that $r(F_1, F_2) = n$, say, two statements must be verified:

(1) *every* red-blue coloring of K_n contains either a red F_1 or a blue F_2 (which shows that $r(F_1, F_2) \leq n$) and

(2) there exists some red-blue coloring of K_{n-1} having neither a red F_1 nor a blue F_2 (which shows that $r(F_1, F_2) \geq n$).

We illustrate this by determining the Ramsey number $r(K_3, K_3)$, which is directly related to the problem in the 1953 Putnam competition mentioned earlier.

Example 11.3 $r(K_3, K_3) = 6$.

Solution. Let there be given a red-blue coloring of K_6. Consider some vertex v_1 of K_6. Since v_1 is incident with five edges, it follows by the Pigeonhole Principle that at least three of these five edges are colored the same, say red. Suppose that $v_1 v_2, v_1 v_3, v_1 v_4$ are red edges, as shown in Figure 11.1. If any of the edges $v_2 v_3, v_2 v_4$ and $v_3 v_4$ is colored red, then we have a red K_3; otherwise, all of these edges are colored blue and a blue K_3 is formed. Hence $r(K_3, K_3) \leq 6$.

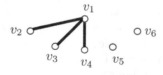

Figure 11.1: Three red edges in K_6

To verify that $r(K_3, K_3) \geq 6$, we must show that there exists a red-blue coloring of K_5 that produces neither a red K_3 nor a blue K_3. Suppose that $V(K_5) = \{v_1, v_2, \ldots v_5\}$. Define a red-blue coloring of K_5 by coloring each edge of the 5-cycle $(v_1, v_2, \ldots, v_5, v_1)$ red and the remaining edges blue, as shown in Figure 11.2, where the red edges are drawn in bold. Since this red-blue coloring produces neither a red K_3 nor a blue K_3, it follows that $r(K_3, K_3) \geq 6$. $\diamond$

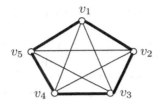

Figure 11.2: A red-blue coloring of K_5
that avoids a red K_3 and a blue K_3

For two nonempty graphs F_1 and F_2, there is an important observation concerning the Ramsey number $r(F_1, F_2)$. Suppose that F_1 has order n_1 and F_2 has order n_2 and that $\max\{n_1, n_2\} = n_1$, say. If we were to color all of the edges of K_{n_1-1} red, then no red F_1 can be produced because K_{n_1-1} doesn't have enough vertices. No blue F_2 can be produced either because no edges of K_{n_1-1} are colored blue. Therefore,

$$r(F_1, F_2) \geq n_1 = \max\{n_1, n_2\}.$$

That is, the Ramsey number $r(F_1, F_2)$ of two graphs F_1 and F_2 is always at least as large as the larger of the orders of F_1 and F_2.

We now determine the Ramsey numbers $r(K_2, K_t)$ for every integer $t \geq 2$.

Example 11.4 *For every integer $t \geq 2$,*

$$r(K_2, K_t) = t.$$

Solution. Since t is the maximum of the orders of K_2 and K_t, it follows that $r(K_2, K_t) \geq t$. Now, let there be given a red-blue coloring of K_t. If any edge of K_t is colored red, then a red K_2 is produced. Otherwise, all edges of K_t are colored blue and a blue K_t is produced. Therefore, $r(K_2, K_t) \leq t$.   ◇

Let F_1 and F_2 be two nonempty graphs and suppose that $r(F_1, F_2) = n$. Therefore, n is the smallest positive integer for which *every* red-blue coloring of K_n produces either a red F_1 or a blue F_2. Since the actual colors used when discussing Ramsey numbers are irrelevant, n is also the smallest positive integer such that every red-blue coloring of K_n produces either a blue F_1 or a red F_2 (or equivalently, a red F_2 or a blue F_1). Therefore,

$$r(F_1, F_2) = r(F_2, F_1);$$

that is, the order of the graphs F_1 and F_2 in $r(F_1, F_2)$ doesn't matter. In particular, $r(K_t, K_2) = t$ for $t \geq 2$.

In Example 11.3, it was shown that $r(K_3, K_3) = 6$. This Ramsey number has a well-known popular interpretation:

How many people must be present at a party to be guaranteed that there are three mutual acquaintances or three mutual strangers?

For any gathering of n people, say, we construct the complete graph K_n whose vertices are the people and where a red edge joins two vertices (people) if the two people are acquaintances and a blue edge joins two vertices (people) if the two people are strangers. That is, the answer to this problem is the Ramsey number $r(K_3, K_3)$, which we have seen is 6. How many people must be present at the party to be guaranteed that there are three mutual acquaintances or four mutual strangers?

Example 11.5 $r(K_3, K_4) = 9$.

Solution. Let there be given a red-blue coloring of $G = K_9$. We show that there is either a red K_3 or a blue K_4. First, observe that it cannot occur that every vertex of K_9 is incident with exactly three red edges; for otherwise, the subgraph of K_9 induced by the red edges of K_9 is 3-regular of order 9, but there is no such graph. Therefore, there are two possibilities.

Case 1. There exists a vertex v_1 that is incident with 4 red edges. Let v_1v_2, v_1v_3, v_1v_4 and v_1v_5 be red edges in K_9. If any two of the vertices v_2, v_3, v_4 and v_5 are joined by a red edge, then a red K_3 is produced; otherwise, every two of the vertices v_2, v_3, v_4 and v_5 are joined by a blue edge, producing a blue K_4.

Case 2. There exists a vertex v_1 that is incident with 6 blue edges. Let $v_1v_2, v_1v_3, v_1v_4, v_1v_5, v_1v_6$ and v_1v_7 be blue edges and let

$$S = \{v_2, v_3, v_4, v_5, v_6, v_7\}.$$

Since $r(K_3, K_3) = 6$, the subgraph $H = G[S] = K_6$ contains either a red K_3 or a blue K_3. If H contains a red K_3, so does K_9. If H contains a blue K_3, then K_9 contains a blue K_4.

Therefore, $r(K_3, K_4) \le 9$. Consider the red-blue coloring of K_8 in which the red and blue subgraphs of K_8 are shown in Figures 11.3(a) and 11.3(b), respectively. That is, the blue subgraph is the graph C_8^2 of C_8 (see Section 5.5). Since there is neither a red K_3 nor a blue K_4, it follows that $r(K_3, K_4) \ge 9$. $\Diamond$

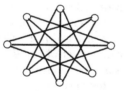

(a) red subgraph of K_8

(b) blue subgraph of K_8

Figure 11.3: A red-blue coloring of K_8

Recall that the complement $\overline{G}$ of a graph G is that graph having the same vertex set as G and such that two vertices are adjacent in $\overline{G}$ if and only if these vertices are not adjacent in G. If G has order n, then $\overline{G}$ has order n as well.

Furthermore, every edge of K_n belongs either to G or to $\overline{G}$. If we think of the edges of G as being colored red and the edges of $\overline{G}$ as being colored blue, then we have a reformulation of the Ramsey number of two graphs. Let F_1 and F_2 be two nonempty graphs. The **Ramsey number** $r(F_1, F_2)$ is the smallest positive integer n such that if G is any graph of order n, then G contains a subgraph isomorphic to F_1 or $\overline{G}$ contains a subgraph isomorphic to F_2. Despite the fact that Ramsey numbers can be studied in terms of graphs and their complements, we will continue study them by means of red-blue colorings of complete graphs.

We know that $r(K_2, K_t) = t$ for every integer $t \geq 2$, $r(K_3, K_3) = 6$ and $r(K_3, K_4) = 9$. The numbers $r(K_s, K_t)$ were the first Ramsey numbers to be studied extensively and, consequently, are often referred to as the **classical Ramsey numbers**. In fact, $r(K_s, K_t)$ is commonly expressed as $r(s, t)$. Despite the fact that the Ramsey numbers $r(K_s, K_t)$ have been studied for decades, relatively few are known for $s, t \geq 3$. Indeed, the only known Ramsey numbers $r(K_s, K_t)$ for $3 \leq s \leq t$ are

$$
\begin{array}{lll}
r(K_3, K_3) = 6 & r(K_3, K_6) = 18 & r(K_3, K_9) = 36 \\
r(K_3, K_4) = 9 & r(K_3, K_7) = 23 & r(K_4, K_4) = 18 \\
r(K_3, K_5) = 14 & r(K_3, K_8) = 28 & r(K_4, K_5) = 25.
\end{array}
$$

In particular, $r(K_5, K_5)$ is not known. However, it is known that

$$43 \leq r(K_5, K_5) \leq 49.$$

For many Ramsey numbers $r(K_s, K_t)$ whose values are unknown, bounds are known that are often far apart. For example,

$$798 \leq r(K_{10}, K_{10}) \leq 12,677.$$

Recall that a set S of vertices in a graph G is independent if every two vertices in S are nonadjacent in G. In particular, if S is a set of s vertices of G such that $G[S] = K_s$, then S is an independent set in the graph $\overline{G}$. Expressed in terms of graphs and their complements, the Ramsey number $r(K_s, K_t)$ is the smallest positive integer n such that for every graph G of order n, either K_s is a subgraph of G or K_t is a subgraph of $\overline{G}$. Equivalently, the Ramsey number $r(K_s, K_t)$ is the smallest positive integer n such that every graph of order n contains either a complete subgraph of order s or an independent set of t vertices. For example, since $r(K_3, K_3) = 6$ and $r(K_3, K_4) = 9$, if G is a graph of order 6, 7 or 8 that does not contain a triangle, then G must contain an independent set of three vertices; while if G is a graph of order 9 that does not contain a complete subgraph of order 4, then G must contain an independent set of three vertices.

We now look at some Ramsey numbers $r(F_1, F_2)$ where F_1 and F_2 are not both complete. Our next example is to determine the Ramsey number $r(P_3, K_3)$. The two graphs P_3 and K_3 are shown in Figure 11.4. How do we begin to determine $r(P_3, K_3)$? Since the maximum of the orders of P_3 and K_3 is 3, it follows that $r(P_3, K_3) \geq 3$. If we color one edge of K_3 red and the other two edges blue, then there is neither a red P_3 nor a blue K_3; that is, we've been able to avoid both a red P_3 and a blue K_3. Therefore, $r(P_3, K_3) \geq 4$.

$$P_3: \qquad K_3:$$

Figure 11.4: Determining $r(P_3, K_3)$

On the other hand, the red-blue coloring of K_4 shown in Figure 11.5 (where, again, a bold edge represents a red edge) also avoids a red P_3 and a blue K_3. So $r(P_3, K_3) \geq 5$. If we have (great) difficulty finding a red-blue coloring of K_5 that avoids both a red P_3 and a blue K_3, then there is reason to suspect that $r(P_3, K_3) = 5$. We now give a formal argument that 5 is, in fact, the Ramsey number of these two graphs.

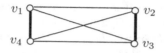

Figure 11.5: A red-blue coloring of K_4
that avoids a red P_3 and a blue K_3

Example 11.6 $r(P_3, K_3) = 5$.

Solution. First we show that $r(P_3, K_3) \geq 5$. As we saw, the red-blue coloring of K_4 shown in Figure 11.5 avoids both a red P_3 and a blue K_3 and so $r(P_3, K_3) \geq 5$.

It remains therefore to show that $r(P_3, K_3) \leq 5$. Let a red-blue coloring of K_5 be given. Consider a vertex v_1 in K_5. If v_1 is incident with two red edges, then a red P_3 is produced. Otherwise, v_1 is incident with at most one red edge. So there are three blue edges incident with v_1, say v_1v_2, v_1v_3 and v_1v_4 are blue edges. If there is a blue edge joining any two of the vertices v_2, v_3 and v_4, a blue K_3 is produced. Otherwise, v_2v_3 and v_3v_4 are red edges, producing a red P_3. Therefore, $r(P_3, K_3) \leq 5$. $\diamond$

Of course, $P_3 = K_{1,2}$ and so $r(K_{1,2}, K_3) = 5$. We now determine $r(K_{1,3}, K_3)$ by providing an argument different from those that we've previously given.

Example 11.7 $r(K_{1,3}, K_3) = 7$.

Solution. First we show that $r(K_{1,3}, K_3) \geq 7$. Consider the red-blue coloring of K_6 shown in Figure 11.6, where again each red edge of K_6 is drawn as a bold edge. Since the red subgraph is $2K_3$ and the blue subgraph is $K_{3,3}$, there is neither a red $K_{1,3}$ nor a blue K_3 in this coloring and so $r(K_{1,3}, K_3) \geq 7$.

Next we show that $r(K_{1,3}, K_3) \leq 7$. Assume, to the contrary, that there is a red-blue coloring of K_7 that produces neither a red $K_{1,3}$ nor a blue K_3. Consider a vertex v_1 in K_7. Then at most two red edges are incident with v_1 and so at least four blue edges are incident with v_1, say v_1v_i ($2 \leq i \leq 5$) are blue edges. If any edge joining two of the vertices in $\{v_2, v_3, v_4, v_5\}$ is colored blue, then we

K_6 :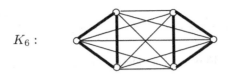

Figure 11.6: A red-blue coloring of K_6
that avoids a red $K_{1,3}$ and a blue K_3

have a blue K_3, which is a contradiction. Hence all edges joining any two of the vertices in $\{v_2, v_3, v_4, v_5\}$ are colored red. In particular, the edges v_2v_3, v_2v_4 and v_2v_5 are colored red and so we have a red $K_{1,3}$, which is a contradiction.   ◇

The two Ramsey numbers that we have just determined are of the type $r(F_1, F_2)$, where F_1 is a tree and F_2 is a complete graph. Remarkably, in the very first issue of the *Journal of Graph Theory* (in 1977), Vašek Chvátal established a simple formula for $r(F_1, F_2)$, where F_1 is *any* tree and F_2 is *any* complete graph. Although the proof is a bit intricate, it is simpler than one might expect for such a general result. Recall, by Theorem 4.9, that if G is a graph such that $\deg v \geq k - 1$ for every vertex v of G and T is a tree of order k, then the graph G contains a subgraph isomorphic to T.

Theorem 11.8 *For every tree T_m of order $m \geq 2$ and every integer $n \geq 2$,*

$$r(T_m, K_n) = (m - 1)(n - 1) + 1.$$

Proof. First, we show that $r(T_m, K_n) \geq (m - 1)(n - 1) + 1$. Let there be given a red-blue coloring of the complete graph $K_{(m-1)(n-1)}$ of order $(m - 1)(n - 1)$ such that the resulting red subgraph is $(n - 1)K_{m-1}$, that is, the red subgraph consists of $n - 1$ copies of K_{m-1}. Since each component of the red subgraph has order $m - 1$, it contains no connected subgraph of order greater than $m - 1$. In particular, there is no red tree of order m. The blue subgraph is then the complete $(n-1)$-partite graph $K_{m-1,m-1,...,m-1}$, where every partite set contains exactly $m - 1$ vertices. There is no blue K_n either. Since this red-blue coloring avoids every red tree of order m and a blue K_n, it follows that

$$r(T_m, K_n) \geq (m - 1)(n - 1) + 1.$$

Next, we show that $r(T_m, K_n) \leq (m-1)(n-1)+1$. We proceed by induction on the order of the complete graph K_n. First, we let $n = 2$ and show that

$$r(T_m, K_2) \leq (m - 1)(2 - 1) + 1 = m.$$

Let there be given a red-blue coloring of K_m. If any edge of K_m is colored blue, then a blue K_2 is produced. Otherwise, every edge of K_m is colored red and a red T_m is produced. Thus $r(T_m, K_2) \leq m$. Therefore, the inequality

$$r(T_m, K_n) \leq (m - 1)(n - 1) + 1$$

holds when $n = 2$. Assume, for every tree T_m of order m and an integer $k \geq 2$, that

$$r(T_m, K_k) \leq (m-1)(k-1) + 1.$$

In particular, this says that every red-blue coloring of $K_{(m-1)(k-1)+1}$ contains either a red T_m or a blue K_k. We show that

$$r(T_m, K_{k+1}) \leq (m-1)k + 1.$$

Let there be given a red-blue coloring of $K_{(m-1)k+1}$. We consider two cases.

Case 1. There exists a vertex v_1 in the complete graph $K_{(m-1)k+1}$ that is incident with $(m-1)(k-1) + 1$ blue edges. Suppose that $v_1 v_i$ is a blue edge for $2 \leq i \leq (m-1)(k-1) + 2$ and let

$$S = \{v_i : 2 \leq i \leq (m-1)(k-1) + 2\}.$$

Consider the subgraph

$$H = G[S] = K_{(m-1)(k-1)+1}.$$

By the induction hypothesis, H contains either a red T_m or a blue K_k. If H contains a red T_m, then so does $K_{(m-1)k+1}$. Otherwise H contains a blue K_k. Since v_1 is joined to every vertex of H by a blue edge, it follows that there is a blue K_{k+1} in $K_{(m-1)k+1}$

Case 2. Every vertex of $K_{(m-1)k+1}$ is incident with at most $(m-1)(k-1)$ blue edges. So every vertex of $K_{(m-1)k+1}$ is incident with at least $m - 1$ red edges. Thus the red subgraph of $K_{(m-1)k+1}$ has minimum degree at least $m - 1$. By Theorem 4.9, this red subgraph contains a red T_m. Therefore, $K_{(m-1)k+1}$ contains a red T_m. ∎

As a consequence of Theorem 11.8, for every positive integer s and every integer $t \geq 2$, it follows that

$$r(K_{1,s}, K_t) = s(t-1) + 1.$$

The next example deals with a Ramsey number $r(F_1, F_2)$, where neither F_1 nor F_2 is complete.

Example 11.9 $r(K_{1,3}, C_4) = 6$.

Proof. Since the red-blue coloring of K_5 shown in Figure 11.7 produces neither a red $K_{1,3}$ nor a blue C_4, it follows that $r(K_{1,3}, C_4) \geq 6$.

It remains to verify that $r(K_{1,3}, C_4) \leq 6$. Let a red-blue coloring of K_6 be given, where we denote the vertices of K_6 by $v_1, v_2, \ldots, v_6$. Since $r(K_3, K_3) = 6$, either a red K_3 or a blue K_3 is produced. We consider these two cases.

Case 1. There is a red K_3 in K_6. We may assume that v_1, v_2 and v_3 are the vertices of a red K_3. If there is no red $K_{1,3}$, then every edge joining a vertex in $\{v_1, v_2, v_3\}$ and a vertex in $\{v_4, v_5, v_6\}$ is colored blue, producing a blue C_4.

Figure 11.7: A red-blue coloring of K_5
that avoids a red $K_{1,3}$ and a blue C_4

Case 2. *There is a blue* K_3 *in* K_6. Assume that v_1, v_2 and v_3 are the vertices
of a blue K_3. If some vertex in $\{v_4, v_5, v_6\}$ is joined to two vertices in $\{v_1, v_2, v_3\}$
by blue edges, then a blue C_4 is produced. If some vertex in $\{v_4, v_5, v_6\}$ is joined
to all three vertices in $\{v_1, v_2, v_3\}$ by red edges, then a red $K_{1,3}$ is produced.
Thus, we may assume that each vertex in $\{v_4, v_5, v_6\}$ is joined to one vertex in
$\{v_1, v_2, v_3\}$ by a blue edge and two vertices in $\{v_1, v_2, v_3\}$ by red edges. If any of
the edges v_4v_5, v_4v_6 and v_5v_6 is red, then there is a red $K_{1,3}$. So we may assume
that v_4, v_5 and v_6 are the vertices of a blue K_3. Then any two blue edges that
join a vertex in $\{v_1, v_2, v_3\}$ and a vertex in $\{v_4, v_5, v_6\}$ lie on a blue C_4.   ◇

Exercises for Section 11.1

11.1 Let F_1 be a graph of order s and F_2 a graph of order t. Prove that
$r(F_1, F_2) \leq r(K_s, K_t)$.

11.2 It was stated that $r(K_4, K_5) = 25$. Show that if G is a graph of order 25
that does not contain K_4 as a subgraph, then G contains five vertices no
two of which are adjacent.

11.3 Show that $r(K_4, K_4) \leq 18$.

11.4 Determine $r(P_3, P_3)$.

11.5 Determine $r(2K_2, P_3)$.

11.6 Determine $r(K_{1,3}, P_3)$.

11.7 Determine $r(2K_2, 2K_2)$.

11.8 Determine $r(2K_2, 3K_2)$.

11.9 Determine $r(K_{1,3}, K_{1,3})$.

11.10 Determine $r(P_3, P_3 \cup P_2)$.

11.11 Determine $r(K_{1,4}, K_{1,4})$.

11.12 Determine $r(P_4, P_4)$.

11.13 Determine $r(C_4, C_4)$.

11.14 Prove that $r(K_{1,3}, C_4) = 6$ by a method different from that used in Example 11.9, namely, by (1) assuming that there is a red-blue coloring of K_6 that has no red $K_{1,3}$ and (2) using the fact that the only 3-regular graphs of order 6 are $K_{3,3}$ and $K_3 \times K_2$.

11.15 Let G be a complete graph of order $r(K_s, K_t) - 1$, where $s, t \geq 2$. Prove that every red-blue coloring of G produces either a red K_{s-1} or a blue K_{t-1}.

11.16 Prove that $r(K_3, K_n) \leq \binom{n+1}{2} = \frac{n^2+n}{2}$ for every integer $n \geq 2$. [Hint: Observe that $\binom{n+1}{2} = \binom{n}{2} + n$.]

11.17 According to Exercise 11.16, every red-blue coloring of a complete graph of order $\binom{n+1}{2}$ results in a red K_3 or a blue K_n. Use this fact and the combinatorial identity $\binom{n+1}{3} + \binom{n+1}{2} = \binom{n+2}{3}$ to obtain a result involving Ramsey numbers. [Note: For integers k and n with $0 \leq k \leq n$, $\binom{n}{k} = \frac{n!}{k!(n-k)!}$ and $\binom{n}{k} = 0$ if $n < k$.]

11.2   Turán's Theorem

For integers $s, t \geq 2$, suppose that $r(K_s, K_t) = n$. Then every graph of order n contains either an independent set of s vertices or a complete subgraph of order t. Therefore, every graph of order n (or more) that fails to contain an independent set of s vertices must contain K_t as a subgraph. There is even a more natural sufficient condition for a graph of a certain order to contain a complete graph of a specified order as a subgraph.

Suppose that G is a graph of order $n \geq 3$ and size m. Then $0 \leq m \leq \binom{n}{2}$. Of course, if $m = \binom{n}{2}$, then $G = K_n$. Consequently, for any graph H of order n, the graph G contains a subgraph isomorphic to H. Therefore, if G has enough edges, then G has a subgraph isomorphic to H. In particular, if G has enough edges, then G contains a triangle. But what is "enough" in this case? Of course, if $n = 3$, then G needs three edges to guarantee that G has a triangle. Let's next answer this question for $n = 4$.

The 4-cycle C_4 in Figure 11.8 certainly doesn't contain a triangle. Indeed, C_4 is the only graph of order 4 and size 4 that doesn't contain a triangle. Consequently, if a graph G of order 4 has four edges, then there is no guarantee that G contains a triangle. There is only one graph of order 4 and size 5 and this graph, denoted by G_1, is shown in Figure 11.8. Since G_1 has a triangle, it follows that every graph of order 4 whose size is at least 5 contains a triangle.

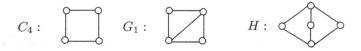

Figure 11.8: The graphs C_4, G_1 and H

Let's next consider graphs of order 5. Notice that the graph $H = K_{2,3}$ of Figure 11.8 has order 5 and size 6 but does not contain a triangle. But what about graphs of order 5 and size 7? Let G be such a graph where $uv \in E(G)$ and x, y and z are the remaining three vertices of G. If G does not contain a triangle, then each vertex in $S = \{x, y, z\}$ is adjacent to at most one of u and v. Furthermore, $G[S]$ contains at most two edges. Thus the size of G is at most 6, a contradiction. That is, every graph of order 5 and size 7 contains a triangle. The graph $H = K_{3,3}$ has order 6 and size 9 and does not contain a triangle. It can be shown that every graph of order 6 and size 10 contains a triangle. Therefore, if G is a graph of order n where $3 \le n \le 6$ and size $m > \lfloor n^2/4 \rfloor$, then G contains a triangle. In 1907 the Dutch mathematician Willem Mantel showed that this is true in general.

Theorem 11.10 *If G is a graph of order $n \ge 3$ and size $m > \lfloor n^2/4 \rfloor$, then G contains a triangle.*

Proof. Suppose that the theorem is false. Then there is a smallest integer n for which the statement is false. So there is some graph G of order n and size $\lfloor n^2/4 \rfloor + 1$ that contains no a triangle.

Let $uv \in E(G)$ and let $v_1, v_2, \ldots, v_{n-2}$ be the remaining vertices of G. Since G contains no triangle, at most one u and v is adjacent to v_i for each i with $1 \le i \le n - 2$. This implies that $\deg u + \deg v \le n$. Let $H = G - u - v$. Then H has order $n - 2$ and size m where

$$m > \lfloor n^2/4 \rfloor + 1 - (\deg u + \deg v - 1)$$

and so

$$m \ge \lfloor n^2/4 \rfloor - n + 2 = \lfloor (n-2)^2/4 \rfloor + 1.$$

Consequently, H contains a triangle, as does G. This is a contradiction. ∎

If n is even in Theorem 11.10, say $n = 2k$, then every graph of order n and size $m > k^2$ contains a triangle. This bound can't be improved because the graph $K_{k,k}$ has order $2k$, size k^2 and contains no triangles. In fact, the graph $K_{k,k}$ is the only graph of order $2k$ and size k^2 containing no triangle. The proof of this fact is similar to that of the preceding result.

Theorem 11.11 *For every integer $k \ge 2$, the only graph of order $2k$ and size k^2 that contains no triangle is $K_{k,k}$.*

Proof. Suppose that the theorem is false. Then there is a smallest integer $k \geq 2$ such that there exists a graph $G \neq K_{k,k}$ having order $2k$ and size k^2 that contains no triangle. Since $K_{2,2}$ is the only graph of order 4 and size 4 that contains no triangle, it follows that $k \geq 3$.

Let uv be an edge of G. Each of the remaining $2k - 2$ vertices different from u and v is adjacent to at most one of u and v. Thus at most $2k - 2$ edges are adjacent to uv. Hence, the graph $H = G - u - v$ contains at least $k^2 - (2k - 1) = (k - 1)^2$ edges. If H contains at least $(k - 1)^2 + 1$ edges, then it follows by Theorem 11.10 that H contains a triangle and so G contains a triangle, which is impossible. Therefore, H must contain exactly $(k - 1)^2$ edges. Since H contains no triangles, $H = K_{k-1,k-1}$. Let U_1 and V_1 be the partite sets of H. Furthermore, every vertex of $V(H) = U_1 \cup V_1$ is adjacent to exactly one of u and v. If u (or v) is adjacent to both a vertex of U_1 and a vertex of V_1, then G contains a triangle, which is impossible. Thus u is adjacent only to the vertices in U_1 or to the vertices in V_1; while v is adjacent only to the vertices in the other set, say u is adjacent to the vertices in V_1 and v is adjacent to the vertices in U_1. Thus $G = K_{k,k}$, whose partite sets are $U_1 \cup \{u\}$ and $V_1 \cup \{v\}$. ∎

Combining Theorems 11.10 and 11.11, we have the following.

Theorem 11.12 *Let G be a graph of order $2k \geq 4$. If the size of G is at least $k^2 + 1$ or if the size of G is k^2 and $G \neq K_{k,k}$, then G contains a triangle.*

By proceeding in the same manner as above for graphs of odd order, the following result can be established.

Theorem 11.13 *Let G be a graph of order $2k + 1 \geq 3$. If the size of G is at least $k^2 + k + 1$ or the size of G is $k^2 + k$ and $G \neq K_{k,k+1}$, then G contains a triangle.*

Theorems 11.10–11.13 can be then combined into a single theorem.

Corollary 11.14 *Let G be a graph of order $n \geq 3$. If the size of G is at least $\left\lfloor \frac{n^2}{4} \right\rfloor + 1$ or the size of G is $\left\lfloor \frac{n^2}{4} \right\rfloor$ and $G \neq K_{\lfloor \frac{n}{2} \rfloor, \lceil \frac{n}{2} \rceil}$, then G contains a triangle.*

For example, according to Corollary 11.14, if G is a graph of order 7 and the size of G is at least 13 or the size of G is 12 and $G \neq K_{3,4}$, then G contains a triangle.

We have been discussing the number of edges required of a graph G of order $n \geq 3$ which guarantees that G contains K_3 as a subgraph. There are more general questions. For example, for integers k and n with $2 \leq k < n$, what is the smallest positive integer m such that every graph of order n and size m contains K_{k+1} as a subgraph?

While the problem for the case $k = 2$ was solved by Mantel, the origin of the more general problem goes back to Paul Turán, a mathematician we encountered earlier. Turán was born on August 28, 1910 in Budapest, Hungary. While a student, he became friends with Paul Erdős, Tibor Gallai, George Szekeres and Esther Klein. Although Turán was to become best known for his work in probabilistic and analytic number theory, what he accomplished in graph theory was to lead to the creation of the area of extremal graph theory.

In 1940, Szekeres wrote a letter to Paul Turán in which he described his unsuccessful attempt to prove a conjecture of William Burnside. After receiving the letter and thinking about it, Turán was led to the following question:

> *What is the maximum number of edges in a graph with n vertices not containing a complete subgraph with k vertices?*

Although Turán found the question interesting, he was mainly concerned with analytic number theory at that time. However, in September 1940 he was placed in a labor camp for the first time. One day, one of his comrades in the labor camp mentioned Turán by name. An officer heard this and recognized the name as that of a mathematician. This officer was able to assign tasks to Turán that were not as physical and which kept Turán outdoors. While this was occurring, the problem he thought of returned to him; however, he had no paper to explore the details of his ideas. Nevertheless, he felt great pleasure dealing with this unusual and beautiful problem. Finally, he obtained a complete solution, which gave Turán a sense of elation. Paul Turán spent some 32 months during World War II in a labor camp. From 1949 he was a professor at Budapest University. He died on September 26, 1976.

We consider Turán's question in the following form: For integers k and n with $2 \le k < n$, what is the maximum size of a graph of order n that fails to contain K_{k+1} as a subgraph? First we show, for any two integers k and n with $2 \le k < n$, that among the graphs of order n which fail to contain K_{k+1} as a subgraph, one of those of maximum size is a k-partite graph. Recall that a graph G is k-partite if $V(G)$ can be partitioned into k independent subsets. Before presenting this result, let's recall some other facts.

If G is a graph of order n and size m, then the First Theorem of Graph Theory tells us that

$$\sum_{v \in V(G)} \deg v = 2m.$$

Let V_1 be a nonempty proper set of $V(G)$ and let $G_1 = G[V_1]$. Suppose that G_1 has size m_1 and that there are m_2 edges joining the vertices in V_1 and vertices not in V_1. Therefore, the sum $\sum_{v \in V_1} \deg_G v$ counts each edge in G_1 twice and each edge joining a vertex of V_1 and a vertex of $V(G) - V_1$ once; that is

$$\sum_{v \in V_1} \deg_G v = 2m_1 + m_2.$$

This is illustrated in Figure 11.9.

Theorem 11.15 *Let k and n be integers with $2 \le k < n$. Among all graphs of order n that do not contain K_{k+1} as a subgraph, at least one of those having maximum size is a k-partite graph.*

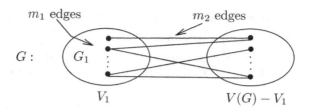

$G:$ G_1 m_1 edges m_2 edges

V_1 $V(G) - V_1$

Figure 11.9: Summing the degrees of the vertices in a subgraph

Proof. Suppose that the result is false. Then there is a smallest integer $k \geq 2$ and an integer $n > k$ such that among the graphs of order n and not containing K_{k+1} as a subgraph, none of those having maximum size m is k-partite. By Corollary 11.14, $k \geq 3$.

Let G be a graph of order n and size m that does not contain K_{k+1} as a subgraph. Therefore, by assumption, G is not a k-partite graph. Let v be a vertex of maximum degree in G, say $\deg v = \Delta$, and let F be the subgraph induced by the neighbors of v in G, that is, $F = G[N(v)]$. Therefore, the order of F is Δ. Suppose that the size of F is s. Since G does not contain K_{k+1} as a subgraph and v is adjacent to every vertex of F, it follows that F does not contain K_k as a subgraph. By assumption, among all graphs of order Δ that do not contain K_k as a subgraph, one of those of maximum size, say s', is a $(k-1)$-partite graph F'. Thus $s' \geq s$.

Define the graph H to be the join of F' and $\overline{K}_{n-\Delta}$, that is,

$$H = F' + \overline{K}_{n-\Delta}.$$

Since F' is a $(k-1)$-partite graph, H is a k-partite graph of order n and size $s' + \Delta(n - \Delta)$ that does not contain K_{k+1} as a subgraph. Therefore, $m > s' + \Delta(n - \Delta)$. Now observe that

$$m \leq s + \sum_{v \in V(G) - V(F)} \deg_G v \leq s + \Delta(n - \Delta) \leq s' + \Delta(n - \Delta) < m,$$

which is a contradiction.   ■

Let $n \geq 3$ be an integer. For each positive integer $k \leq n$, let $t_1, t_2, \ldots, t_k$ be k integers such that

$$n = t_1 + t_2 + \ldots + t_k, \ 1 \leq t_1 \leq t_2 \leq \ldots \leq t_k, \text{ and } t_k - t_1 \leq 1.$$

For every two integers k and n with $1 \leq k \leq n$, the integers $t_1, t_2, \ldots, t_k$ are unique. For example, if $n = 11$ and $k = 3$, then t_1, t_2, t_3 is $3, 4, 4$; while if $n = 14$ and $k = 6$, then $t_1, t_2, \ldots, t_6$ is $2, 2, 2, 2, 3, 3$. The complete k-partite graph $K_{t_1, t_2, \ldots, t_k}$ is the **Turán graph** $T_{n,k}$. Thus the Turán graph $T_{n,k}$ is the complete k-partite graph of order n, the cardinalities of whose partite sets differ by at most 1. The cardinality of each partite set of $T_{n,k}$ is either $\lfloor n/k \rfloor$ or $\lceil n/k \rceil$. If n/k is an integer, then $\lfloor n/k \rfloor = \lceil n/k \rceil$; while if n/k is not an integer and r is the remainder when n is divided by k, then exactly r of the partite sets of $T_{n,k}$ have cardinality $\lceil n/k \rceil$.

Theorem 11.16 *Let k and n be integers with $1 \le k \le n$ and $n \ge 3$. Among all k-partite graphs of order n, the Turán graph $T_{n,k}$ is the unique graph of maximum size.*

Proof. Among all k-partite graphs of order n, let G be one of maximum size. Suppose that the partite sets of G are $V_1, V_2, \ldots, V_k$, where $|V_i| = n_i$ $(1 \le i \le k)$ and $1 \le n_1 \le n_2 \le \ldots \le n_k$. Certainly,

$$G = K_{n_1,n_2,\ldots,n_k}.$$

Suppose that $n_k - n_1 \ge 2$. Let $H = K_{n_1+1,n_2,\ldots,n_k-1}$, that is, H can be considered to be the complete k-partite graph of order n obtained from G by moving a vertex v in the largest partite set V_k of G to the smallest partite set V_1. This transfer results in a loss of n_1 edges in G that are incident with v and a gain of $n_k - 1$ new edges in H but with no other changes. Since $n_k - n_1 \ge 2$, it follows that H has more edges then G. This contradiction implies that $n_k - n_1 \le 1$ and that G is isomorphic to the Turán graph $T_{n,k}$.   ∎

More generally, we have the following result of Turán.

Theorem 11.17 (**Turán's Theorem**) *Let k and n be integers with $2 \le k < n$. Among all graphs of order $n \ge 3$ that do not contain K_{k+1} as a subgraph, the Turán graph $T_{n,k}$ is the unique graph of maximum size.*

We mentioned earlier that for a given positive integer n and a graph F of order at most n, there is a smallest positive integer m such that every graph of order n and size m contains a subgraph isomorphic to F. Turán's theorem determines the exact value of m when $F = K_{k+1}$ for every pair k, n of integers with $2 \le k < n$.

We now look at another example of this type, namely, $F = C_n$ for $n \ge 3$. That is, we are investigating the question: How many edges must a graph of order $n \ge 3$ have to be certain that it is Hamiltonian? If $n = 3$, then the only Hamiltonian graph is K_3. For $n = 4$, four edges is not enough as the graph G_1 of Figure 11.10 shows. We have seen that there is only one graph of order 4 and size 5 (the graph G_2 of Figure 11.10) and this graph is Hamiltonian. For $n = 5$, the situation is a bit more complicated. However, the graph G_3 of order 5 and size 7 is not Hamiltonian, so the number of edges needed for a graph G of order 5 to be Hamiltonian is certainly at least 8. If turns out that 8 is the answer. Consequently, if G is a graph of order n where $3 \le n \le 5$ and size at least m, where $m = \binom{n-1}{2} + 2$, then G is Hamiltonian. This is the correct number for *every* integer $n \ge 3$. Before showing this, we need to recall a couple of things. First, by Theorem 6.6, if G is a graph of order $n \ge 3$ such that $\deg u + \deg v \ge n$ for every pair u, v of nonadjacent vertices of G, then G is Hamiltonian. Also for every integer $k \ge 2$,

$$\binom{k}{2} = 1 + 2 + \ldots + (k - 1).$$

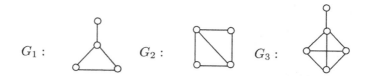

G_1 : G_2 : G_3 :

Figure 11.10: Investigating the maximum size of a non-Hamiltonian graph

Theorem 11.18 *Every graph of order $n \geq 3$ and size at least $\binom{n-1}{2} + 2$ is Hamiltonian.*

Proof. Assume that the statement is false. Then there is a smallest positive integer n for which there exists a graph G of order n and size $\binom{n-1}{2} + 2$ that is not Hamiltonian. Because the result is true for graphs of orders 3, 4 and 5, it follows that $n \geq 6$. Since G is not Hamiltonian, certainly, G is not complete.

Let u and v be any two nonadjacent vertices of G. Then the size of $H = G - u - v$ is $\binom{n-1}{2} + 2 - \deg u - \deg v$. Since the order of H is $n - 2$, the size of H cannot exceed $\binom{n-2}{2}$ and so

$$\binom{n-1}{2} + 2 - \deg u - \deg v \leq \binom{n-2}{2}.$$

Therefore,

$$\begin{aligned}
\deg u + \deg v \;\geq\; & \binom{n-1}{2} - \binom{n-2}{2} + 2 \\
= \;& [1 + 2 + \ldots + (n-2)] - [1 + 2 + \ldots + (n-3)] + 2 = n.
\end{aligned}$$

By Theorem 6.6, G is Hamiltonian, which is a contradiction. ∎

Exercises for Section 11.2

11.18 Prove Theorem 11.13: *Let G be a graph of order $2k + 1 \geq 3$. If the size of G is at least $k^2 + k + 1$ or the size of G is $k^2 + k$ and $G \neq K_{k,k+1}$, then G contains a triangle.*

11.19 (a) There is a unique graph G of order 10 and maximum size that fails to contain K_4 as a subgraph. What is G and what is its size?

   (b) What is the smallest possible integer m such that every graph of order 10 and size m contains K_4 as a subgraph.

11.20 Prove that every graph of order $n \geq 4$ and size at least $2n - 3$ contains a subdivision of $K_4 - e$ as a subgraph.

11.21 Determine the Turán graph $T_{n,k}$ when

(a) $n = 5$ and $k = 1$

(b) $n = 7$ and $k = 2$

(c) $n = 6$ and $k = 3$

(d) $n = 6$ and $k = 4$

(e) $n = k = 5$

11.22 Let $n \geq 3$ be an integer. What is the smallest positive integer m for which every graph G of order n and size m

(a) contains P_3 as a subgraph?

(b) contains P_3 as an induced subgraph?

11.23 Let $n \geq 2$ be an integer. What is the smallest positive integer m for which every graph G of order n and size m contains a vertex of degree at least k (where $1 \leq k < n$)?

11.24 Let $n \geq 2$ be an even integer. What is the smallest positive integer m for which every graph G of order n and size m contains a 1-factor?

11.25 Let $n \geq 2$ be an integer. What is the smallest positive integer m for which every graph G of order n and size m contains a Hamiltonian path. [Hint: Use Theorem 11.18].

11.26 Let $n \geq 4$ be an integer. What is the smallest positive integer m for which every graph G of order n and size m contains a cycle of each of the lengths $3, 4, \cdots, n$. [Hint: Consider Exercise 11.25.]

11.27 Use mathematical induction to prove Theorem 11.18: *Every graph of order $n \geq 3$ and size at least $\binom{n-1}{2} + 2$ is Hamiltonian.*

11.3 Exploration: Modified Ramsey Numbers

Recall for graphs F and H that the Ramsey number $r(F, H)$ is the smallest positive integer n such that for every red-blue coloring of K_n there is either a red F or a blue H. For example, we saw that $r(K_{1,3}, K_3) = 7$; that is, for every red-blue coloring of K_7, there is either a red $K_{1,3}$ or a blue K_3. Furthermore, the fact that $r(K_{1,3}, K_3) = 7$ implies that there is a red-blue coloring of K_6 that produces neither a red $K_{1,3}$ nor a blue K_3. In fact, the red-blue coloring of K_6 for which the red graph is $2K_3$ (and the blue graph is $K_{3,3}$) has this property. Since $r(F, H) = r(H, F)$ for all pairs F, H of graphs, it therefore follows that $r(K_3, K_{1,3}) = 7$, implying that every red-blue coloring of K_7 produces either a red K_3 or a blue $K_{1,3}$.

Suppose that we are given an edge coloring of a graph G (where adjacent edges may be colored the same). If the edges of some subgraph F of G are colored the same, then this is called a **monochromatic F**. In fact, if F and H are graphs such that $F \cong H$, then $r(F, H) = r(H, F) = r(F, F)$ may be stated as the smallest positive integer n such that if each edge of K_n is colored with one of two colors, then a monochromatic F results. This leads to the following definition.

For two graphs F and H, the **monochromatic Ramsey number $mr(F, H)$** is the smallest positive integer n such that if each edge of K_n is colored with one of two colors, then a monochromatic F or a monochromatic H results. Certainly,

$$mr(F, H) = mr(H, F)$$

for every two graphs F and H. Also,

$$mr(F, H) \leq r(F, H).$$

Furthermore, if $F \cong H$, then $mr(F, H) = r(F, H)$ and if $F \subseteq H$, then $mr(F, H) = r(F, F)$. Since $r(K_3, K_{1,3}) = 7$, it follows that $mr(K_3, K_{1,3}) \leq 7$.

Example 11.19 $mr(K_3, K_{1,3}) = 6$.

Solution. The red-blue coloring of K_5 for which the red subgraph of K_5 is C_5 (and so the blue subgraph is C_5 as well) has neither a monochromatic K_3 nor a monochromatic $K_{1,3}$. Therefore, $mr(K_3, K_{1,3}) \geq 6$.

Next, let there be given a red-blue coloring of K_6. Since $r(K_3, K_3) = 6$ by Example 11.3, it follows that K_6 has a red K_3 or a blue K_3, that is, a monochromatic K_3 and so $mr(K_3, K_{1,3}) = 6$. $\Diamond$

Let the vertex set of the complete graph K_n be a set of n distinct positive integers. A coloring of the edges of K_n is called a **minimum coloring** if two edges ij and $k\ell$ are colored the same if and only if $\min\{i, j\} = \min\{k, \ell\}$; while a coloring of the edges of K_n is called a **maximum coloring** if two edges ij and $k\ell$ are colored the same if and only if $\max\{i, j\} = \max\{k, \ell\}$.

If, in an edge-colored graph G, the edges of a subgraph F in G are colored differently, then F is called a **rainbow F**. A coloring of the edges of the graph K_5, with vertex set $\{1, 2, 3, 4, 5\}$, is shown in Figure 11.11. This graph contains a monochromatic triangle and a rainbow triangle, as well as a triangle with a minimum coloring and a triangle with a maximum coloring.

Paul Erdős and Richard Rado showed for a sufficiently large integer n and a complete graph of order n whose vertices are $1, 2, \ldots, n$ and whose edges are colored from the set of positive integers that there must be a complete subgraph of prescribed order that is monochromatic or rainbow or has a minimum or maximum coloring.

Theorem 11.20 *For every positive integer k, there exists a positive integer n such that if each edge of K_n with vertex set $\{1, 2, \ldots, n\}$ is colored from the set of positive integers, then there is a complete subgraph of order k that is either monochromatic or rainbow or has a minimum or maximum coloring.*

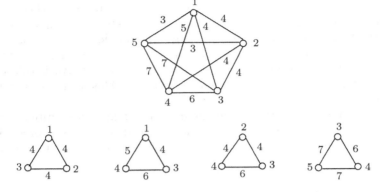

Figure 11.11: A coloring of K_5

Arie Bialostocki and William Voxman defined, for a nonempty graph F, the **rainbow Ramsey number** $RR(F)$ of F as the smallest positive integer n such that if each edge of the complete graph K_n is colored from any number of colors, then either a monochromatic F or a rainbow F is produced. Rainbow Ramsey numbers are not defined for all graphs, however.

Theorem 11.21 *Let F be a graph without isolated vertices. The rainbow Ramsey number $RR(F)$ is defined if and only if F is a forest.*

Proof. First, assume that F is not a forest. Then F contains a cycle C, of length $k \geq 3$ say. Let n be an integer with $n \geq k$. Let the vertex set of a complete graph K_n be $\{1, 2, \ldots, n\}$ and for $1 \leq i < j \leq n$, assign the color i to the edge ij. This is then a minimum coloring of K_n. We may assume that $C = (v_1, v_2, \ldots, v_k, v_1)$, where $\{v_1, v_2, \ldots, v_k\} \subseteq \{1, 2, \ldots, n\}$ and

$$\min\{v_1, v_2, \ldots, v_k\} = v_1.$$

Consequently, the edges v_1v_2 and v_1v_k are colored v_1 and the edge v_2v_3 is not colored v_1. That is, with this coloring of edges of K_n, no cycle has a monochromatic or a rainbow coloring. Therefore, if F is any graph containing a cycle, then $RR(F)$ is not defined.

Next, we verify the converse. Let F be a forest, say of order k. By Theorem 11.20, there exists a positive integer n such that if the edges of K_n are colored from the set of positive integers in any way whatsoever, then there is a complete subgraph G of order k that is either monochromatic or rainbow or has a minimum or maximum coloring. Since G contains subgraphs that are isomorphic to F, if there is either a monochromatic G or a rainbow G, then there is either a monochromatic F or a rainbow F.

Suppose then that G has either a minimum coloring or a maximum coloring. Assume, without loss of generality, that G has a minimum coloring. We show in this case that G contains a rainbow F. If F is disconnected, then edges can be added to F to produce a tree T of order k. Select some vertex r of T as the root of T. Suppose that $V(G) = \{v_1, v_2, \ldots, v_k\}$, where

$$v_1 < v_2 < \ldots < v_k.$$

Label the vertices of T in the order $v_k, v_{k-1}, v_{k-2}, \ldots, v_1$, according to the distances of the vertices of T from r. That is, the root r is labeled v_k, some vertex adjacent to r is labeled v_{k-1} and so on, up to a vertex of T farthest from r, which is labeled v_1. (See Figure 11.12 for an illustration of this when $k = 8$.)

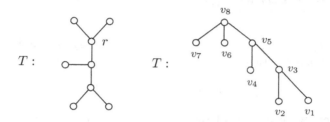

Figure 11.12: A step in the proof of Theorem 11.21

The minimum coloring of G then assigns the colors $v_1, v_2, \ldots, v_{k-1}$ to the edges of T. Since this produces a rainbow T, there is a rainbow F. Hence in any case the rainbow Ramsey number $RR(F)$ is defined. ∎

Let's consider an example of this.

Example 11.22 $RR(K_{1,3}) = 6$.

Solution. Since the red-blue coloring of K_5 shown in Figure 11.13 has neither a monochromatic $K_{1,3}$ nor a rainbow $K_{1,3}$, it follows that $RR(K_{1,3}) \geq 6$. It remains to show that $RR(K_{1,3}) \leq 6$. Let there be given a coloring of the edges of K_6 with no monochromatic $K_{1,3}$. Consider a vertex v of K_6. Since v is incident with five edges and at most two of these edges can be colored the same, there are three edges incident with v that are colored differently, producing a rainbow $K_{1,3}$. ◇

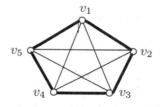

Figure 11.13: A red-blue coloring of K_5

Rainbow Ramsey numbers for a single graph (a forest) were extended in the following way to rainbow Ramsey numbers for two graphs. For two nonempty graphs F_1 and F_2, the **rainbow Ramsey number** $RR(F_1, F_2)$ is defined as the smallest positive integer n such that if each edge of K_n is colored from any number of colors, then there is either a monochromatic F_1 or a rainbow F_2. In view of Theorem 11.21, it wouldn't be expected that $RR(F_1, F_2)$ is defined for every pair F_1, F_2 of nonempty graphs. The conditions under which $RR(F_1, F_2)$ is defined is a consequence of a result of Erdös and Rado.

Theorem 11.23 *Let F_1 and F_2 be two graphs without isolated vertices. The rainbow Ramsey number $RR(F_1, F_2)$ exists if and only if F_1 is a star or F_2 is a forest.*

If F_1 and F_2 are nonempty graphs of orders n_1 and n_2, respectively, for which $RR(F_1, F_2)$ is defined, then

$$RR(F_1, F_2) \geq \max\{n_1, n_2\}.$$

Let's consider two examples of these rainbow Ramsey numbers involving the path P_3 of order 3 and the graph F shown in Figure 11.14.

Figure 11.14: The graphs F and P_3

Example 11.24 $RR(F, P_3) = 4$.

Solution. Since P_3 is a forest, it follows by Theorem 11.23 that $RR(F, P_3)$ exists. Because the order of F is 4, $RR(F, P_3) \geq 4$. First, we color each edge of K_4 from any number of colors. Suppose that there is no rainbow P_3. Since every two adjacent edges of K_4 must be colored the same, it follows that every two edges of K_4 are colored the same. Consequently, there is a monochromatic F and so $RR(F, P_3) = 4$.   ◇

We now reverse the order of the graphs F and P_3.

Example 11.25 $RR(P_3, F) = 5$.

Solution. Since $P_3 = K_{1,2}$ is a star, it follows by Theorem 11.23 that $RR(F, P_3)$ exists. In the coloring of the edges of K_4 shown in Figure 11.15 using colors 1, 2, 3, there is neither a monochromatic P_3 nor a rainbow F and so $RR(P_3, F) \geq 5$.

Let there be given a coloring of the edges of K_5 and suppose that there is no monochromatic P_3. Let the vertices of K_5 be v, v_1, v_2, v_3, v_4. Since there is no monochromatic P_3, we may assume that the edge vv_i is colored i for $i = 1, 2, 3, 4$ (see Figure 11.16).

K_4 :

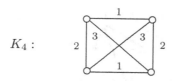

Figure 11.15: A coloring of K_4 that avoids
a monochromatic P_3 and a rainbow F

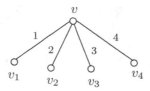

Figure 11.16: Coloring the edges vv_i $(i = 1, 2, 3, 4)$ in K_5

Since K_5 contains no monochromatic P_3, the edge v_1v_2 is not colored 1 or 2. However, regardless of whether v_1v_2 is colored 3, 4 or some other color, a rainbow F is produced. Therefore, $RR(P_3, F) = 5$. ◇

Examples 11.24 and 11.25 illustrate that, unlike Ramsey numbers, even if $RR(F_1, F_2)$ and $RR(F_2, F_1)$ are both defined, then these rainbow Ramsey numbers need not be equal.

Let F_1 and F_2 be two graphs without isolated vertices, where F_2 has size m. For an integer $k \geq m$, the k-**rainbow Ramsey number** $RR_k(F_1, F_2)$ is the smallest positive integer n such that if the edges of K_n are colored in any manner whatsoever from the set $\{1, 2, \ldots, k\}$, then either a monochromatic F_1 or a rainbow F_2 is produced. It turns out that $RR_k(F_1, F_2)$ exists for every two graphs F_1 and F_2 without isolated vertices, where F_2 has size m, and for any integer $k \geq m$. Thus, even though $r(K_3, K_3) = 6$ and $RR(K_3, K_3)$ is not defined, $RR_k(K_3, K_3)$ is defined for all $k \geq 3$. (See Exercise 11.39.)

As an illustration, we compute $RR_3(2K_2, K_3)$. Observe that the only connected graphs F not containing $2K_2$ are those where every two edges are adjacent, which is then either a star or K_3.

Example 11.26 $RR_3(2K_2, K_3) = 6$.

Solution. The coloring of K_5 with three colors, say red, blue and green, shown in Figure 11.17 (where bold edges are red, standard edges are blue and dashed edges are green) has no monochromatic $2K_2$. Any rainbow K_3 must contain a red edge and therefore the vertex v_1. However, v_1 is incident only with red edges and so any K_3 containing v_1 has two red edges. Thus there is no rainbow K_3, implying that $RR_3(2K_2, K_3) \geq 6$.

To show that $RR_3(2K_2, K_3) \leq 6$, let there be given a coloring of K_6 with three colors, say red, blue and green. Since K_6 has 15 edges, at least five edges of

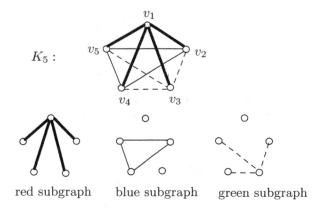

Figure 11.17: A red-blue-green coloring of K_5

K_6 are colored the same. If there are six or more edges of K_6 colored the same, then there are two nonadjacent edges that are colored the same and we have a monochromatic $2K_2$. Thus we may assume that for each color, exactly five edges in K_6 are colored with this color and that these five edges are incident with a common vertex. Suppose that the five red edges are incident with a vertex v and the five blue edges are incident with a vertex u. Then a contradiction is obtained for the edge uv, as it is colored both red and blue.   ◊

Exercises for Section 11.3

11.28 Determine $mr(C_3, C_4)$.

11.29 Determine $mr(K_{1,4}, P_4)$.

11.30 Determine $RR(K_{1,4})$.

11.31 For the tree T in Figure 11.18, determine $RR(T)$.

T :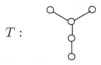

Figure 11.18: The tree T in Exercise 11.29

11.32 (a) Determine $RR(P_3, 2K_2)$.

     (b) Determine $RR(2K_2, P_3)$.

11.33 Determine $RR(K_2, mK_2)$.

11.34 Determine $RR(C_n, P_2)$.

11.35 Determine $RR(C_n, P_3)$.

11.36 Determine $RR(K_{1,n}, P_3)$.

11.37 Let F_1 be a subgraph of a graph G_1 and F_2 be a subgraph of a graph G_2 such that $RR(F_1, F_2)$ and $RR(G_1, G_2)$ both exist. Prove that $RR(F_1, F_2) \leq RR(G_1, G_2)$.

11.38 Show that $RR(K_{1,3}, 3K_2) \geq 7$.

11.39 It is known that $RR_3(K_3, K_3) = 11$. Show that $RR_3(K_3, K_3) \geq 11$.

11.4 Excursion: Erdős Numbers

Within a day or two after September 20, 1996, the news appeared on computer screens of mathematicians around the world that Paul Erdős had died while attending a graph theory workshop at the Banach Center in Warsaw, Poland. Thus ended the life of a unique and most unusual mathematician whose accomplishments were notable by any standards.

Paul Erdős was born on March 26, 1913 of Jewish parents in Budapest, Hungary. Shortly before Erdős was born, his two older sisters died of scarlet fever. In 1914 World War I started in Europe. Erdős' father was captured by the Russians and was taken as a prisoner of war to Siberia. He was to remain there for six years. There is little doubt that Erdős was a child prodigy as far as mathematics was concerned. He showed interest in and uncanny ability with numbers as early as 4 years old, when he discovered negative numbers on his own. He could multiply 4-digit numbers in his head.

Erdős' mother and father were both mathematics teachers and were responsible for his early education. At age 16, his father introduced him to infinite series and set theory, subjects that remained among his favorites throughout his life.

While in high school, Erdős was an ardent problem solver. As a winner of national mathematical competitions, he (along with Paul Turán and Tibor Gallai) was admitted to Pázmány University in Budapest. Erdős and his friends would journey to the hills of Budapest to discuss a variety of topics. Mathematics was always the main subject of their conversations, however.

By age 19, Erdős had essentially completed his Ph.D. in number theory under the direction of Leopold Fejér (although Fejér was well known for his work in analysis). Erdős gave an elegant proof of a theorem of Pafnuty Lvovich Chebyshev (the renowned Russian mathematician who made important contributions to number theory, probability theory and approximation theory): *For every integer $n \geq 2$, there is a prime between n and $2n$.* Indeed, Erdős showed the existence of prime numbers between n and $2n$ belonging to certain arithmetic progressions.

We have often mentioned how graph theory has been influenced by the accomplishments of Hungarian mathematicians. Indeed, the master problem-solver and mathematics educator George Pólya (1887 - 1985) was also born in Budapest. Pólya was the author of the famous book *How to Solve It*, which sold over a million copies (and for whose English language version he had great difficulty finding a publisher). Pólya gave several reasons why Hungarians were so influential in mathematics: (1) there were mathematical journals for high school students that stimulated their interest in mathematics, (2) there were mathematical competitions for students and (3) there was Leopold Fejér, who was responsible for attracting many young people to mathematics.

When Erdős graduated in 1934, he was already considered a leading number theorist. He went to Manchester in Britain on a 4-year fellowship. While there, he became a frequent traveler, visiting universities to do research. With World War II on the horizon in Europe, he left for the United States, not to return to Europe for another decade. He held a fellowship at the Institute for Advanced Study during 1938-39. Erdős felt this was his best year mathematically. During that year the subject of probabilistic number theory was born. He also solved an outstanding unsolved problem in dimension theory. Despite these and other major accomplishments, Erdős' fellowship was not renewed. He survived on small loans from colleagues.

Erdős received a research instructorship from Purdue University in 1943, only to be unemployed again in 1945. None of these ever slowed his research accomplishments, however. In 1946 his work with others led to the initiation of extremal graph theory.

While all of this was going on, World War II was in progress and Erdős anguished over the fate of his parents and friends. In fact, Erdős' father died of a heart attack in 1942 and his mother fell into depression. However, his mother survived the war, as did his close friends Paul Turán and Tibor Gallai. Vera Sós, a student of Gallai, also survived. Later she married Turán and became a collaborator of Erdős.

In 1948, Erdős met the young mathematician Atle Selberg at the Institute for Advanced Study and this led to independent proofs of the Prime Number Theorem: For a positive integer n, let $\pi(n)$ denote the number of primes less than n.

$$\text{Then } \pi(n) \sim \frac{n}{\ln n}, \text{ that is, } \lim_{n \to \infty} \frac{\pi(n)}{\frac{n}{\ln n}} = 1.$$

This was conjectured to be true by the famous mathematician Carl Friedrich Gauss in 1791, but it was not proved until 1896 when Jacques Hadamard and Charles de la Vallée Poussin gave independent proofs (using complex analysis). In 1949, Erdős and Selberg gave independent elementary proofs (that is, not using modern complex analysis methods). In 1950 Selberg was a recipient of the Fields Medal, the mathematical equivalent of a Nobel Prize.

Although most mathematicians are associated with the places they have worked and lived, for much of his life Erdős had no job and no home. He traveled from university to university, country to country, continent to continent, visiting one mathematician after another to discuss, pose and solve research problems. He worked with the famous and not-so-famous, with established mathematicians and with students. During much of his life, he traveled with his mother. After she died in 1971 at the age of 91, he traveled on his own. Since he had little interest in material things, his mathematics friends, of whom there were many, took care of or assisted him with the everyday items, such as clothing, food and money.

Erdős was a problem solver, often working on several challenging problems at the same time. Although he was not one to develop theory, the problems he worked on frequently led to theory developed by others. These problems were primarily in the areas of combinatorics, graph theory and number theory. He was not satisfied with simply solving problems, however. He sought proofs that provided insight as to why the result was true. Although a person of limited wealth, he enjoyed offering monetary incentives for others to solve these problems. For solutions to some problems, he would sometimes offer thousands of dollars. He was not good at keeping mathematical secrets though. If there was a problem he found intriguing, he would tell those with whom he came in contract. One anecdote of this type occurred in the 1980s. A conjecture had been made at Western Michigan University just minutes before Erdős arrived to give a colloquium talk. When he heard about the conjecture, he mentioned it to the audience and offered $5 for a proof or a counterexample.

Erdős authored or co-authored some 1500 papers, the largest number by any mathematician. Because he traveled so much and encountered so many mathematicians, he had a large number of co-authors - more than 500. This is contrary to the way many mathematicians work, especially the early mathematicians, who did research on their own. To Erdős, working on research was a social event, to be done with others. There are more than 4600 individuals who never co-authored a paper with Erdős but who did write a paper with a co-author of his. Such occurrences inspired the definition of **Erdős numbers**. Only Paul Erdős has Erdős number 0. Any mathematician who co-authored a paper with Erdős has Erdős number 1. Any mathematician who does not have Erdős number 1 but who co-authored a paper with someone who has Erdős number 1 has Erdős number 2. More generally, for an integer $k \geq 3$, a mathematician has Erdős number k if the mathematician does not have Erdős number less than k but has co-authored a paper with a mathematician having Erdős number $k - 1$.

Erdős numbers can be considered from another point of view. The **collaboration graph** G has the set of all mathematicians as its vertex set and two vertices (mathematicians) are adjacent if they have co-authored a paper (possibly with other co-authors). The Erdős number of a mathematician (vertex) is the distance from that vertex to the Erdős vertex in the collaboration graph. A very small subgraph of this graph is shown in Figure 11.19. Not only is the Erdős number a dynamic concept (its value varies with time), so too is the collaboration graph.

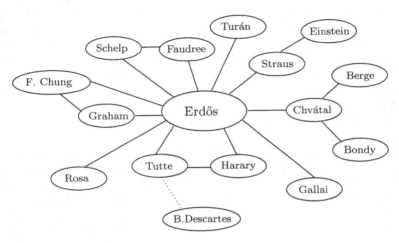

Figure 11.19: A subgraph of the collaboration graph

For example, the mathematician Ernst Straus wrote separate papers with both Erdős and Albert Einstein, so Straus has Erdős number 1. Since Einstein never wrote a paper with Erdős, his Erdős number is 2. Indeed, Erdős and Einstein met on only one occasion, at which time they discussed religion (not mathematics or physics). As of this writing, the web site developed by Jerrold Grossman of Oakland University, contains a multitude of information on Erdős numbers: http://www.oakland.edu/enp/.

Augustin-Louis Cauchy, Leonhard Euler, Arthur Cayley, Paul Erdős. These are the four mathematicians who are credited with authoring or co-authoring the largest number of mathematical papers. Curiously, the last three of these famous mathematicians have connections to graph theory, as we have seen. Indeed, the last three of these four individuals had numerous and important connections to graph theory.

There are other related graphs and numbers that can be associated with individuals in certain occupations. Perhaps the best known of these occurs in what is called the **Kevin Bacon Game**, named for the movie actor Kevin Bacon. In this case, the associated graph has movie actors as its vertices and two actors (vertices) are adjacent if the two individuals appeared in the same feature movie. The **Kevin Bacon number** of an actor is the distance in this graph from that actor (vertex) to the Kevin Bacon vertex. Therefore, only Kevin Bacon has Kevin Bacon Number 0. Tom Hanks has Kevin Bacon number 1 since he acted with Kevin Bacon in the movie *Apollo* 13, while Sarah Jessica Parker has Kevin Bacon number 1 since she appeared in the movie *Footloose* with Kevin Bacon. On the other hand, Cary Grant has Kevin Bacon number

2 as he never appeared in a movie with Kevin Bacon; however, Cary Grant appeared with Walter Matthau in *Charade* and Walter Matthau appeared with Kevin Bacon in the movie *JFK*. As of this writing, the interactive web site http://oracleofbacon.org allows one to determine the Kevin Bacon numbers of many movie actors.

Erdős numbers and Kevin Bacon numbers have their origins in a 1967 experiment of the psychologist Stanley Milgram (1933-1984) who tracked chains of acquaintances in the United States. Because people know people, who know people, etc., Milgram claimed that on the average, every two people could be connected by a path of length 6, which gave rise to the phrase: *six degrees of separation*.

There is also a play and a movie with the title *Six Degrees of Separation*, written by John Guare and starring Will Smith. The story is based on the real-life story of David Hampton, a con man who convinced a number of people in the 1980s that he was the son of movie actor Sidney Poitier.

Actually in recent years Erdős numbers and Kevin Bacon numbers have been extended to **Erdős-Bacon numbers**. People who have acted in feature films *and* who have co-authored mathematical papers have the possibility of having a defined Erdős-Bacon number, which is the sum of their Erdős and Kevin Bacon numbers. For example, the physicist Brian Greene appeared in the movie *Frequency* with John Di Benedetto, who was in *Sleepers* with Kevin Bacon. Also, Greene wrote a paper with Shing-Tung Yau, who wrote a paper with Ronald Graham who has Erdős number 1. Thus the Erdős-Bacon number of Greene is $2 + 3 = 5$.

Ronald Graham has made major contributions to many areas of graph theory, to many subjects in mathematics and to the mathematical community. Ronald Graham did undergraduate work in electrical engineering at the University of Chicago and in physics at the University of Alaska at Fairbanks. As an undergraduate Graham supported himself as a circus performer and worked for Cirque du Soleil. Graham received his Ph.D. in 1962 from the University of California at Berkeley under the direction of Derrick Lehmer. Graham went to Bell Laboratories where he worked for 37 years. In 1999 he became the Irwin and Joan Jacobs Professor of Computer Science and Engineering at the University of California at San Diego.

In 2003 Ronald Graham became president of the Mathematical Association of America, thereby becoming only the sixth person to hold that position as well as president of the American Mathematical Society, a position he held during 1993-1995.

Among his non-mathematical accomplishments, Graham is a skilled juggler and at one time served as president of the International Juggler's Association. Graham has lectured often and in many places, including at Walt Disney World.

Chapter 12

Distance

12.1 The Center of a Graph

We've mentioned a number of times of how a graph can be used to model the street system of a town. Of course, as a town grows in size, so too does the graph that models it. As a reminder, we see in Figure 12.1 the street system of a town T and a graph G_T that models it.

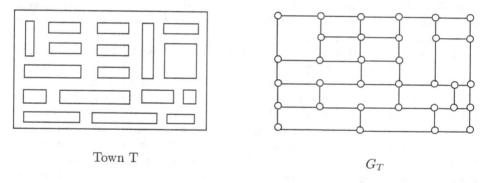

Town T G_T

Figure 12.1: Town T and a graph modeling town T

As a town grows into a city, new questions arise. For example, when a town is small, it might be appropriate to rely on and pay for the services of the fire department of a neighboring city. However, when a town reaches a certain size (and is able to afford it), it becomes necessary for that town to have its own fire department. Assuming that the decision has been made by the town to build its own firehouse, we now have another question: Where in the town should we build it? Let's assume that we decide to build the firehouse at some street intersection in the town. This, however, does not answer our question. Of course, the main reason for building the firehouse is so that *all* citizens of the town are protected in the event of a fire. Consequently, no location in the town should be too far from this new firehouse. We see that answering our question concerns distances in town T and, therefore, distances in the graph G_T as well.

Let's review the definition of distance in a graph. For two vertices u and v in a graph G, the **distance** $d(u, v)$ from u to v is the length of a shortest $u - v$ path in G. A $u - v$ path of length $d(u, v)$ is called a $u - v$ **geodesic**. In order for $d(u, v)$ to be defined for all pairs u, v of vertices in G, the graph G must be connected. We therefore assume that G is a connected graph. The term distance that we just defined satisfies all four of the following properties in any connected graph G.

1. $d(u, v) \geq 0$ for all $u, v \in V(G)$.

2. $d(u, v) = 0$ if and only if $u = v$.

3. $d(u, v) = d(v, u)$ for all $u, v \in V(G)$ [**the symmetric property**].

4. $d(u, w) \leq d(u, v) + d(v, w)$ for all $u, v, w \in V(G)$ [**the triangle inequality**].

That a connected graph satisfies all four of these properties should be clear, with the possible exception of property 4 (the triangle inequality), which we now verify. Let P_1 be a $u - v$ geodesic and P_2 a $v - w$ geodesic in the graph G. The path P_1 followed by P_2 produces a $u - w$ walk of length $d(u, v) + d(v, w)$. By Theorem 1.6, G contains a $u - w$ path whose length is at most $d(u, v) + d(v, w)$. Therefore, $d(u, w) \leq d(u, v) + d(v, w)$. Since the distance d satisfies property 2 (the symmetric property), we can refer to the distance *between* two vertices rather than the distance *from* one vertex *to* another.

The fact that the distance d satisfies properties 1–4 means that d is a **metric** and $(V(G), d)$ is a **metric space**. It is ordinarily very useful when a distance is a metric as this concept has been studied widely. There are many concepts involving connected graphs that are defined in terms of distance and which are valuable in providing information about these graphs.

For a vertex v in a connected graph G, the **eccentricity** $e(v)$ of v is the distance between v and a vertex farthest from v in G. The minimum eccentricity among the vertices of G is its **radius** and the maximum eccentricity is its **diameter**, which are denoted by $\text{rad}(G)$ and $\text{diam}(G)$, respectively. A vertex v in G is a **central vertex** if $e(v) = \text{rad}(G)$ and the subgraph induced by the central vertices of G is the **center** $\text{Cen}(G)$ of G. If every vertex of G is a central vertex, then $\text{Cen}(G) = G$ and G is called **self-centered**. For example, if $G = C_n$ where $n \geq 3$, then G is self-centered.

To illustrate the concepts we have just presented, consider the graph H of Figure 12.2, where each vertex is labeled by its eccentricity. Since the smallest eccentricity is 2, $\text{rad}(H) = 2$. Because the largest eccentricity is 4, $\text{diam}(H) = 4$. The center of H is also shown in Figure 12.2.

There are a number of observations that can be made about the graph H of Figure 12.2. We have already mentioned that $\text{rad}(H) = 2$ and $\text{diam}(H) = 4$. The terms "radius" and "diameter" are familiar because of circles, where, of course,

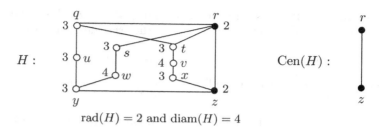

$$\text{rad}(H) = 2 \text{ and } \text{diam}(H) = 4$$

Figure 12.2: The eccentricities of the vertices of a graph

the diameter is always twice the radius. This fact together with the knowledge that $\text{diam}(H) = 2\,\text{rad}(H)$ for the graph H of Figure 12.2 might reasonably suggest that $\text{diam}(G) = 2\,\text{rad}(G)$ for *every* connected graph G. Such is not the case, however. Figure 12.3 shows three graphs G_2, G_3 and G_4, each of which has radius 2, where $\text{diam}(G_k) = k$ for $k = 2, 3, 4$. There is, therefore, no identity that relates the radius and the diameter of a graph. As we now show, Figure 12.3 illustrates the only possible diameters for a graph having radius 2.

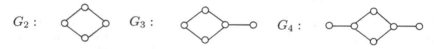

Figure 12.3: Three graphs having radius 2

Theorem 12.1 *For every nontrivial connected graph G,*

$$\text{rad}(G) \leq \text{diam}(G) \leq 2\,\text{rad}(G).$$

Proof. The inequality $\text{rad}(G) \leq \text{diam}(G)$ is immediate since the smallest eccentricity cannot exceed the largest eccentricity. Let u and v be two vertices such that $d(u, v) = \text{diam}(G)$ and let w be a central vertex of G. Therefore, $e(w) = \text{rad}(G)$. Hence the distance between w and any other vertex of G is at most $\text{rad}(G)$. By the triangle inequality,

$$\text{diam}(G) = d(u, v) \leq d(u, w) + d(w, v) \leq \text{rad}(G) + \text{rad}(G) = 2\,\text{rad}(G). \qquad \blacksquare$$

Another observation about the graph H in Figure 12.2 is that the eccentricities of every two adjacent vertices differ by at most 1. This statement too is true for all connected graphs.

Theorem 12.2 *For every two adjacent vertices u and v in a connected graph,*

$$|e(u) - e(v)| \leq 1.$$

Proof. Assume, without loss of generality, that $e(u) \geq e(v)$. Let x be a vertex that is farthest from u. So $d(u, x) = e(u)$. By the triangle inequality,

$$e(u) = d(u, x) \leq d(u, v) + d(v, x) \leq 1 + e(v).$$

Hence $e(u) \leq 1 + e(v)$, which implies that $0 \leq e(u) - e(v) \leq 1$. Therefore, $|e(u) - e(v)| \leq 1$. ∎

In much the same way, the following can be proved (see Exercise 12.10).

Theorem 12.3 *Let u and v be adjacent vertices in a connected graph G. Then*

$$|d(u, x) - d(v, x)| \leq 1$$

for every vertex x of G.

Returning once again to the graph H of Figure 12.2, we see that $\mathrm{Cen}(H) = K_2$. This brings up a natural question. Which graphs can be the center of some graph? Stephen Hedetniemi showed that "every graph" is the answer to this question.

Theorem 12.4 *Every graph is the center of some graph.*

Proof. Let G be a graph. We show that G is the center of some graph. First, add two new vertices u and v to G and join them to every vertex of G but not to each other. Next, we add two other vertices u_1 and v_1, where we join u_1 to u and join v_1 to v. The resulting graph is denoted by F (see Figure 12.4).

Figure 12.4: The graph F in the proof of Theorem 12.4

Since $e(u_1) = e(v_1) = 4$, $e(u) = e(v) = 3$ and $e_F(x) = 2$ for every vertex x in G, it follows that $V(G)$ is the set of central vertices of F and so $\mathrm{Cen}(F) = G$. ∎

Stephen Hedetniemi was born on February 7, 1939 in Washington, D.C. His father, who at one time worked for U.S. Supreme Court Justice Frank Murphy, encouraged his four sons and daughter to pursue higher education. Hedetniemi became an undergraduate at the University of Michigan and majored in mathematics but became interested in the new field of computer science. While an undergraduate at the University of Michigan, he took a course in graph theory from a professor who was teaching it for the first time: Frank Harary. In 1966, Hedetniemi received his Ph.D. with co-supervisors Frank Harary and John Holland. Holland, a computer scientist, would later become the founder of the field of genetic algorithms.

After working in the computer science departments at the Universities of Iowa, Virginia and Oregon, Hedetniemi went to Clemson University in 1982, where he taught until he retired. While he contributed not only to graph theory but to the areas of algorithms, computation theory, combinatorial optimization and parallel processing, Hedetniemi is known for initiating several areas of study within graph theory. The area for which he is probably best known will be visited in the next chapter.

Although every graph can be the center of some connected graph, there are some restrictions as to where the center of a graph G can be located in G.

Theorem 12.5 *The center of every connected graph G is a subgraph of some block of G.*

Proof. Assume, to the contrary, that G is a connected graph whose center $\text{Cen}(G)$ is *not* a subgraph of a single block of G. Then there is a cut-vertex v of G such that $G - v$ contains two components G_1 and G_2, each of which contains vertices of $\text{Cen}(G)$. Let u be a vertex of G such that $d(u, v) = e(v)$ and let P_1 be a $u - v$ geodesic in G. At least one of G_1 and G_2 contains no vertices of P_1, say G_2 contains no vertices of P_1. Let w be a central vertex of G that belongs to G_2 and let P_2 be a $v - w$ geodesic. Then P_1 followed by P_2 produces a $u - w$ geodesic, whose length is greater than that of P_1. Hence $e(w) > e(v)$, which contradicts the fact that w is a central vertex of G.   ∎

The graph G_T of Figure 12.1 (which, recall, models town T in that figure) is shown again in Figure 12.5, where in this case, every vertex is labeled with its eccentricity. We asked earlier where a firehouse should be built so that no location in the town is too far from the firehouse. We now see that an appropriate answer is for the firehouse to be built at any of the three intersections that correspond to central vertices (vertices having eccentricity 5) in G_T.

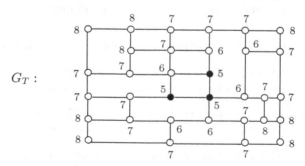

$G_T :$

Figure 12.5: The eccentricities of the vertices of G_T

Exercises for Section 12.1

12.1 Find the radius and diameter of the graph G in Figure 12.6. What is the center of G?

G :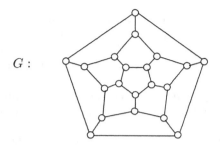

Figure 12.6: The graph in Exercise 12.1

12.2 (a) Find the radius and diameter of C_n for $n \geq 3$.

    (b) Find the radius and diameter of P_n for $n \geq 3$. What is the center of P_n?

    (c) Find the radius and diameter of Q_n for $n \geq 2$.

12.3 Find the radius and diameter of $K_{s,t}$ for $1 \leq s \leq t$. What is the center of $K_{s,t}$?

12.4 Find the radius and diameter of the Petersen graph PG. What is the center of PG?

12.5 Give an example of a connected graph G such that $\mathrm{Cen}(G)$ is disconnected.

12.6 Show that every graph of order n is the center of some graph of order $2n$.

12.7 (a) Prove that if G is a connected graph with $\mathrm{diam}(G) \geq 3$, then $\mathrm{diam}(\overline{G}) \leq 3$.

    (b) Give an example of a connected graph G with $\mathrm{diam}(G) = \mathrm{diam}(\overline{G}) = 3$.

12.8 Prove that for each pair a, b of positive integers with $a \leq b \leq 2a$, there exists a graph G with $\mathrm{rad}(G) = a$ and $\mathrm{diam}(G) = b$.

12.9 Prove the following generalization of Theorem 12.2: For every two vertices u and v in a connected graph, $|e(u) - e(v)| \leq d(u, v)$.

12.10 (a) Prove Theorem 12.3: *Let u and v be adjacent vertices in a connected graph G. Then $|d(u, x) - d(v, x)| \leq 1$ for every vertex x of G.*

    (b) Let G be a connected graph and suppose that $d(u, x) = k$ for some $u, x \in V(G)$. Show that if v is a neighbor of u, then $d(v, x)$ is $k - 1$, k or $k + 1$.

12.11 Let G be a connected graph and k an integer with $\mathrm{rad}(G) < k < \mathrm{diam}(G)$. Use Theorem 12.2 to prove that there is a vertex v of G with $e(v) = k$.

12.12 Prove that if T is a tree of order $n \geq 3$, then $\Delta(T) + \mathrm{diam}(T) \leq n + 1$.

12.13 (a) Let T be a tree of order $n \geq 3$ and let T' be the tree obtained from T by deleting the end-vertices of T. Prove that $\mathrm{Cen}(T) = \mathrm{Cen}(T')$.

(b) Prove that the center of a tree T is either K_1 or K_2.

(c) Prove that if the center of a tree T is K_1, then $\mathrm{diam}(T) = 2\,\mathrm{rad}(T)$.

12.14 Show that for every pair r, s of positive integers, there exists a positive integer n such that for every connected graph G of order n, either $\Delta(G) \geq r$ or $\mathrm{diam}(G) \geq s$.

12.2   Distant Vertices

If we were to find ourselves at a certain location in some town (such as in town T of Figure 12.1) and ask for a location in the town that is farthest from where we are, then this is the same question as: For a given vertex u in a connected graph G, what is a vertex v in G that is farthest from u? Of course, what we're seeking is a vertex v such that $d(u, v) = e(u)$. Depending on where u is located in G, the distance between u and v might be as small as $\mathrm{rad}(G)$, as large as $\mathrm{diam}(G)$ or some number between these two.

A vertex v in a connected graph G is called a **peripheral vertex** if $e(v) = \mathrm{diam}(G)$. Thus, in certain sense, a peripheral vertex is opposite to a central vertex. The subgraph of G induced by its peripheral vertices is the **periphery** $\mathrm{Per}(G)$ of G. For the graph H of Figure 12.2, which is redrawn in Figure 12.7, the periphery of H is shown in Figure 12.7.

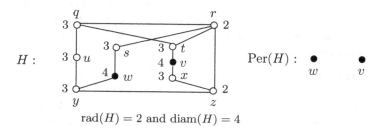

$$\mathrm{rad}(H) = 2 \text{ and } \mathrm{diam}(H) = 4$$

Figure 12.7: The eccentricities of the vertices of a graph

The periphery of the graph H of Figure 12.7 is $2K_1$ (that is, it consists of two isolated vertices) and so it is disconnected. Is the periphery of *every* graph disconnected? The answer is no, as the graph F of Figure 12.8 shows. Each vertex of F is labeled with its eccentricity. Since $\mathrm{diam}(F) = 3$, it follows that $\mathrm{Per}(F) = C_6$, which is connected. In fact, if $G = C_n$, where then $n \geq 3$, it follows that $\mathrm{Per}(G) = C_n$. Could it be then that as with centers, *every* graph is

the periphery of some graph? Halina Bielak and Maciej Syslo showed that the
answer to this question is no.

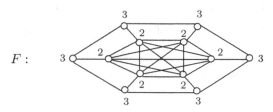

Figure 12.8: A graph F with $\text{Per}(F) = C_6$

Theorem 12.6 *A nontrivial graph G is the periphery of some graph if and only
if every vertex of G has eccentricity 1 or no vertex of G has eccentricity 1.*

Proof. Assume first that every vertex of G has eccentricity 1 or no vertex of G
has eccentricity 1. If every vertex of G has eccentricity 1, then G is complete and
$\text{Per}(G) = G$. Now assume that *no* vertex of G has eccentricity 1. This implies
that for every vertex u of G, there is a vertex v in G that is not adjacent to u.
Let H be the graph obtained by adding a new vertex w and joining w to every
vertex of G. Then $e_H(w) = 1$. Since $e_H(x) = 2$ for every vertex x of G, it follows
that every vertex of G is a peripheral vertex of H and so $\text{Per}(H) = G$.

For the converse, let G be a graph that contains some vertices of eccentricity
1 and some vertices whose eccentricity is not 1. Assume, to the contrary, that
there exists a graph H such that $\text{Per}(H) = G$. Necessarily, G is a proper induced
subgraph of H. Then there exists an integer $k \geq 2$ such that $e_H(v) = k$ for every
vertex v of G, while $e_H(v) < k$ for every vertex v of H that is not in G. Let
x be a vertex of G such that $e_G(x) = 1$ and let w be a vertex of H such that
$d(x, w) = e_H(x) = k \geq 2$. Since w is not adjacent to x, it follows that w is not
in G. However, $d(w, x) = k$ and so $e_H(w) \geq k$, contradicting the fact that w is
not in the periphery of H.   ∎

According to Theorem 12.6 then, no star of order 3 or more is the periphery
of any graph. For a given vertex u in a connected graph G, we have discussed
seeking a vertex v such that $d(u, v) = e(u)$, that is, v is a vertex that is farthest
from u. Such a vertex v is called an **eccentric vertex of** u. A vertex v is an
eccentric vertex of the graph G if v is an eccentric vertex of some vertex of
G. In other words, a vertex v is an eccentric vertex of G if v is farthest from
some vertex of G.

Consider the graph G of Figure 12.9, where each vertex is labeled with its
eccentricity. For example, $e(u) = 3$. Since $d(u, v) = 3$, it follows that v is an
eccentric vertex of u. Because there is a $u - v$ path of length 3 in G, there is
certainly a $v - u$ path of length 3 in G. This does not mean, however, that u is
an eccentric vertex of v as there may be a vertex farther from v than u is. This
only implies therefore that $e(v) \geq 3$. In fact, $e(v) = 4$ and so u is *not* an eccentric

vertex of v, although w *is* an eccentric vertex of v. More generally, if a vertex y is an eccentric vertex of a vertex x in a connected graph, then $e(y) \geq e(x)$.

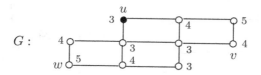

Figure 12.9: An eccentric vertex in G

If a vertex x in a connected graph G is a peripheral vertex of G, then, as we have seen, $e(x) = \text{diam}(G)$. Necessarily then, there exists a vertex y such that $d(x, y) = e(x) = \text{diam}(G)$. This also implies, however, that $d(x, y) = e(y) = \text{diam}(G)$ and that y is a peripheral vertex of G as well. Therefore, every peripheral vertex of G is an eccentric vertex. The converse is not true, however. We saw that the vertex v in the graph G of Figure 12.9 is an eccentric vertex of G but that v is not a peripheral vertex of G.

Consider next the graph H shown in Figure 12.10, where $\text{rad}(H) = 2$ and $\text{diam}(H) = 4$. Since q and r are peripheral vertices (the *only* peripheral vertices of H), they are also eccentric vertices of H. The vertices x and z are also eccentric vertices of each other; while t and u are both eccentric vertices of x and z. Furthermore, w and y are eccentric vertices of each other; while s and v are both eccentric vertices of w and y. That is, every vertex of H is an eccentric vertex.

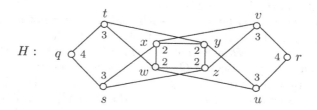

Figure 12.10: A graph each of whose vertices is an eccentric vertex

If every vertex of some graph G has the same eccentricity (and is therefore a peripheral vertex), then certainly every vertex of G is an eccentric vertex. However, the graph H of Figure 12.10 shows that every vertex of a graph can be an eccentric vertex without all the eccentricities being the same.

A connected graph G is an **eccentric graph** if every vertex of G is an eccentric vertex. Therefore, the graph H of Figure 12.10 is an eccentric graph, as is every graph all of whose vertices have the same eccentricity. Ordinarily, however, only some of the vertices of a graph are eccentric.

Let G be a connected graph. The **eccentric subgraph** $\text{Ecc}(G)$ of G is the subgraph of G induced by the set of eccentric vertices of G. For example, a connected graph F and its eccentric subgraph are shown in Figure 12.11. If every vertex of a graph G is an eccentric vertex, then $\text{Ecc}(G) = G$.

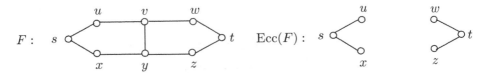

Figure 12.11: A graph and its eccentric subgraph

In the graph F of Figure 12.11, $\text{Ecc}(F) = 2P_3$. This brings up the question: Which graphs are eccentric subgraphs of some graph? Perhaps surprisingly, this question has the same answer as the question: Which graphs are the peripheries of some graph?

Theorem 12.7 *A nontrivial graph G is the eccentric subgraph of some graph if and only if every vertex of G has eccentricity 1 or no vertex of G has eccentricity 1.*

Proof. Assume, first, that every vertex of a graph G has eccentricity 1 or no vertex of G has eccentricity 1. If every vertex of G has eccentricity 1, then G is complete and G is an eccentric graph. Thus $\text{Ecc}(G) = G$. Next, assume that no vertex of G has eccentricity 1. Let H be the graph obtained by adding a new vertex w and joining w to every vertex of G. Since every vertex v in H that belongs to G is an eccentric vertex of w but w is not an eccentric vertex of any vertex of H, it follows that $\text{Ecc}(H) = G$.

For the converse, let G be a graph such that some but not all of its vertices have eccentricity 1. Then G is connected. Assume, to the contrary, that there exists a connected graph H such that $\text{Ecc}(H) = G$. Let u be a vertex of G that is adjacent to all other vertices of G and let v be an eccentric vertex of u in H. Since all eccentric vertices of H belong to G, it follows that v is in G. However, v is adjacent to u; so $e_H(u) = 1$, which implies that u is adjacent all other vertices in H and that all vertices of H that are not in G also belong to $\text{Ecc}(H)$. This is a contradiction.   ∎

If v is an eccentric vertex of a vertex u in a connected graph G, then no vertex of G is farther from u than v is. In particular, if w is a neighbor of v, then $d(u, w) \le d(u, v)$. However, a vertex can have this particular property without being an eccentric vertex of u.

A vertex v in a connected graph G is a **boundary vertex of a vertex** u if $d(u, w) \le d(u, v)$ for each neighbor w of v; while a vertex v is a **boundary vertex of the graph** G if v is a boundary vertex of some vertex of G.

We have mentioned that in a connected graph, every peripheral vertex is an eccentric vertex, but not conversely. Also, every eccentric vertex is a boundary vertex, but a boundary vertex need not be an eccentric vertex. Consider the graph G in Figure 12.12. The vertex z is an eccentric vertex of the vertex w, which in turn is a boundary vertex of the vertex s. However, z is not a peripheral vertex of G and w is not an eccentric vertex of G.

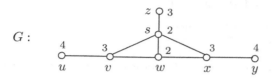

Figure 12.12: Peripheral, eccentric and boundary vertices in a graph

While the distance from a vertex u of a graph G to an eccentric vertex v of u attains the absolute maximum $\max_{w \in V(G)}\{d(u,w)\}$, the distance from u to a boundary vertex v of u attains the local maximum $\max_{w \in N[v]}\{d(u,w)\}$. Equivalently, a vertex v is a boundary vertex of u if no $u - v$ geodesic can be extended at v to a longer geodesic. Intuitively, beginning at u, a boundary vertex of u is reached when, locally, it is not possible to proceed farther from u.

There are certain vertices in a nontrivial connected graph that *cannot* be boundary vertices.

Theorem 12.8 *No cut-vertex is a boundary vertex of any connected graph.*

Proof. Assume, to the contrary, that there exists a connected graph G and a cut-vertex v of G such that v is a boundary vertex of some vertex u in G. Let G_1 be component of $G - v$ that contains u and let G_2 be another component of $G - v$. If w is a neighbor of v that belongs to G_2, then $d(w,u) = d(u,v) + 1$, which contradicts our assumption that v is a boundary vertex of u. ∎

Since no cut-vertex can be a boundary vertex, no cut-vertex can be an eccentric vertex or a peripheral vertex either. There are certain vertices, however, that must be boundary vertices.

A vertex v in a graph G is called a **complete vertex** (or an **extreme** or **simplicial vertex**) if the subgraph of G induced by the neighbors of v is complete. In particular, every end-vertex is complete. Therefore, if v is a complete vertex and u is a neighbor of v, then $d(w,u) = d(w,v) = 1$ for every $w \in N(v)$. Thus v is a boundary vertex of u. A complete vertex v is not only a boundary vertex of each neighbor of v, it is a boundary vertex of every vertex different from v.

Theorem 12.9 *Let G be a connected graph. A vertex v of G is a boundary vertex of every vertex distinct from v if and only if v is a complete vertex of G.*

Proof. First, let v be a complete vertex in G and let u be a vertex distinct from v. Also, let $(u = v_0, v_1, \cdots, v_k = v)$ be a $u - v$ geodesic and let w be a neighbor of v. If $w = v_{k-1}$, then $d(u,w) < d(u,v)$. So we may assume that $w \neq v_{k-1}$. Since v is complete, $wv_{k-1} \in E(G)$ and $(u = v_0, v_1, \cdots, v_{k-1}, w)$ is a $u - w$ path in G, implying that $d(u,w) \leq d(u,v)$. Hence v is a boundary vertex of u.

For the converse, let v be a vertex of G that is not a complete vertex. Then there exist nonadjacent vertices $u, w \in N(v)$. Since $d(u, w) > d(u, v)$, it follows that v is not a boundary vertex of u.   ∎

We now present a result which deals with a question that is opposite to that considered in Theorem 12.9.

Theorem 12.10 *Let G be nontrivial connected graph and let u be a vertex of G. Every vertex distinct from u is a boundary vertex of u if and only if $e(u) = 1$.*

Proof. Assume first that $e(u) = 1$ and let v be a vertex of G distinct from u. Let w be a neighbor of v. Then $d(u, w) \le 1$ and $d(u, v) = 1$. Hence v is a boundary vertex of u. For the converse, assume, to the contrary, that every vertex of G different from u is a boundary vertex of u but $e(u) \neq 1$. Then there exists a vertex x in G such that $d(u, x) = 2$. Let (u, y, x) be a $u - x$ geodesic in G. Then $d(u, y) = 1$ and x is a neighbor of y but $d(u, x) = 2 > 1 = d(u, y)$. Thus y is not a boundary vertex of u, which is a contradiction.   ∎

There are certain vertices in a connected graph G that have a close connection with boundary vertices. Let x and z be two distinct vertices in G. A vertex y distinct from x and z is said to lie **between** x and z if

$$d(x, z) = d(x, y) + d(y, z),$$

that is, the triangle inequality becomes an equality. A vertex v is an **interior vertex** of G if for every vertex u distinct from v, there exists a vertex w such that v lies between u and w. The **interior** Int(G) of G is the subgraph of G induced by interior vertices. For example, for the graph G of Figure 12.13 (which is also shown in Figure 12.12), the vertices s, v and x are the interior vertices of G and so Int$(G) = P_3$, as shown in Figure 12.13.

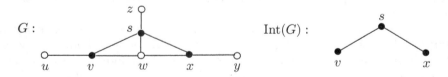

Figure 12.13: The interior of a graph

We now see that the interior vertices are precisely those vertices that are *not* boundary vertices.

Theorem 12.11 *Let G be a connected graph. A vertex v is a boundary vertex of G if and only if v is not an interior vertex of G.*

Proof. Let v be a boundary vertex of G, say v is a boundary vertex of the vertex u. Assume, to the contrary, that v is also an interior vertex of G. Since v is an interior vertex of G, there exists a vertex w distinct from u and v such that v lies between u and w. Let

$$P = (u = v_1, v_2, \cdots, v = v_j, v_{j+1}, \ldots, v_k = w)$$

be a $u - v$ path, where $1 < j < k$. However, $v_{j+1} \in N(v)$ and $d(u, v_{j+1}) = d(u, v) + 1$, a contradiction.

For the converse, let v be a vertex that is not an interior vertex of G. Hence there exists some vertex u such that for every vertex w distinct from u and v, the vertex v does not lie between u and w. Let $x \in N(v)$. Then

$$d(u, x) \leq d(u, v) + d(v, x) = d(u, v) + 1.$$

Since v does not lie between u and x, this inequality is strict and so $d(u, x) \leq d(u, v)$, that is, v is a boundary vertex of u. ∎

Exercises for Section 12.2

12.15 What is the periphery of the graph G of Figure 12.6 in Exercise 12.1.

12.16 What is the periphery of P_n for $n \geq 2$?

12.17 What is the periphery of $K_{s,t}$ for $1 \leq s \leq t$?

12.18 What is the periphery of the Petersen graph?

12.19 Give an example of a connected graph G and a vertex v of G such that (1) v does not belong to the center of G, (2) v does not belong to the periphery of G and (3) v is neither adjacent to a vertex in the center nor adjacent to a vertex in the periphery of G.

12.20 Prove or disprove: There exists a connected graph whose center and periphery are distinct but not disjoint.

12.21 Let G be a connected graph for which some but not all vertices of G have eccentricity 1. Does there exist a connected graph H such that $\text{Per}(H) = G$, where every vertex of H has eccentricity 2 or 3?

12.22 Let G be a connected graph of order $n \geq 3$ that is not complete. Prove that G is the periphery of some graph if and only if $\Delta(G) \leq n - 2$.

12.23 Show that a connected graph G of diameter 2 is the periphery of some graph if and only if G is self-centered.

12.24 Show that for every integer $n \geq 3$, there is exactly one tree of order n that is not the periphery of some graph.

12.25 If a graph G is the eccentric subgraph of a graph H, does it follows that G is the periphery of H?

12.26 For the graph G of Figure 12.14, determine

(a) the set of peripheral vertices of G,

(b) the set of eccentric vertices of G,

(c) the set of boundary vertices of G,

(d) the periphery, eccentric subgraph and boundary of G.

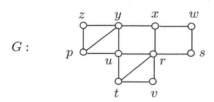

G :

Figure 12.14: The graph in Exercise 12.26

12.27 Give an example of a connected graph G and a set $S = \{v_1, v_2, v_3, v_4\}$ of vertices of G such that v_{i+1} is an eccentric vertex of v_i for $i = 1, 2, 3$ but no vertex of S is an eccentric vertex of any other vertex of S.

12.28 Let F be a nontrivial connected graph with no vertices of eccentricity 1 and let $G = F + K_k$, where $k = 1, 2$.

(a) Prove that the boundary of G is F if $k = 1$.

(b) Prove that the boundary of G is G itself if $k = 2$.

12.29 For each of the graphs G_i, $i = 1, 2$, in Figure 12.15, show that the boundary of G_i is G_i.

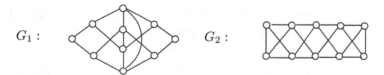

G_1 : G_2 :

Figure 12.15: Graphs in Exercise 12.29

12.30 Show that for every positive integer k, there exists a connected graph G and an eccentric vertex v of G such that $\operatorname{diam}(G) - e(v) \geq k$.

12.31 Prove that for every graph G, there exists a connected graph H such that $\operatorname{Cen}(H) = \operatorname{Int}(H) = G$.

12.3 Excursion: Locating Numbers

Suppose that a certain facility consists of five rooms R_1, R_2, R_3, R_4, R_5 (shown in Figure 12.16). The distance between rooms R_1 and R_3 is 2 and the distance between R_2 and R_4 is also 2. The distance between all other pairs of distinct rooms is 1. The distance between a room and itself is 0. A certain (red) sensor is placed in one of the rooms. If an unauthorized individual should enter a room, then the sensor is able to detect the distance from the room with the red sensor to the room containing the intruder. Suppose, for example, that the sensor is placed in R_1. If an intruder enters room R_3, then the sensor alerts us that an intruder has entered a room at distance 2 from R_1; that is, the intruder is in R_3 since R_3 is the only room at distance 2 from R_1. If the intruder is in R_1, then the sensor indicates that an intruder has entered a room at distance 0 from R_1; that is, the intruder is in R_1. However, if the intruder is in any of the other three rooms, then the sensor tells us that there is an intruder in a room at distance 1 from R_1. But with this information, we cannot determine the precise room containing the intruder. In fact, there is *no* room in which the (red) sensor can be placed to identify the exact location of an intruder in every instance.

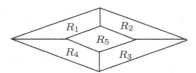

Figure 12.16: A facility consisting of five rooms

On the other hand, if we place the red sensor in R_1 and a blue sensor in R_2 and an intruder enters R_5, say, then the red sensor in R_1 tells us that there is an intruder in a room at distance 1 from R_1, while the blue sensor tells us that the intruder is in a room at distance 1 from R_2, that is, the ordered pair $(1, 1)$ is produced for R_4. Since these ordered pairs are distinct for all rooms, the minimum number of sensors required to detect the exact location of an intruder is 2. Care must be taken, however, as to where the two sensors are placed. For example, we cannot place sensors in R_1 and R_3 since, in this case, the ordered pairs for R_2, R_4 and R_5 are all $(1, 1)$, and we cannot determine the precise location of a possible intruder.

The facility that we have just described can be modeled by the graph of Figure 12.17, whose vertices are the rooms and such that two vertices in this graph are adjacent if the corresponding two rooms are adjacent. This gives rise to a problem involving graphs.

Let G be a connected graph. For an ordered set $W = \{w_1, w_2, \cdots, w_k\}$ of vertices of G and a vertex v of G, the **locating code** (or simply the **code**) of v

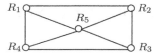

Figure 12.17: A graph modeling a facility with five rooms

with respect to W is the k-vector

$$c_W(v) = (d(v, w_1), d(v, w_2), \cdots, d(v, w_k)).$$

The set W is a **locating set** (also called a **resolving set**) for G if distinct vertices have distinct codes. A locating set containing a minimum number of vertices is a **minimum locating set** (or **metric basis**) for G. The **location number** $\mathrm{loc}(G)$ of G (also called the **metric dimension**) is the number of vertices in a minimum locating set for G. For example, consider the graph G shown in Figure 12.18, which you will notice is isomorphic to the graph of Figure 12.17. The ordered set $W_1 = \{v_1, v_3\}$ is not a locating set for G since $c_{W_1}(v_2) = (1, 1) = c_{W_1}(v_4)$, that is, G contains two vertices with the same code with respect to W_1. On the other hand, $W_2 = \{v_1, v_2, v_5\}$ is a locating set for G since the codes for the vertices of G with respect to W_2 are

$$c_{W_2}(v_1) = (0, 1, 1), \quad c_{W_2}(v_2) = (1, 0, 1), \quad c_{W_2}(v_3) = (2, 1, 1),$$
$$c_{W_2}(v_4) = (1, 2, 1), \quad c_{W_2}(v_5) = (1, 1, 0).$$

However, W_2 is not a minimum locating set for G since $W_3 = \{v_1, v_2\}$ is also a locating set. The codes for the vertices of G with respect to W_3 are

$$c_{W_3}(v_1) = (0, 1), \quad c_{W_3}(v_2) = (1, 0), \quad c_{W_3}(v_3) = (2, 1),$$
$$c_{W_3}(v_4) = (1, 2), \quad c_{W_3}(v_5) = (1, 1).$$

Since no single vertex constitutes a locating set for G, it follows that W_3 is a minimum locating set for this graph G and so $\mathrm{loc}(G) = 2$.

$G:$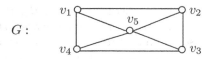

Figure 12.18: Resolving sets in a graph G

Peter Slater described the usefulness of these ideas in connection with U.S. sonar and Coast Guard Loran (long range aids to navigation) stations. We can think of a locating set of a connected graph G as a set W of vertices in G so that each vertex in G is uniquely determined by its distances to the vertices of W.

For every ordered set $W = \{w_1, w_2, \ldots, w_k\}$ of vertices in a connected graph G of order $n \geq 2$, the only vertex of G whose code with respect to W has 0 in its ith coordinate is w_i. So the vertices of W necessarily have distinct codes. Since only vertices of G that are not in W have coordinates all of which are positive, it is only these vertices that need to be examined to determine if their codes are distinct. This implies that the locating number of G is at most $n - 1$. In fact, for every connected graph G of order $n \geq 2$,

$$1 \leq \text{loc}(G) \leq n - 1.$$

Only one connected graph of order $n \geq 2$ has locating number 1.

Theorem 12.12 *A connected graph G of order n has locating number 1 if and only if $G = P_n$.*

Proof. Let $P_n = (v_1, v_2, \ldots, v_n)$. Since $d(v_i, v_1) = i - 1$ for $1 \leq i \leq n$, it follows that $\{v_1\}$ is a minimum locating set of P_n and so $\text{loc}(P_n) = 1$. For the converse, assume that G is a connected graph of order n with locating number 1 and let $W = \{w\}$ be a minimum locating set for G. For each vertex v of G, $c_W(v) = d(v, w)$ is a nonnegative integer less than n. Since the codes of the vertices of G with respect to W are distinct, there exists a vertex u of G such that $d(u, w) = n - 1$. Consequently, the diameter of G is $n - 1$. This implies that $G = P_n$. ∎

At the other extreme, only one connected graph of order $n \geq 2$ has locating number $n - 1$.

Theorem 12.13 *A connected graph G of order $n \geq 2$ has locating number $n-1$ if and only if $G = K_n$.*

Proof. Assume first that $G = K_n$ and let W be a minimum locating set for G. If $u \notin W$, then every coordinate of $c_W(u)$ is 1. Therefore, every minimum locating set for G must contain all but one vertex of G and so $\text{loc}(G) = n - 1$. For the converse, assume that $G \neq K_n$. Then G contains two vertices u and v with $d(u, v) = 2$. Let (u, x, v) be a $u - v$ geodesic in G and let $W = V(G) - \{x, v\}$. Since $d(v, u) = 2$ and $d(x, u) = 1$, it follows that $c_W(x) \neq c_W(v)$ and so W is a locating set. Therefore, $\text{loc}(G) \leq n - 2$. ∎

As we mentioned, if G is a connected graph of order $n \geq 2$, then $1 \leq \text{loc}(G) \leq n-1$. Furthermore, we know exactly the graphs for which the two extreme values are attainable. If, in addition to the order of G, we also know the diameter and maximum degree of G, then bounds for the locating number of G can be improved.

Theorem 12.14 *Let G be a nontrivial connected graph of order $n \geq 2$, diameter d and maximum degree Δ. Then*

$$\lceil \log_3 (\Delta + 1) \rceil \leq \text{loc}(G) \leq n - d. \tag{12.1}$$

Proof. First, we establish the upper bound. Let u and v be vertices of G for which $d(u, v) = d$ and let $(u = v_0, v_1, \ldots, v_d = v)$ be a $u - v$ geodesic. Let

$$W = V(G) - \{v_1, v_2, \ldots, v_d\}.$$

Since $u \in W$ and $d(v_i, u) = i$ for $1 \le i \le d$, it follows that W is a locating set of cardinality $n - d$ for G. Thus $\mathrm{loc}(G) \le n - d$.

Next, we consider the lower bound. Let $\mathrm{loc}(G) = k$ and let $v \in V(G)$ with $\deg v = \Delta$. Moreover, let $N(v)$ be the neighborhood of v and let $W = \{w_1, w_2, \cdots, w_k\}$ be a locating set of G. Observe that if $u \in N(v)$, then for each i, $1 \le i \le k$, the distance $d(u, w_i)$ is one of the numbers $d(v, w_i)$, $d(v, w_i) + 1$ or $d(v, w_i) - 1$ by Theorem 12.3. Moreover, since W is a locating set, $c_W(u) \ne c_W(v)$ for all $u \in N(v)$. Thus there are three possible numbers for each of the k coordinates of $c_W(u)$. On the other hand, since it cannot occur that $d(u, w_i) = d(v, w_i)$ for all i ($1 \le i \le k$), it follows that there are at most $3^k - 1$ distinct codes of the vertices in $N(v)$ with respect to W. Therefore, $|N(v)| = \Delta \le 3^k - 1$, which implies that

$$\mathrm{loc}(G) = k \ge \log_3 (\Delta + 1).$$

Since $\mathrm{loc}(G)$ is an integer, $\mathrm{loc}(G) \ge \lceil \log_3 (\Delta + 1) \rceil$.   ∎

Since the path P_n, $n \ge 2$, has maximum degree $\Delta = 2$ and diameter $d = n-1$, the inequalities in (12.1) say that

$$\lceil \log_3 (2 + 1) \rceil = 1 \le \mathrm{loc}(P_n) \le 1 = n - (n - 1)$$

and both bounds are sharp for P_n. Furthermore, the complete graph K_n has diameter 1 and locating number $n - 1$, so the upper bound is attainable in this case. Each of the two bounds is sharp for other graphs as well.

Let G be a connected graph of order $n \ge 2$. Two vertices u and v of G are **distance similar** if $d(u, x) = d(v, x)$ for all $x \in V(G) - \{u, v\}$. Two distinct vertices u and v are therefore distance similar if either

(1) $uv \notin E(G)$ and $N(u) = N(v)$ or

(2) $uv \in E(G)$ and $N[u] = N[v]$.

Distance similarity is an equivalence relation on $V(G)$ (see Exercise 12.36). Consequently, $V(G)$ can be partitioned into k distinct distance similar equivalence classes, say $V_1, V_2, \cdots, V_k$. For each integer i ($1 \le i \le k$), the set V_i is either independent in G or induces a complete subgraph of G. Necessarily, each locating set of G either contains all or all but one vertex in each equivalence class V_i. Therefore,

$$\mathrm{loc}(G) \ge n - k.$$

Consider the graph G of Figure 12.19, which has order $n = 11$. Three of the distance similar equivalence classes are $V_1 = \{u_1, w_1\}$, $V_2 = \{u_2, w_2\}$ and $V_3 = \{u_3, w_3\}$. Each of the remaining five classes consists of a single vertex. Thus there are $k = 8$ equivalence classes and so $\mathrm{loc}(G) \ge n - k = 3$. We may assume that any locating set for G contains w_1, w_2 and w_3. In fact, $W = \{w_1, w_2, w_3\}$ *is* a locating set and consequently is a minimum locating set. Therefore, $\mathrm{loc}(G) = n - k$. The codes for the vertices of $V(G) - W$ with respect to W are

$$c_W(u_1) = (1,3,4), \quad c_W(u_2) = (3,2,3), \quad c_W(u_3) = (4,3,1),$$
$$c_W(u_4) = (1,2,3), \quad c_W(u_5) = (1,3,3), \quad c_W(u_6) = (2,1,2),$$
$$c_W(u_7) = (2,2,2), \quad c_W(u_8) = (3,2,1).$$

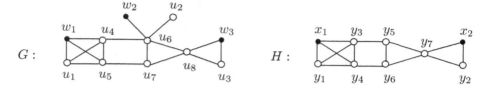

Figure 12.19: The graphs G and H

Next, consider the graph H of order $n = 9$ of Figure 12.19. In this graph, $V_1 = \{x_1, y_1\}$ and $V_2 = \{x_2, y_2\}$ are distance similar equivalence classes, while each of the remaining classes consists of a single vertex. Thus, there are $k = 7$ equivalence classes and so $\text{loc}(H) \geq n - k = 2$. Let $W = \{x_1, x_2\}$. The codes for the vertices of $V(G) - W$ with respect to W are

$$c_W(y_1) = (1,4), \quad c_W(y_2) = (4,2), \quad c_W(y_3) = (1,3), \quad c_W(y_4) = (1,3),$$
$$c_W(y_5) = (2,2), \quad c_W(y_6) = (2,2), \quad c_W(y_7) = (3,1).$$

Since $c_W(y_3) = c_W(y_4)$ and $c_W(y_5) = c_W(y_6)$, the set W is not a locating set for H. On the other hand, if $W' = \{x_1, x_2, y_3\}$, then W' is a locating set and so $\text{loc}(H) = 3$. Therefore, $\text{loc}(H) > n - k$.

Exercises for Section 12.3

12.32 Determine $\text{loc}(C_n)$ for $n \geq 3$.

12.33 Determine $\text{loc}(K_{s,t})$ for $1 \leq s \leq t$.

12.34 (a) Give an example of a connected graph of order $n \geq 3$ (different from P_n and K_n) with diameter d such that $\text{loc}(G) = n - d$.

   (b) Give an example of a connected graph of order $n \geq 3$ (different from P_n and K_n) with maximum degree Δ such that $\text{loc}(G) = \lceil \log_3 (\Delta + 1) \rceil$.

12.35 Prove that for each pair k, n of integers with $1 \leq k \leq n - 1$, there exists a connected graph G of order n with $\text{loc}(G) = k$.

12.36 Let G be a connected graph. Show that distance similarity is an equivalence relation on $V(G)$.

12.37 Determine the locating number of the graph G of Figure 12.20. Find a minimum locating set W for G and indicate the code of each vertex of G with respect to W.

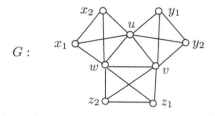

Figure 12.20: The graph G in Exercise 12.37

12.4   Excursion: Detour and Directed Distance

While the standard distance $d(u, v)$ from a vertex u to a vertex v in a connected graph G is the length of a shortest $u - v$ path in G, it is by no means the only definition of distance. For two vertices u and v in a connected graph G of order n, the **detour distance** $D(u, v)$ from u to v is defined as the length of a *longest* $u - v$ path in G. A $u - v$ path of length $D(u, v)$ is called a $u - v$ **detour**. For example, for the graph G of Figure 12.21 $d(u, v) = 3$ while $D(u, v) = 8$. A $u - v$ detour (drawn in bold) is also shown in that figure.

$G :$

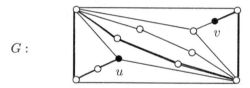

Figure 12.21: Illustrating detour distance

As with standard distance, detour distance is also a metric on the vertex set of every connected graph.

Theorem 12.15 *Detour distance is a metric on the vertex set of every connected graph.*

Proof. Let G be a connected graph. Since (1) $D(u, v) \geq 0$, (2) $D(u, v) = 0$ if and only if $u = v$ and (3) $D(u, v) = D(v, u)$ for every pair u, v of vertices of G, it remains only to show that detour distance satisfies the triangle inequality.

Let u, v and w be any three vertices of G. Since the inequality $D(u, w) \leq D(u, v) + D(v, w)$ holds if any two of these three vertices are the same vertex, we assume that u, v and w are distinct. Let P be a $u - w$ detour in G of length $k = D(u, w)$. We consider two cases.

Case 1. v lies on P. Let P_1 be the $u - v$ subpath of P and let P_2 be the $v - w$ subpath of P. Suppose that the length of P_1 is s and the length of P_2 is

t. So $s + t = k$. Therefore,

$$D(u, w) = k = s + t \le D(u, v) + D(v, w).$$

Case 2. v does not lie on P. Since G is connected, there is a shortest path Q from v to a vertex of P. Suppose that Q is a $v - x$ path. Thus x lies on P but no other vertex of Q lies on P. Let r be the length of Q. Thus $r > 0$ (see Figure 12.22).

Figure 12.22: A step in the proof of Case 2

Let the $u - x$ subpath P' of P have length a and the $x - w$ subpath P'' of P have length b. Then $a \ge 0$ and $b \ge 0$. Therefore, $D(u, v) \ge a + r$ and $D(v, w) \ge b + r$. So

$$D(u, w) = k = a + b < (a + r) + (b + r) \le D(u, v) + D(v, w),$$

establishing the triangle inequality. ∎

The detour eccentricity, detour radius and detour diameter are defined as expected. Let G be a connected graph and let v be a vertex of G. The **detour eccentricity** $e_D(v)$ of v is the maximum detour distance from v to a vertex of G. The minimum detour eccentricity among the vertices of G is the **detour radius** $\text{rad}_D(G)$ of G and the maximum detour eccentricity is its **detour diameter** $\text{diam}_D(G)$. There are upper and lower bounds for $\text{diam}_D(G)$ that are analogues (as are the proofs) to those given for the standard diameter of G in Theorem 12.1.

Theorem 12.16 *For every nontrivial connected graph G,*

$$\text{rad}_D(G) \le \text{diam}_D(G) \le 2\text{rad}_D(G).$$

Proof. The definitions of $\text{rad}_D(G)$ and $\text{diam}_D(G)$ give the inequality $\text{rad}_D(G) \le \text{diam}_D(G)$. Now let u and v be two vertices of G such that $D(u, v) = \text{diam}_D(G)$ and let w be a vertex of G such that $e_D(w) = \text{rad}_D(G)$. Since detour distance is a metric on $V(G)$,

$$\text{diam}_D(G) = D(u, v) \le D(u, w) + D(w, v) \le 2\text{rad}_D(G),$$

as desired. ∎

Every pair a, b of positive integers can be realized as the detour radius and detour diameter, respectively, of some connected graph provided $a \le b \le 2a$.

Theorem 12.17 *For each pair a, b of positive integers with $a \leq b \leq 2a$, there exists a connected graph G with*

$$\text{rad}_D(G) = a \text{ and } \text{diam}_D(G) = b.$$

Proof. For $a = b = k \geq 1$, the complete graph K_{k+1} has the desired property. For $a < b \leq 2a$, let G be the graph of order $b + 1$ obtained by identifying a vertex v of K_{a+1} and a vertex of K_{b-a+1} (see Figure 12.23 for $a = 3$ and $b = 5$). Since $b \leq 2a$, it follows that $b - a + 1 \leq a + 1$. Thus $e_D(v) = a$. Since there is a Hamiltonian path in G with initial vertex x for every vertex $x \in V(G) - \{v\}$, it follows that $e_D(x) = b$. Hence $\text{rad}_D(G) = a$ and $\text{diam}_D(G) = b$.   ∎

$$G: \quad$$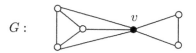

Figure 12.23: The graph G of Theorem 12.17 for $a = 2$ and $b = 5$

For integers a and b with $a < b \leq 2a$, each vertex in the graph G in the proof of Theorem 12.17 has eccentricity a or b. So unlike ordinary eccentricity, if k is an integer such that $\text{rad}_D(G) < k < \text{diam}_D(G)$, there may not be a vertex x of G such that $e_D(x) = k$.

We have now seen two definitions of distance in connected graphs, both of which are metrics on the vertex set of the graph. We now turn to digraphs.

Let D be a connected digraph. For two vertices u and v of D, recall that the **directed distance** $\vec{d}(u, v)$ from u to v is the length of a shortest directed $u - v$ path in D. Actually, when we refer to a $u - v$ path in a digraph D, we always refer to a directed $u - v$ path and when we refer to the distance from a vertex u to a vertex v in D, we mean the directed distance from u to v. A $u - v$ path of length $\vec{d}(u, v)$ is a $u - v$ **geodesic**. The fact that D is connected does not guarantee the existence of a $u - v$ path in D. For this reason, when discussing distance in digraphs, we ordinarily assume that D is strong. Typically, we take D to be a strong *oriented* graph. For example, for the strong tournament T of order 3 in Figure 12.24, $\vec{d}(u, v) = 1$. Since $\vec{d}(v, u) = 2$, however, it follows that $\vec{d}(u, v) \neq \vec{d}(v, u)$, that is, this distance is not symmetric and is therefore not a metric on the vertex set of a nontrivial strong oriented graph.

$$T: \quad$$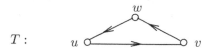

Figure 12.24: Directed distance in strong digraphs

Directed distance does satisfy the triangle inequality however. Let u, v and w be vertices of a strong digraph D. Let P_1 be a $u - v$ geodesic and P_2 a $v - w$

geodesic in D. Then P_1 followed by P_2 produces a $u - w$ walk W of length $\vec{d}(u, v) + \vec{d}(v, w)$. By Theorem 7.2, D contains a $u - w$ path whose length is at most the length of W and so

$$\vec{d}(u, w) \leq \vec{d}(u, v) + \vec{d}(v, w).$$

For directed distance, the eccentricities of the vertices of a strong digraph are defined as expected. Let D be a strong digraph and let v be a vertex of D. The **eccentricity** $e(v)$ of v is the length of a longest geodesic from v to a vertex of D. The minimum eccentricity among the vertices of D is its **radius** rad(D) and the maximum eccentricity is its **diameter** diam(D).

To illustrate these concepts, let r and d be positive integers with $r \leq d$ and let C_{r+1} be the directed $(r + 1)$-cycle, say $C_{r+1} = (u_1, u_2, \ldots, u_{r+1}, u_1)$. Notice that $e(u_i) = r$ for $1 \leq i \leq r + 1$. Next, let $P_d = (v_1, v_2, \ldots, v_d)$ be a path of order d. The digraph D is constructed from C_{r+1} and P_d by joining every vertex of C_{r+1} to v_i for $1 \leq i \leq d - r + 1$ and joining v_d to every vertex of C_{r+1} (see Figure 12.25). Thus

$$
\begin{aligned}
e(u_i) &= r \quad \text{for } 1 \leq i \leq r + 1 \\
e(v_1) &= d \\
e(v_i) &= d - i + 2 \quad \text{for } 2 \leq i \leq d - r + 1 \\
e(v_i) &= r \quad \text{for } d - r + 2 \leq i \leq d.
\end{aligned}
$$

In particular, this shows that the diameter of D can be more than twice its radius and so for directed distance, there are no theorems analogous to Theorems 12.1 and 12.2 in this case.

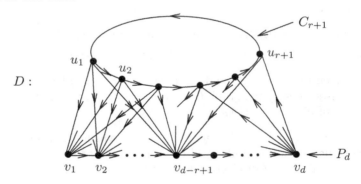

Figure 12.25: A digraph D with rad(D) $= r$ and diam(D) $= d$

Exercises for Section 12.4

12.38 (a) Give an example of a connected graph G of order 5 or more such that $D(u, v) = d(u, v)$ for every pair u, v of vertices of G.

(b) Determine all connected graphs G for which $D(u,v) = d(u,v)$ for every pair u, v of vertices of G.

12.39 Give an example of a connected graph G and a positive constant K such that $D(u,v) = d(u,v) + K$ for every pair u, v of distinct vertices of G.

12.40 Give an example of a connected graph G having the property that for every pair u, v of distinct vertices of G, each $u - v$ geodesic and $u - v$ detour have only u and v in common.

12.41 Determine $\mathrm{rad}_D(G)$ and $\mathrm{diam}_D(G)$ for $G = K_n, C_n, Q_n$ $(n \geq 3)$ and for $G = K_{r,s}$ $(2 \leq r \leq s)$.

12.42 Prove that $|e_D(u) - e_D(v)| \leq D(u,v)$ for every two vertices u and v in a connected graph D.

12.43 Prove that the detour center of every connected graph G lies in a single block of G.

12.44 Let G be a connected graph and define $\overline{d}(u,v) = d(u,v) + D(u,v)$. Is $\overline{d}$ a metric on $V(G)$?

12.45 Let G be a connected graph and let d' be any metric on $V(G)$.

(a) Define $\mathrm{rad}_{d'}(G)$ and $\mathrm{diam}_{d'}(G)$.

(b) Prove or disprove: $\mathrm{rad}_{d'}(G) \leq \mathrm{diam}_{d'}(G) \leq 2\,\mathrm{rad}_{d'}(G)$.

12.46 Let G be a connected graph with cut-vertices and let $v, w \in V(G)$. Prove or disprove:

(a) If $e_D(v) = \mathrm{rad}_D(G)$, then v is a cut-vertex of G.

(b) If $e_D(w) = \mathrm{diam}_D(G)$, then w is not a cut-vertex of G.

12.47 For a given integer $n \geq 3$, find all integers k with $1 \leq k \leq n - 1$ for which there exists a connected graph G of order n such that $D(u,v) = k$ for every pair u, v of distinct vertices of G.

12.48 Give an example of an integer $n \geq 4$ for which there exist two non-isomorphic graphs G_1 and G_2 of order n such that $d(u,v) + D(u,v) = n$ for every pair u, v of distinct vertices of both G_1 and G_2.

12.49 Prove that for every pair a, b of integers with $1 \leq a \leq b$, there exists

(a) a connected graph F for which $\mathrm{rad}(F) = a$ and $\mathrm{rad}_D(F) = b$;

(b) a connected graph H for which $\mathrm{diam}(H) = a$ and $\mathrm{diam}_D(H) = b$.

12.50 For vertices u and v in a strong digraph D, define $d_s(u,v) = \overrightarrow{d}(u,v) + \overrightarrow{d}(v,u)$.

(a) Is this distance a metric on $V(D)$?

(b) Ask and answer a question of your own concerning this concept.

12.51 For vertices u and v in a strong digraph D, define $d_p(u,v) = \vec{d}(u,v) \cdot \vec{d}(v,u)$. Is this distance a metric on $V(D)$?

12.5 Exploration: Channel Assignment

Radio waves, which are electromagnetic waves propagated by antennas, have different frequencies. When a radio receiver is tuned to a particular frequency, a specific signal can be accessed. In the United States, it is the responsibility of the Federal Communications Commission (FCC) to decide which frequencies are used for which purposes. It is also the FCC that licenses specific frequencies to radio stations as well as call letters for the stations. AM (amplitude modulated) radio is in a band of 550 kHz (kilohertz) to 1700 kHz which means that AM radio broadcasts in a frequency band of 550,000 to 1,700,000 cycles per second. The first radio broadcasts occurred around 1906. Frequency allocation for AM radio began in the 1920s. Because radio technology was not highly developed during that period, low frequencies for AM radio were appropriate then.

Perhaps the major inventor in the early days of radio was Edwin Armstrong (1890-1954). It was Armstrong who in 1933 developed the complete FM (frequency modulated) system. All FM radio stations transmit radio waves in a band of frequencies between 88 MHz (megahertz) and 108 MHz, that is, the transmitter of an FM radio station oscillates at an assigned frequency between 88,000,000 and 108,000,000 cycles per second. Only FM radio stations are permitted to use these frequencies. Certain frequencies above and below these are reserved for television stations. For example, the band 54 MHz to 88 MHz is for channels 2 through 6, while the band 174 MHz to 220 MHz is for channels 7 through 13.

The FM radio frequency band, which, as we said, begins at 88.0 MHz and ends at 108.0 MHz, is divided into 100 channels, each having a width of 0.2 MHz (or 200 kHz). The frequency that is identified with an FM radio station is the midpoint of its 200 kHz channel. For example, in the state of Michigan, the FM radio station WVTI is located near the city of Holland and broadcasts on the frequency 96.1 MHz, while the FM radio station WFAT located in the city of Portage broadcasts on 96.5 MHz. It is not uncommon for radio stations to give themselves names (actually nicknames). For example, the station WVTI calls itself "the new I-96" and the nickname of WFAT is "the fat one." Each FM radio station is assigned a station class that depends on a number of factors, including its antenna height and the effective radiated power of its signal. Five common station classes A, B1, B, C1 and C are described in Figure 12.26. For each class, the maximum Effective Radiated Power (ERP) of the signal, measured in kilowatts (kW), and the maximum antenna Height Above Average Terrain

(HAAT) of the station, measured in meters (m), are indicated in Figure 12.26 as well. The station classes for FM radio stations WVTI and WFAT are shown in Figure 12.27.

FM Station Class	Maximum ERP (kW)	Maximum HAAT (m)
Class A	6.0	100
Class B1	25.0	100
Class B	50.0	150
Class C1	100.0	299
Class C	100.0	600

Figure 12.26: Examples of FM station classes

	WVTI The Fat One	WFAT The New I-96
ERP (kW)	3.6	50
HAAT (m)	79	150
Station class	A	B

Figure 12.27: ERP and HAAT of two stations

Channels assigned to FM radio stations depend not only on the effective radiated power of their signals and the heights of their antennas but also on their distances from other stations. In particular, two stations that share the same channel (called co-channel stations) must be separated by at least 115 kilometers (or 71 miles); however, the required separation depends on the classes of the two stations. Two channels are first-adjacent, or simply adjacent, if their frequencies differ by 200 kHz, that is, if they are consecutive on the FM radio dial. For example, FM stations on channels 105.7 MHz and 105.9 MHz are adjacent. The distance between two radio stations on adjacent channels must be at least 72 kilometers (or 45 miles). Again, this distance varies according to the classes of the two stations. Moreover, the distance between two radio stations whose channels differ by 400 kHz or 600 kHz (so-called second- or third-adjacent channels) must be at least 31 kilometers. The actual minimum distance between stations on such channels is shown in Figure 12.28.

Since the FM station WFAT is a Class A station broadcasting on channel 96.5 MHz and WVTI is a Class B station broadcasting on channel 96.1 MHz, they are second-adjacent station and the distance between them is required to be at least 69 kilometers. This condition is met, though just barely.

In general then, FM radio stations located within a certain proximity of one another must be assigned distinct channels. The nearer two stations are to each other, the greater the difference must be in their assigned channels. The task of efficiently allocating channels to transmitters is called the **Channel Assignment Problem**.

The use of graph theory to study the Channel Assignment Problem dates back to at least 1970. In 1980 William Hale provided a model of the Channel Assignment Problem. Most often the Channel Assignment Problem has been

	Co-Channels	First-Adjacent Channels	Second- or Third-Adjacent Channels
A to A	115	72	31
A to B1	143	96	48
A to B	178	113	69
A to C1	200	133	75
A to C	226	165	95
B1 to B1	175	114	50
B1 to B	211	145	71
B1 to C1	233	161	77
B1 to C	259	193	105
B to B	241	169	74
B to C1	270	195	79
B to C	274	217	105
C1 to C1	245	177	82
C1 to C	270	209	105
C to C	290	241	105

Figure 12.28: Required distance (in kilometers) between FM radio stations

modeled as a graph coloring problem, where (1) the transmitters are the vertices of a graph, (2) two vertices (transmitters) are adjacent if they are sufficiency close to each other, (3) the colors of the vertices are the channels assigned to the transmitters and (4) some sort of minimum separation rule is stipulated, that is, for every pair of colors, there is a minimum allowable distance between two distinct vertices assigned these colors.

We consider one of these models that was inspired by the Channel Assignment Problem. For a connected graph G of order n and an integer k with $1 \leq k \leq \text{diam}(G)$, a **radio k-coloring** of G is a function $c : V(G) \to \mathbf{N}$ for which

$$d(u, v) + |c(u) - c(v)| \geq 1 + k$$

for every two distinct vertices u and v of G. For $k = 2$, a radio 2-coloring then requires that

$$d(u, v) + |c(u) - c(v)| \geq 3$$

for every two distinct vertices u and v of G. This says that

(1) the colors assigned to adjacent vertices must differ by at least 2,

(2) the colors assigned to vertices whose distance is 2 must differ and

(3) there is no restriction on colors assigned to vertices whose distance is 3 or more.

A radio 2-coloring of a graph H is shown in Figure 12.29.

The **value** $\text{rc}_k(c)$ of a radio k-coloring c of a connected graph G is the maximum color assigned to a vertex of G, while the **radio k-chromatic number** $\text{rc}_k(G)$ of G is $\min\{\text{rc}_k(c)\}$ over all radio k-colorings c of G. A radio k-coloring c of G is a **minimum radio k-coloring** if $\text{rc}_k(c) = \text{rc}_k(G)$. For any minimum radio k-coloring c of a connected graph, there are necessarily vertices u and v such that $c(u) = 1$ and $c(v) = \text{rc}_k(c)$.

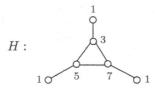

Figure 12.29: A radio 2-coloring of a graph

For the radio 2-coloring c of the graph H of Figure 12.29, $\text{rc}_2(c) = 7$. However, the radio 2-chromatic number of H is not 7. A radio 2-coloring having value 6 is shown in Figure 12.30. In fact, $\text{rc}_2(H) = 6$.

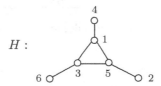

Figure 12.30: A minimum radio 2-coloring of a graph

Example 12.18 *For the graph H of Figure* 12.29, $\text{rc}_2(H) = 6$.

Solution. The radio 2-coloring given in Figure 12.30 has a value of 6 and so $\text{rc}_2(H) \leq 6$. Assume, to the contrary, that there is a radio 2-coloring of H using only the colors 1, 2, 3, 4, 5. Since the vertices of degree 3 in H form a triangle, they must be colored 1, 3, 5. Let y be the vertex colored 3 and let v be the end-vertex of H that is adjacent to y (see Figure 12.31). Since $|c(v) - c(y)| \geq 2$ and $1 \leq c(v) \leq 5$, it follows that either $c(v) = 1$ or $c(v) = 5$. However, there is a contradiction in both cases since $d(v, x) = d(v, w) = 2$, implying that $c(v) \neq 1$ and $c(v) \neq 5$.   ◇

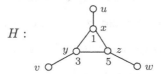

Figure 12.31: A step in showing $\text{rc}_2(H) = 6$ in Example 12.18

There are some values of k for which radio k-colorings are well-studied parameters. For $k = 1$, a radio 1-coloring of a connected graph G requires that

$$d(u, v) + |c(u) - c(v)| \geq 2$$

for every two distinct vertices u and v of G. This says that different colors must be assigned to adjacent vertices and there is no restriction on vertices whose distance is 2 or more. However, this is the standard definition of a proper vertex coloring and shows that radio k-coloring is a generalization of standard vertex coloring.

Radio 2-colorings have also been studied and are also referred to as **labelings at distance 2** and **L(2, 1)-labelings**. For connected graphs of diameter d, a radio d-coloring of G is also called a **radio labeling**. More specifically, a **radio labeling** of a connected graph G is a function $c : V(G) \rightarrow \mathbf{N}$ with the property that

$$d(u, v) + |c(u) - c(v)| \geq 1 + \operatorname{diam}(G)$$

for every two distinct vertices u and v of G. Since $d(u, v) \leq \operatorname{diam}(G)$ for every two vertices u and v of G, it follows that no two vertices are colored (or labeled) the same in a radio labeling.

For a radio labeling c of a connected graph G, the **value** $\operatorname{rn}(c)$ of c is the maximum label (or color) assigned to a vertex of G, while the **radio number** $\operatorname{rn}(G)$ of the graph G is the minimum value of a radio labeling of G. A radio labeling c with $\operatorname{rn}(c) = \operatorname{rn}(G)$ is a **minimum radio labeling**. To illustrate these concepts, we consider the following example.

Example 12.19 *Determine* $\operatorname{rn}(G)$ *for the the graph G of Figure 12.32(a).*

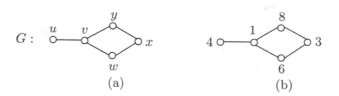

$$G:$$

(a)   (b)

Figure 12.32: A radio labeling of a graph

Solution. Since $\operatorname{diam}(G) = 3$, it follows that in any radio labeling of G, the labels of every two adjacent vertices must differ by at least 3 and the labels of every two vertices whose distance is 2 must differ by at least 2. The colors of two vertices can differ by exactly 1 only if their distance is 3. Thus the labeling of G given in Figure 12.32(b) is a radio labeling. Consequently, $\operatorname{rn}(G) \leq 8$. On the other hand, $\operatorname{rn}(G) \neq 7$, for assume, to the contrary, that there is a radio labeling c of G with $\operatorname{rn}(c) = \operatorname{rn}(G) = 7$. Since exactly two of the integers 2, 3, 4, 5, 6 are not used in this labeling, either three consecutive integers in $\{1, 2, \cdots, 7\}$ are labels for the vertices of G or two pairs of consecutive integers are labels, both of which are impossible since u and v are the only two vertices of G whose distance is 3. Therefore, $\operatorname{rn}(G) = 8$ and the labeling given in Figure 12.32(b) is a minimum radio labeling.   ◇

Let G be a connected graph with $V(G) = \{v_1, v_2, \cdots, v_n\}$ and $d = \operatorname{diam}(G)$. Let c be a radio labeling of G. The **complementary labeling** $\bar{c}$ of c is defined by

$$\bar{c}(v_i) = (\operatorname{rn}(c) + 1) - c(v_i)$$

for all i with $1 \leq i \leq n$. The complementary labeling of the labeling of the graph G in Figure 12.32(b) is shown in Figure 12.33.

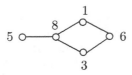

Figure 12.33: illustrating a complementary labeling

Since $|\overline{c}(v_i) - \overline{c}(v_j)| = |c(v_i) - c(v_j)|$ for all i, j with $1 \leq i < j \leq n$, we have the following observation.

Theorem 12.20 *Let G be a connected graph. If c is a radio labeling of G, then $\overline{c}$ is also a radio labeling of G.*

Hence if c is a minimum radio labeling of a connected graph G, then so too is $\overline{c}$ and $\mathrm{rn}(\overline{c}) = \mathrm{rn}(c)$.

Exercises for Section 12.5

12.52 Show that if there exists a radio labeling of a graph G having value k, then there exists a radio labeling of G having value $k + 1$.

12.53 Determine $\mathrm{rn}(K_n)$ for $n \geq 3$.

12.54 Determine $\mathrm{rc}_k(K_{r,s})$, where $1 \leq r \leq s$, for $k = 1, 2$.

12.55 Determine $\mathrm{rc}_k(P_n)$ for $3 \leq n \leq 7$ and $1 \leq k \leq n - 1$.

12.56 Determine $\mathrm{rc}_k(C_n)$ for $3 \leq n \leq 7$ and $1 \leq k \leq n/2$.

12.6   Exploration: Distance Between Graphs

Two graphs G and H are, of course, either isomorphic or they are not. For two graphs G and H, we often ask (and are satisfied with the answer to) the question:

Are G and H isomorphic?

Since the answer is obvious if G and H have different orders or different sizes, the question is only interesting if G and H have the same order and same size. Once the question is answered, one way or the other, we probably just go on to consider other questions. However, if the answer is *no*, that is, if G and H are not isomorphic, other questions may occur to us. For example, showing that G and H are not isomorphic may have been quite easy (as the graphs were clearly different) or extraordinarily difficult (as the graphs were strikingly similar). This suggests the problem of comparing two graphs, at least two graphs of the same order and same size. That is, how close to being isomorphic are two non-isomorphic graphs? There are several ways of answering this. We look at one of these.

Let G and H be two graphs of order n and size m for positive integers n and m, where then $1 \leq m \leq \binom{n}{2}$. We define a distance $d(G, H)$ between them, called the rotation distance. If $G \cong H$, then define $d(G, H) = 0$. Suppose then that $G \not\cong H$. We say that G can be **transformed into H by an edge rotation** (or G can be **rotated** into H) if G contains distinct vertices u, v and w such that $uv \in E(G)$, $uw \notin E(G)$ and $H \cong G - uv + uw$. For example, the graph G of Figure 12.34 can be rotated into H but G cannot be rotated into F.

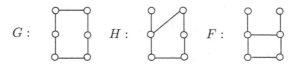

Figure 12.34: Edge rotations

For two graphs G and H of the same order and same size, the **rotation distance** $d(G, H)$ between G and H is defined as the smallest nonnegative integer k for which there exists a sequence G_0, G_1, ..., G_k of graphs such that $G_0 \cong G$, $G_k \cong H$ and G_i can be rotated into G_{i+1} for $i = 0, 1, \ldots, k - 1$. Thus for the graphs G, H and F of Figure 12.34, $d(G, H) = 1$ and $d(G, F) = 2$. For the graphs G' and H' of Figure 12.35, $d(G', H') = 3$.

Figure 12.35: Graphs G' and H' with rotation distance 3

The rotation distance is a metric on the set of all graphs having a fixed order and fixed size and provides a measure of how close two graphs are to being isomorphic - the smaller the distance, the closer the graphs are to being isomorphic.

Another concept occurs quite naturally when discussing this distance. For two nonempty graphs G_1 and G_2 (not necessarily having the same order or same size), a graph G is called a **greatest common subgraph** of G_1 and G_2 if G is a graph of maximum size that is isomorphic to both an edge-induced subgraph of G_1 and an edge-induced subgraph of G_2. The graphs G_1 and G_2 of Figure 12.36 have three distinct greatest common subgraphs, namely G, G' and G''.

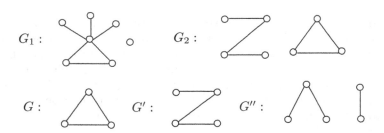

Figure 12.36: Greatest common subgraphs

There is an upper bound for the rotation distance between two graphs in terms of the size of a greatest common subgraph of these graphs.

Theorem 12.21 *Let G and H be graphs of order n and size m for positive integers n and m and let F be a greatest common subgraph of G and H, where F has size s. Then*

$$d(G, H) \le 2(m - s).$$

Proof. If $s = m$, then $G = H$ and $d(G, H) = 0$. Hence we may assume that $1 \le s < m$. Let G^* and H^* be edge-induced subgraphs of G and H, respectively, such that $G^* \cong H^* \cong F$. Furthermore, assume that $V(G) = V(H) = \{v_1, v_2, \ldots, v_n\}$ and that the subgraphs G^* and H^* are identically labeled. Since $G \not\cong H$, the graph G contains an edge $v_i v_j$ that is not in H and H contains an edge $v_p v_q$ that is not in G. Suppose that $\{v_i, v_j\} \cap \{v_p, v_q\} \ne \emptyset$, say $v_j = v_p$. Then G can be rotated into $G_1 = G - v_i v_j + v_j v_q$ and $d(G, G_1) = 1$. Next, suppose that $\{v_i, v_j\} \cap \{v_p, v_q\} = \emptyset$.

Suppose that at least one of v_i and v_j is not adjacent in G to at least one of v_p and v_q, say $v_i v_p \notin E(G)$. Then G can be rotated into $G' = G - v_i v_j + v_i v_p$ and G' can be rotated into $G'' = G' - v_i v_p + v_p v_q$ and so $d(G, G'') \le 2$.

If, on the other hand, each of v_i and v_j is adjacent to both v_p and v_q, then G can be rotated into $G_1 = G - v_i v_p + v_p v_q$ and G_1 can be rotated into $G_2 = G_1 - v_i v_j + v_i v_p$ and so $d(G, G_2) \le 2$.

In any case, G can be transformed into $H' = G - v_i v_j + v_p v_q$ by at most two rotations and so $d(G, H') \le 2$. The graphs H' and H have $s + 1$ edges in common. Continuing in this manner, we have $d(G, H) \le 2(m - s)$.   ∎

How a collection of graphs, all of the same order and same size, are related to each other in terms of rotation can itself be modeled by a graph. Let $S = \{G_1, G_2, \ldots, G_k\}$ be such a set. Then the **rotation distance graph** $D(S)$ of S has S as its vertex set and vertices (graphs) G_i and G_j are adjacent if $d(G_i, G_j) = 1$. The distance graph $D(S)$ is shown for the set $S = \{G_1, G_2, G_3, G_4\}$ of Figure 12.37.

A graph G is **a rotation distance graph** if $G \cong D(S)$ for some set S of graphs. Consequently, the graph $G = K_4 - e$ of Figure 12.37 is a rotation distance graph.

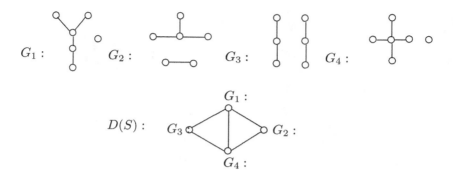

Figure 12.37: A rotation distance graph

Exercises for Section 12.6

12.57 For each positive integer k, show, with justification, that there exist two graphs G and H such that $d(G, H) = k$.

12.58 Give an example of two graphs G and H that have a unique greatest common subgraph.

12.59 For each positive integer k, give an example of two graphs G and H that have exactly k greatest common subgraphs.

12.60 Show that the bound in Theorem 12.21 is sharp.

12.61 Show that K_3 is a rotation distance graph.

12.62 Show that C_4 is a rotation distance graph.

12.63 Define another distance d' between graphs of a fixed order and fixed size and give an example of two graphs G and H for which $d'(G, H)$ does not equal the rotation distance between these graphs.

Chapter 13

Domination

13.1   The Domination Number of a Graph

For a vertex v of a graph G, recall that a neighbor of v is a vertex adjacent to v in G. Also, the neighborhood (or **open neighborhood**) $N(v)$ of v is the set of neighbors of v. The **closed neighborhood** $N[v]$ is defined as $N[v] = N(v) \cup \{v\}$. A vertex v in a graph G is said to **dominate** itself and each of its neighbors, that is, v dominates the vertices in its closed neighborhood $N[v]$. Therefore, v dominates $1 + \deg v$ vertices of G.

A set S of vertices of G is a **dominating set** of G if every vertex of G is dominated by some vertex in S. Equivalently, a set S of vertices of G is a dominating set of G if every vertex in $V(G) - S$ is adjacent to some vertex in S. Consider the graph G of Figure 13.1. The sets $S_1 = \{u, v, w\}$ and $S_2 = \{u_1, u_4, v_1, v_4\}$, indicated by solid vertices, are both dominating sets in G.

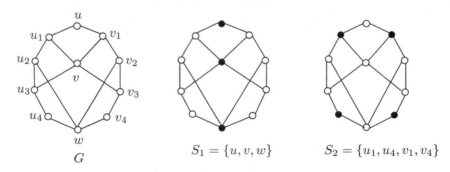

$$S_1 = \{u, v, w\} \qquad S_2 = \{u_1, u_4, v_1, v_4\}$$

Figure 13.1: Two dominating sets in a graph G

A **minimum dominating set** in a graph G is a dominating set of minimum cardinality. The cardinality of a minimum dominating set in G is called the

domination number of G and is denoted by $\gamma(G)$. (The notation used to denote the domination number of a graph is the same as that used for its genus. However, this is the common notation in both instances. There should be no confusion as domination and genus will not occur in the same discussion.)

The topic of domination began with Claude Berge in 1958 and Oystein Ore in 1962, with Ore actually using that term. However, it wasn't until 1977, following an article by Ernie Cockayne and Stephen Hedetniemi, that domination became an area of study by many. In 1998, a book devoted to this subject was written by Teresa Haynes, Hedetniemi and Peter Slater. Well over 2000 articles have been written on domination.

Since the vertex set of a graph is always a dominating set, the domination number is defined for every graph. If G is a graph of order n, then $1 \le \gamma(G) \le n$. A graph G of order n has domination number 1 if and only if G contains a vertex v of degree $n-1$, in which case $\{v\}$ is a minimum dominating set; while $\gamma(G) = n$ if and only if $G = \overline{K}_n$, in which case $V(G)$ is the unique (minimum) dominating set.

Let's return to the graph G of Figure 13.1. We saw that the set $S_1 = \{u, v, w\}$ is a dominating set for G. Therefore, $\gamma(G) \le 3$. To show that the domination number of G is actually 3, it is required to show that there is no dominating set with two vertices. Notice that the order of G is 11 and that the degree of every vertex of G is at most 4. This means that no vertex can dominate more than 5 vertices and that every two vertices dominate at most 10 vertices. That is, $\gamma(G) > 2$ and so $\gamma(G) = 3$.

Let's look at a practical example involving domination. Figure 13.2 shows a portion of a city, consisting of six city blocks, determined by three horizontal streets and four vertical streets. A security protection agency has been retained to watch over the street intersections. A security guard stationed at an intersection can observe the intersection where he or she is located as well as all intersections up to one block away in straight line view from this intersection. The question is: What is the minimum number of security officers needed to guard all 12 intersections? Figure 13.2 shows four intersections where security guards can be placed (labeled by SG) so that all 12 intersections are under observation.

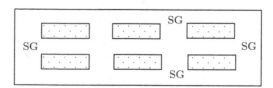

Figure 13.2: A city map

This situation can be modeled by the graph G of Figure 13.3. The graph G is actually the Cartesian product $P_3 \times P_4$, which is a bipartite graph. The street intersections are the vertices of G and two vertices are adjacent if the vertices represent intersections on the same street at opposite ends of a city block. Looking for the smallest number of security guards in the city of Figure 13.2 is the same problem as seeking the domination number of the graph G in Figure 13.3. The solid vertices in Figure 13.3 correspond to the placement of security officers in Figure 13.2.

G :

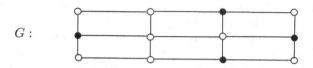

Figure 13.3: A graph modeling a city map

Example 13.1 *For the graph G in Figure* 13.3, $\gamma(G) = 4$.

Solution. Since the four solid vertices in Figure 13.3 form a dominating set of G, it follows that $\gamma(G) \leq 4$. To verify that $\gamma(G) \geq 4$, it is necessary to show that there is no dominating set with three vertices in G.

The graph G has 12 vertices, two of which have degree 4 and six have degree 3. The remaining four vertices have degree 2. Therefore, there are two vertices that dominate five vertices each and six vertices that dominate four vertices each. Conceivably, then, there is some set of three vertices that together dominate all 12 vertices of G. However, we have already noticed that G is bipartite and so its vertices can be colored with two colors, say red (R) and blue (B). Without loss of generality, we can assume that the vertices of G are colored as in Figure 13.4. Notice that the neighbors of each vertex have a color that is different from the color assigned to this vertex.

G :

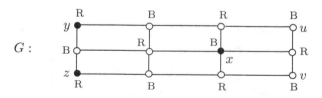

Figure 13.4: The graph $P_3 \times P_4$

Assume, to the contrary, that G has a dominating set S containing three vertices. At least two vertices of S are colored the same. If all three vertices of S are colored the same, say red, then only three of the six red vertices will be dominated. Therefore, exactly two vertices of S are colored the same, say red, with the third vertex colored blue. If the blue vertex of S has degree at most 3, then it can dominate at most three red vertices and S dominates at most five red vertices of G, which is impossible. Hence S must contain x (see Figure 13.4) as its only blue vertex. Since y and z are the only two red vertices not dominated by x, it follows that $S = \{x, y, z\}$. However, u and v are not dominated by any vertex of S, which cannot occur. Therefore, $\gamma(G) = 4$.   ◇

Showing that $\gamma(P_3 \times P_4) = 4$ illustrates the general procedure for establishing the domination number of a graph. To show that $\gamma(G) = k$, say, for some graph G, we need to find a dominating set for G with k vertices (which shows that $\gamma(G) \le k$) and, in addition, we must verify that every dominating set of G must contain at least k vertices (which shows that $\gamma(G) \ge k$).

As we have mentioned, a graph G of order n has domination number 1 if and only if G contains a vertex v of degree $n - 1$. Thus all complete graphs and all stars have domination number 1. The next example provides the domination numbers of graphs belonging to another familiar class.

Example 13.2 $\gamma(C_n) = \lceil n/3 \rceil$ *for* $n \ge 3$.

Solution. First, write $n = 3q + r$, where $0 \le r \le 2$. Since C_n is 2-regular, every vertex of C_n dominates exactly three vertices. Therefore, any q vertices of C_n dominate at most $3q$ vertices of C_n. If $r = 0$, then this says that $\gamma(C_n) \ge q = \lceil n/3 \rceil$. If $r = 1$ or $r = 2$, then $\gamma(C_n) \ge q + 1 = \lceil n/3 \rceil$.

We now show that $\gamma(C_n) \le \lceil n/3 \rceil$, where $n = 3q + r$. Suppose first that $r = 0$. Let S be the set consisting of any vertex v of C_n and every third vertex of C_n begining with v as we proceed cyclically about C_n in some direction. Then every vertex of C_n is dominated by exactly one vertex of S. Since S contains exactly q vertices, $\gamma(C_n) \le q = \lceil n/3 \rceil$. Next suppose that $r = 1$ or $r = 2$. Now let S be the set consisting of any vertex v of C_n and every third vertex of C_n begining with v as we proceed cyclically about C_n in some direction until we have a total of $q + 1$ vertices (see Figure 13.5). Then, every vertex of C_n is dominated by at least one vertex of S. So S is a dominating set of C_n and $\gamma(C_n) \le q + 1 = \lceil n/3 \rceil$. In both cases, $\gamma(C_n) = \lceil n/3 \rceil$.   ◇

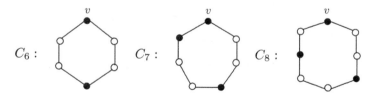

Figure 13.5: A minimum dominating set in C_n for $6 \le n \le 8$

In the following result, both a lower bound and an upper bound are established for the domination number of a graph, each in terms of the order and the maximum degree of the graph.

Theorem 13.3 *If G is a graph of order n, then*

$$\frac{n}{1 + \Delta(G)} \le \gamma(G) \le n - \Delta(G).$$

Proof. We have already mentioned that each vertex v in a graph G of order n dominates $1 + \deg v$ vertices. If v is chosen so that $\deg v = \Delta(G)$, then v dominates $1 + \Delta(G)$ vertices, that is, v dominates all but $n - (1 + \Delta(G))$ vertices of G. Since each of the $n - (1 + \Delta(G))$ vertices not dominated by v dominates itself, there is certainly a dominating set for G with $n - (1 + \Delta(G)) + 1 = n - \Delta(G)$ vertices. So

$$\gamma(G) \leq n - \Delta(G).$$

Next, suppose that $\gamma(G) = k$. Let $S = \{v_1, v_2, \ldots, v_k\}$ be a minimum dominating set for G. Since v_i dominates $1 + \deg v_i$ vertices of G for $1 \leq i \leq k$ and the vertices of S dominate all n vertices of G,

$$\sum_{i=1}^{k}(1 + \deg v_i) \geq n.$$

However, $1 + \deg v_i \leq 1 + \Delta(G)$ for $1 \leq i \leq k$. So

$$n \leq \sum_{i=1}^{k}(1 + \deg v_i) \leq k(1 + \Delta(G))$$

and $k(1 + \Delta(G)) \geq n$. Since $\gamma(G) = k$, it follows that

$$\gamma(G) \geq \frac{n}{1 + \Delta(G)},$$

as desired. ∎

Both bounds for $\gamma(G)$ in Theorem 13.3 are sharp. For positive integers r and n with $r \leq n - 2$, let G consist of $n - r$ components, one of which is the star $K_{1,r}$ and each remaining component is an isolated vertex, that is, $G = K_{1,r} \cup (n - r - 1)K_1$. The central vertex of $K_{1,r}$ dominates all vertices of $K_{1,r}$ and each isolated vertex can only be dominated by itself. Thus $\gamma(G) = n - r = n - \Delta(G)$.

We show that the lower bound is sharp even for regular graphs. The graph $G = \overline{K}_n$ consists of n isolated vertices; so G is 0-regular and $\Delta(G) = 0$. Thus

$$\gamma(G) = n = \frac{n}{1 + \Delta(G)}.$$

For n even, let $n = 2k$, where $k \geq 1$ and let $G = kK_2$, that is, G consists of k components, all isomorphic to K_2. Thus G is 1-regular and $\Delta(G) = 1$. The two vertices of each component are dominated by either vertex in the component and so

$$\gamma(G) = k = \frac{n}{2} = \frac{n}{1 + \Delta(G)}.$$

We have seen that $\gamma(C_n) = \lceil n/3 \rceil$. So if $n = 3k$ for some $k \geq 1$, then $\gamma(C_n) = k$. Since C_n is 2-regular, $\Delta(G) = 2$ and

$$\gamma(C_n) = k = \frac{n}{3} = \frac{n}{1 + \Delta(G)}.$$

Let's turn to the 3-regular graph G of order 20 shown in Figure 13.6. Since $\{u_0, u_4, v_0, v_4, w_0, w_4\}$ is a dominating set of G, it follows that $\gamma(G) \leq 6$. Next, let $U = \{u_1, u_2, \ldots, u_5\}$ and consider the induced subgraph $F = G[U]$ of G. Since each vertex of G dominates four vertices and $|U| = 5$, at least two vertices from $U \cup \{u_0\}$ are needed to dominate the vertices of U. Applying this argument to the other two subgraphs isomorphic to F in G, we see that $\gamma(G) \geq 3 \cdot 2 = 6$. Hence for the graph G of Figure 13.6,

$$\gamma(G) = 6 > 5 = \frac{20}{1+3} = \frac{n}{1+\Delta(G)}.$$

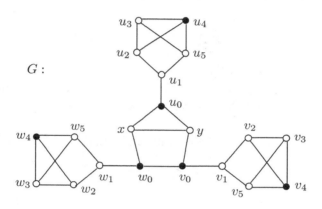

Figure 13.6: A 3-regular graph of order 20

Let S be a dominating set in a graph G. Then, of course, every vertex of G is dominated by at least one vertex of S. It may occur that the vertices dominated by some vertex v of S are also dominated by some other vertex of S. In this case, v is not needed to dominate the vertices of G, that is, $S - \{v\}$ is also a dominating set for G. In other words, v can be deleted from S and the remaining set is a dominating set. We can continue to delete vertices from S in this manner until we have found a subset S' of S such that S' is a dominating set for G and no proper subset of S' is a dominating set. This does not mean that S' is a minimum dominating set, however.

If S is a dominating set of a graph G and no proper subset of S is a dominating set of G, then S is called a **minimal dominating set**. Every minimum dominating set is minimal, but the converse is not true in general. For example, consider the graph C_8 in Figure 13.7. The set $S = \{v_1, v_2, v_5, v_6\}$ is a dominating set. If we were to delete any vertex from S, however, then the resulting set is not a dominating set, that is, S is a minimal dominating set. Since $\gamma(C_8) = \lceil 8/3 \rceil = 3$, it follows that S is not a minimum dominating set of C_8. The dominating set S_2 in the graph G of Figure 13.1 is also a minimal dominating set that is not a minimum dominating set.

Since each isolated vertex in a graph G can only be dominated by itself,

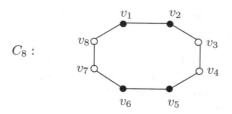

C_8 :

Figure 13.7: A minimal dominating set in C_8

every dominating set in G must contain its isolated vertices. For graphs without isolated vertices, however, there are always two disjoint dominating sets.

Theorem 13.4 *Let G be a graph without isolated vertices. If S is a minimal dominating set of G, then $V(G) - S$ is a dominating set of G.*

Proof. We show that $V(G) - S$ is a dominating set of G. Let $v \in V(G)$. If $v \in V(G) - S$, then v is dominated by itself. Thus we may assume that $v \notin V(G) - S$ and so $v \in S$. We show that v is dominated by some vertex in $V(G) - S$. Assume, to the contrary, that v is not dominated by any vertex in $V(G) - S$. Therefore, v is not adjacent to any vertex in $V(G) - S$. Since S is a dominating set of G, each vertex in $V(G) - S$ is adjacent to some vertex in S different from v. Thus, each vertex in $V(G) - S$ is dominated by some vertex in $S - \{v\}$. On the other hand, G has no isolated vertices and so v is not an isolated vertex of G. Since v is not adjacent to any vertex in $V(G) - S$, the vertex v must be adjacent to some vertex in $S - \{v\}$. Thus v is dominated by some vertex in $S - \{v\}$. Therefore, $S - \{v\}$ is a dominating set of G, which contradicts the fact that S is a minimal dominating set of G. ∎

For graphs without isolated vertices, we now present an upper bound for the domination number of a graph in terms of its order.

Corollary 13.5 *If G is a graph of order n without isolated vertices, then*

$$\gamma(G) \le \frac{n}{2}.$$

Proof. Let S be a minimum dominating set of G. By Theorem 13.4, $V(G) - S$ is also a dominating set of G. Since $|S| + |V(G) - S| = n$ and $|S| \le |V(G) - S|$, it follows that $\gamma(G) = |S| \le n/2$. ∎

We have seen that a vertex u dominates a vertex v in a graph if either $u = v$ or v is a neighbor of u. However, there are a number of variations of domination. We consider one of the best known of these. In this variation, a vertex u dominates a vertex v only if v is a neighbor of u. (In this context, a vertex does not dominate itself.) This type of domination is called **total domination**. A set S of vertices in a graph G is a **total dominating set** of G if every vertex of G is adjacent to at least one vertex of S. Therefore, a graph G has a total dominating set if and only if G contains no isolated vertices. Furthermore, if S is a total dominating set of G, then the subgraph $G[S]$ induced by S contains no isolated vertices. The minimum cardinality of a total dominating set is the **total domination number** $\gamma_t(G)$ of G. A total dominating set of cardinality $\gamma_t(G)$ is a **minimum total dominating set** for G. For example, for the graph G of Figure 13.1, which is redrawn in Figure 13.8, the set $S = \{u_1, v, w, v_4\}$ is a minimum total dominating set of G and so $\gamma_t(G) = 4$.

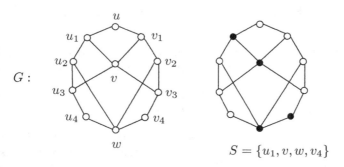

$$S = \{u_1, v, w, v_4\}$$

Figure 13.8: A minimum total dominating set in a graph

In Example 13.2, we saw that $\gamma(C_n) = \lceil n/3 \rceil$ for $n \geq 3$. We now determine $\gamma_t(C_n)$.

Example 13.6 *For $n \geq 3$,*

$$\gamma_t(C_n) = \begin{cases} \left\lceil \frac{n}{2} \right\rceil & \text{if } n \not\equiv 2 \ (\text{mod } 4) \\ \frac{n+2}{2} & \text{if } n \equiv 2 \ (\text{mod } 4). \end{cases}$$

Solution. Let $C_n = (v_1, v_2, \cdots, v_n, v_{n+1} = v_1)$. Since C_n is 2-regular, every vertex of C_n dominates exactly two vertices of C_n in this case, namely its two neightbors. Therefore, $\gamma_t(C_n) \geq \lceil n/2 \rceil$. Suppose first that $n \not\equiv 2 \ (\text{mod } 4)$. We consider three cases.

Case 1. $n \equiv 0 \ (\text{mod } 4)$. Thus $n = 4k \geq 4$. Then $\{v_1, v_2, v_5, v_6, \cdots, v_{4k-3}, v_{4k-2}\}$ is a total dominating set of $2k = n/2$ vertices.

Case 2. $n \equiv 1 \ (\text{mod } 4)$. Thus $n = 4k + 1 \geq 5$. Then $\{v_1, v_2, v_5, v_6, \cdots, v_{4k-3}, v_{4k-2}, v_{4k+1}\}$ is a total dominating set of $2k + 1 = \lceil n/2 \rceil$ vertices.

Case 3. $n \equiv 3 \ (\text{mod } 4)$. Thus $n = 4k + 3 \geq 3$. Then $\{v_1, v_2, v_5, v_6, \cdots, v_{4k+1}, v_{4k+2}\}$ is a total dominating set of $2k + 2 = \lceil n/2 \rceil$ vertices.

Therefore, if $n \not\equiv 2 \ (\text{mod } 4)$, then $\gamma_t(C_n) = \lceil n/2 \rceil$. It remains to show that $\gamma_t(C_n) = (n+2)/2$ if $n \equiv 2 \ (\text{mod } 4)$. Then $n = 4k + 2$ for some positive integer k. Since $\{v_1, v_2, v_5, v_6, \cdots, v_{4k+1}, v_{4k+2}\}$ is a total dominating set with $2k + 2 = (n + 2)/2$ vertices, it follows that $\gamma_t(C_n) \leq (n + 2)/2$. Therefore, $\gamma_t(C_n) = n/2$

or $\gamma_t(C_n) = (n+2)/2$. Assume, to the contrary, that $\gamma_t(C_n) = n/2 = 2k+1$. Let S be a minimum total dominating set of C_n.

We make some observations about the set S. First, let v_i, v_{i+1}, v_{i+2}, v_{i+3} ($1 \le i \le n$) be four consecutive vertices on C_n, where the addition in the subscripts is performed modulo n. The vertex v_{i+1} can be dominated only by v_i or v_{i+2}, while v_{i+2} can be dominated only by v_{i+1} or v_{i+3}. Let

$$S_i = S \cap \{v_i, v_{i+1}, v_{i+2}, v_{i+3}\}.$$

From our observation, $|S_i| \ge 2$ and there exist $v_r, v_t \in S_i$, where r is even and t is odd. In particular, there exists an integer j with $1 \le j \le n$ such that $v_j, v_{j+1} \notin S$. This implies that $v_{j-2}, v_{j-1}, v_{j+2}, v_{j+3} \in S$. Let

$$S' = V(G) - \{v_{j-2}, v_{j-1}, \cdots, v_{j+3}\}.$$

Then $|S'| = (4k+2)-6 = 4(k-1)$. Hence S contains at least $2(k-1)$ vertices of S' and 4 vertices of $\{v_{j-2}, v_{j-1}, \cdots, v_{j+3}\}$. However then, $|S| \ge 2(k-1)+4 = 2k+2$, a contradiction. Therefore, $\gamma_t(C_n) = (n+2)/2$ if $n \equiv 2 \pmod 4$. ∎

Minimum total dominating sets are indicated for C_n, $6 \le n \le 9$, in Figure 13.9.

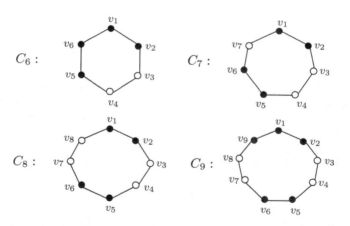

Figure 13.9: Minimum total dominating sets for C_n, $6 \le n \le 9$

Since every total dominating set of a graph G contains at least two vertices,

$$2 \le \gamma_t(G) \le n \qquad (13.1)$$

for every graph G of order n containing no isolated vertices. The lower bound in (13.1) can be attained if G is a star or a double star, while the upper bound in (13.1) is attainable if $G = kK_2$ for a positive integer k.

There are bounds for the total domination number of a graph (without isolated vertices) in terms of of its domination number.

Theorem 13.7 *For every graph G containing no isolated vertices,*

$$\gamma(G) \leq \gamma_t(G) \leq 2\gamma(G).$$

Proof. Since every total dominating set in G is also a dominating set, $\gamma(G) \leq$
$\gamma_t(G)$. It remains to show that $\gamma_t(G) \leq 2\gamma(G)$. Let $S = \{v_1, v_2, \ldots, v_k\}$ be a
minimum dominating set of G. Each vertex in $V(G) - S$ is therefore adjacent to
some vertex of S. Since G contains no isolated vertices, each open neighborhood
$N(v_i)$ is nonempty. Now let $u_i \in N(v_i)$ $(1 \leq i \leq k)$ and let $S' = \{u_1, u_2, \ldots, u_k\}$.
Thus the vertices of S are dominated by the vertices of S'. Therefore, the set
$S \cup S'$ is a total dominating set of G. Hence $\gamma_t(G) \leq |S \cup S'| \leq 2|S| = 2\gamma(G)$. ∎

Both bounds in Theorem 13.7 are sharp. For example, the domination number and total domination number of every double star is 2. For the graph G of
Figure 13.10, let $S = \{v_1, v_2, v_3, v_4, v_5\}$ and $S' = \{u_1, u_2, u_3, u_4, u_5\}$. Then S
is a minimum dominating set of G, while $S \cup S'$ is a minimum total dominating
set. Therefore, $\gamma(G) = 5$ and $\gamma_t(G) = 10$.

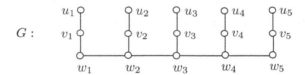

Figure 13.10: A graph G with $\gamma_t(G) = 2\gamma(G)$

Corollary 13.5 states that the domination number of a graph G of order n
having no isolated vertices is at most $n/2$. For total domination, Ernie Cockayne,
Robyn Dawes and Stephen Hedetniemi showed that the best upper bound of this
type is $2n/3$.

Theorem 13.8 *If G is a connected graph of order $n \geq 3$, then*

$$\gamma_t(G) \leq \frac{2n}{3}.$$

The graph G of Figure 13.10 also illustrates the sharpness of the bound in
Theorem 13.8.

Exercises for Section 13.1

13.1 For the graph G of Figure 13.11, determine

   (a) the domination number of G,

   (b) the total domination number of G.

13.2 For the graph G of Figure 13.12, determine

$G:$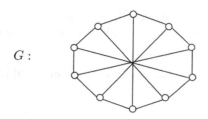

Figure 13.11: The graph G in Exercise 13.1

(a) the domination number of G,

(b) the total domination number of G.

$G:$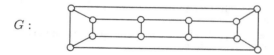

Figure 13.12: The graph in Exercise 13.2

13.3 Determine the domination number and the total domination number for each of the following graphs.

(1) K_n, $n \geq 2$, (2) P_n, $n \geq 2$, (3) $K_{s,t}$, (4) Q_3, (5) the Petersen graph.

13.4 For each positive integer n, show that there exists a connected graph G of order n such that

(a) $\gamma(G) = n - \Delta(G)$, (b) $\gamma(G) = n/(1 + \Delta(G))$.

13.5 Give an example of

(a) a graph G of some order n, without isolated vertices and with $\Delta(G) \leq n - 2$ and containing a minimum dominating set S such that for each v in S, there is *no* w in $V(G) - S$ such that $N(w) \cap S = \{v\}$.

(b) a graph G of odd order $n \geq 9$ without isolated vertices having the maximum possible domination number.

13.6 Prove for each pair k, n of integers with $1 \leq k \leq n$ that there exists a graph G of order n with $\gamma(G) = k$.

13.7 Prove for each pair k, n of integers with $1 \leq k \leq n/2$ that there exists a connected graph G of order n with $\gamma(G) = k$.

13.8 Give an example of a minimal dominating set that is not a minimum dominating set in each of the following graphs

(a) P_9, (b) the graph G in Figure 13.11 (see Exercise 13.1).

13.9 Prove that if every two vertices in a dominating set S of a graph G are not adjacent, then S is necessarily a minimal dominating set but not necessarily a minimum dominating set.

13.10 Show that there exists a graph G and a minimal dominating set S of G such that $|S| - \gamma(G) \geq 2$.

13.11 (a) Prove that if G is a graph of order $n \geq 2$, then $3 \leq \gamma(G) + \gamma(\overline{G}) \leq n + 1$.

(b) Show for each integer $n \geq 2$ that there exists a graph G of order n such that $\gamma(G) + \gamma(\overline{G}) = 3$.

(c) Show for each integer $n \geq 2$ that there exists a graph G of order n such that $\gamma(G) + \gamma(\overline{G}) = n + 1$.

13.12 Prove that if G is a graph with $\gamma(G) \geq 3$ and $\gamma(\overline{G}) \geq 3$, then diam $G = 2$. [Hint: First show that G must be connected.]

13.13 For each integer $k \geq 2$, give an example of a connected graph G for which $\gamma(G) = \gamma_t(G) = k$.

13.14 For each positive integer k, give an example of a connected graph G for which $\gamma(G) = k$ and $\gamma_t(G) = 2k$.

13.15 Give an example of a connected graph G for which $\gamma_t(G) = 1.5\gamma(G)$.

13.16 For each integer $n \geq 3$ with $n \equiv 0 \pmod{3}$, give an example of a connected graph G of order n for which $\gamma_t(G) = 2n/3$.

13.2   Exploration: Stratification

We have seen examples where the vertex set of a graph has been divided into classes in some manner. This might be as fundamental as separating the vertices into even and odd vertices or perhaps distinguishing the vertices that are cut-vertices from those that are not. The best known example of this, however, occurs with graph coloring, when the vertex set of a graph is partitioned into independent sets ·in some manner.

A graph G whose vertex set has been partitioned in some manner is referred to as a **stratified graph**. If $V(G)$ is partitioned into k subsets, say $V_1, V_2, \ldots, V_k$, then G is a k-**stratified graph** and these subsets are called the **strata** or the **color classes** of G. Unlike vertex coloring, no condition is placed on the subsets V_i, $1 \leq i \leq k$. If G is 2-stratified, then we commonly color the vertices of one color class red and those of the other color class blue. For a given graph G, a partition of the vertices of G (that is, a coloring of the vertices of G) is called a **stratification** of G (or a k-**stratification** of G if the partition is into k subsets).

For example, for the graph G of Figure 13.13, two 2-stratifications G_1 and G_2 of G are shown, where the solid vertices represent red vertices and the open vertices represent blue vertices.

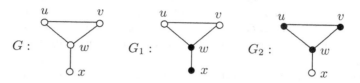

Figure 13.13: Two 2-stratifications of a graph

Two k-stratified graphs G and H are **isomorphic** if there exists a bijective function $\phi : V(G) \to V(H)$ such that

(1) u and v are adjacent in G if and only if $\phi(u)$ and $\phi(v)$ are adjacent in H and

(2) x and $\phi(x)$ are colored the same for all $x \in V(G)$.

The function ϕ is then called a **color-preserving isomorphism**.

In this context, a **red-blue coloring** of a graph G is an assignment of the colors red and blue to the vertices of G, one color to each vertex of G. In a red-blue coloring of G, it may occur that every vertex is red or that every vertex is blue. If there is at least one vertex of each color, then the red-blue coloring produces a 2-stratification of G.

The study of stratified graphs was initiated in the 1990s by Reza Rashidi and Naveed Sherwani when it was observed that it was desirable to use graphs whose vertex sets are partitioned into classes in the design of algorithms to solve multilayer routing problems that occur when transistors are being assembled in Very Large Scale Integrated (VLSI) circuit chips.

There is a close connection between domination in graphs and stratification of graphs, in particular 2-stratification of graphs. Let F be a 2-stratified graph. Therefore, F contains one or more red vertices and one or more blue vertices. One of the blue vertices of F is selected as the **root** of F, which we denote by v. Another 2-stratification G_3 of the graph G of Figure 13.13 is shown in Figure 13.14. Since G_3 contains two nonsimilar blue vertices, we distinguish between these, according to which blue vertex is selected as the root. We denote these 2-stratified rooted graphs by F' and F''.

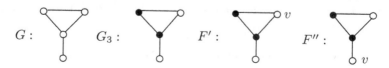

Figure 13.14: 2-stratified rooted graphs

Now, let F be a 2-stratified rooted graph, that is, a 2-stratified graph in which some blue vertex v of F has been designated as the root. By an F-**coloring** of a graph G, we mean a red-blue coloring of G such that for every blue vertex w of G, there is a copy of F in G with v at w. That is, for every blue vertex w of G, there exists a 2-stratified subgraph G' of G containing w and a color-preserving isomorphism α from F to G' such that $\alpha(v) = w$. The red-blue coloring of G in which every vertex is colored red is vacuously an F-coloring for every 2-stratified rooted graph F.

For example, for the 2-stratified rooted graph F' of Figure 13.15 and the graph G of the same figure, an F'-coloring is given. This is illustrated when the root v of F' is placed at the blue vertex u_2 of G.

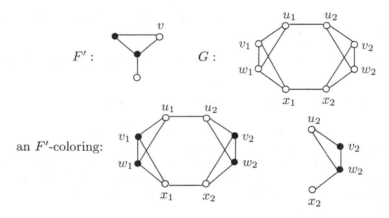

Figure 13.15: An F'-coloring of a graph G

Let F be a 2-stratified graph rooted at some blue vertex v and let G be a graph. The set of red vertices in an F-coloring of G is an F-**dominating set** of G. An F-dominating set of minimum cardinality is a **minimum F-dominating set** and the number of vertices in minimum F-dominating set is the F-**domination number** $\gamma_F(G)$ of G. Note that $\gamma_F(G)$ is defined for *every* graph G, even if G contains no subgraph isomorphic to the (uncolored) graph F, since the red-blue coloring of G in which every vertex is colored red is an F-coloring of G. An F-coloring of G in which there are $\gamma_F(G)$ red vertices is called a **minimum F-coloring**.

Let's see what $\gamma_F(G)$ means for some small connected 2-stratified rooted graphs F. Of course, the simplest example is when F is a 2-stratified K_2 (see Figure 13.16). For this 2-stratified graph F, the F-domination number of a graph is a familiar parameter.

Theorem 13.9 *Let F be the 2-stratified K_2. For every graph G,*

$$\gamma_F(G) = \gamma(G).$$

$$F:$$

Figure 13.16: The 2-stratified K_2

Proof. Let G be a graph and let there be given a minimum F-coloring of G. This implies that every blue vertex of G is adjacent to a red vertex of G. Hence the red vertices form a dominating set for G and so $\gamma(G) \leq \gamma_F(G)$. Next, consider a minimum dominating set S and color all of the vertices in S red and the remaining vertices blue. Since every blue vertex is adjacent to at least one red vertex, this red-blue coloring is an F-coloring of G. Hence $\gamma_F(G) \leq \gamma(G)$ and so $\gamma_F(G) = \gamma(G)$. ∎

By Theorem 13.9, ordinary domination can be considered as F-domination for an appropriately chosen 2-stratified rooted graph F. We now turn to 2-stratified rooted graphs P_3. Actually, there are five possibilities in this case, all of which are shown in Figure 13.17.

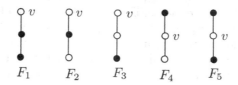

Figure 13.17: The 2-stratified rooted graphs P_3

The values of the five domination parameters associated with the 2-stratified rooted graphs F_i in Figure 13.17 are given for the graph G of Figure 13.18. A minimum F_i-dominating set for this graph G is shown in Figure 13.18 for $i = 1, 2, 3, 4, 5$.

For the 2-stratified rooted graph F_1 of Figure 13.17, the corresponding domination parameter is also a familiar one.

Theorem 13.10 *Let F_1 be the 2-stratified rooted graph shown in Figure 13.17. If G is a graph without isolated vertices, then the F_1-domination number of G is the total domination number of G, that is,*

$$\gamma_{F_1}(G) = \gamma_t(G).$$

Proof. Since G has no isolated vertices, G has an total dominating set. Let S be a minimum total dominating set in G. Color the vertices of S red and color the remaining vertices of G blue. Now let v be a blue vertex of G. Since $v \notin S$, the vertex v is adjacent to a vertex u in S, that is, v is adjacent to a red vertex u. Since u must be adjacent to a vertex w in S distinct from v, it follows that v

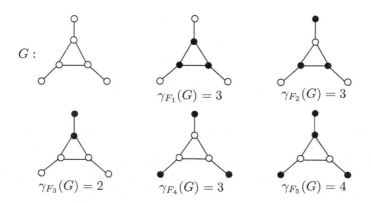

$$G:$$

$$\gamma_{F_1}(G) = 3 \qquad \gamma_{F_2}(G) = 3$$

$$\gamma_{F_3}(G) = 2 \qquad \gamma_{F_4}(G) = 3 \qquad \gamma_{F_5}(G) = 4$$

Figure 13.18: A minimum F_i-dominating set ($1 \le i \le 5$) in a graph G

is a root of a copy of F_1. Therefore, this red-blue coloring of G is an F_1-coloring and so $\gamma_{F_1}(G) \le \gamma_t(G)$.

Next, we show that $\gamma_t(G) \le \gamma_{F_1}(G)$. Among all minimum F_1-colorings of G, let c be one for which the subgraph induced by the red vertices contains a minimum number of isolated vertices. Let R_c be the set of red vertices of G colored by c. Thus $|R_c| = \gamma_{F_1}(G)$. Since every blue vertex v in G is adjacent to a red vertex, R_c is a dominating set in G. We claim that every red vertex in R_c is also adjacent to a red vertex. Assume, to the contrary, that there is a red vertex $u \in R_c$ that is adjacent only to blue vertices. Let v be a neighbor of u. Then v belongs to a copy of F rooted at v. Thus, v must be adjacent to a red vertex w which itself is adjacent to some other red vertex, which implies that $u \ne w$. Interchanging the colors of u and v produces a new γ_F-coloring of G having fewer isolated vertices in the subgraph induced by its red vertices, contradicting the choice of c. Hence, as claimed, every red vertex in R_c is adjacent to some other red vertex. Therefore, R_c is a total dominating set of G. This implies that$\gamma_t(G) \le |R_c| = \gamma_{F_1}(G)$. Consequently, $\gamma_t(G) = \gamma_F(G)$. ∎

While the total domination number is defined only for graphs without isolated vertices, the F_1-domination number is defined for all graphs. By Theorem 13.10, for graphs where both of these domination numbers are defined, the values are the same.

For the 2-stratified rooted graph F_2 of Figure 13.17, we once agian have a familiar parameter. We do not include the proof in this case.

Theorem 13.11 *For every connected graph G of order 3 or more,*

$$\gamma_{F_2}(G) = \gamma(G).$$

If F is a 2-stratified rooted graph and G is a graph of order n containing no subgraph isomorphic to the (uncolored) graph F, then surely, $\gamma_F(G) = n$. The converse of this statement is not true, however. Consider the 2-stratified rooted graph F_3 of Figure 13.17. Certainly, the star $K_{1,n-1}$, $n \geq 3$, contains many subgraphs isomorphic to P_3; indeed, it contains $\binom{n-1}{2}$ such subgraphs. However, there is no F_3-coloring of $K_{1,n-1}$ in which any vertex can be colored blue. Therefore, $\gamma_{F_3}(K_{1,n-1}) = n$.

Exercises for Section 13.2

13.17 For the graph G of Figure 13.19, determine $\gamma_{F_i}(G)$ for each 2-stratified graph F_i in Figure 13.17.

$G:$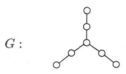

Figure 13.19: The graph G in Exercise 13.17

13.18 For the Petersen graph PG, determine $\gamma_{F_i}(PG)$ for each 2-stratified graph F_i in Figure 13.17.

13.19 Consider the 2-stratified graph F_3 in Figure 13.17.

(a) Give an example of a graph of at least 3 such that $\gamma_{F_3}(G) = 1$.

(b) Prove that if G is a graph with $\gamma_{F_3}(G) = 1$, then diam$(G) \leq 4$.

(c) Prove that if G is a bipartite graph, then $\gamma_{F_3}(G) \geq 2$.

13.20 Study F_4-domination and F_5-domination for the 2-stratified graphs F_4 and F_5 in Figure 13.17.

13.21 Choose a 2-stratified rooted graph F and several graphs G of your own. Determine the F-domination number of each such graph G.

13.3 Exploration: Lights Out

On a certain floor of a business building, a firm occupies three offices A, B and C located in a row. Each office has a large ceiling light and a light button which, when pressed, reverses the light in that office (on to off or off to on) as well as the light in each adjacent office. So if we begin the day, as in Figure 13.20(a), with all lights off and push the light button in the central office B, then we arrive at the situation in Figure 13.20(b), where all lights are on.

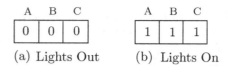

(a) Lights Out   (b) Lights On

Figure 13.20: Lights Out and Lights On

Each light arrangement of the three offices can be represented by an ordered triple (a, b, c) or abc, where a, b and c can be 0 or 1, with 0 meaning that the light is off in the particular office and 1 meaning that the light is on. The eight possibilities are shown in Figure 13.21.

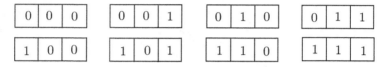

Figure 13.21: The possible light arrangements of the three offices

This situation can be represented by a graph G of order 8, whose vertices are the ordered triples abc, where $a, b, c \in \{0, 1\}$. If we can change from one light arrangement to another by pressing a single light button, then we draw an edge between the two vertices representing these arrangements. The graph G is shown in Figure 13.22. You might notice that G is the 3-cube Q_3. The graph Q_3 of Figure 13.22 shows that, beginning with lights out in all three offices, we can obtain any light pattern we desire, although it may require pressing as many as three buttons.

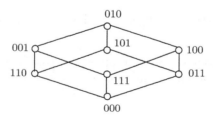

Figure 13.22: The graph Q_3

The situation that we have just described can be interpreted in terms of graphs from the beginning. Consider the graph G in Figure 13.23, where the vertices are drawn as solid vertices, indicating that all lights are on. If we "press" the solid vertex in the middle, this causes all lights to go out and we obtain the graph H.

So the general situation might go something like this. Let G be a connected graph where there is a light as well as a light button at each vertex. For each vertex, the light at that vertex is either on or off. When the light button at that

Figure 13.23: The graphs G and H

vertex is pressed, it reverses the light (changing it from on to off or off to on) not only at that vertex but at all vertices adjacent to that vertex. There is a variety of questions that can be asked here but our chief question concerns the following:

The Lights Out Puzzle: *Let G be a graph. If all vertex lights of G are on, does there exist a collection of light buttons which when pressed will turn out all vertex lights? If so, what is the smallest number of light buttons in such a collection?*

There is an electronic game called Lights Out marketed by Tiger Electronics of Hasbro, Inc. that gave rise to the more general graph theory puzzle mentioned above. Earlier manufactured as a cube, the current Lights Out game is played on a grid with multi-colored LEDs and digitized sound. Indeed, there are interactive web sites where various versions of the game can be played.

In terms of the Lights Out Puzzle on a graph G mentioned above, what we are asking is whether there is a dominating set of G such that every vertex of G is dominated by an odd number of vertices. The following (possibly unexpected) result of Klaus Sutner says that this game always has a solution.

Theorem 13.12 *If G is a connected graph all of whose vertex lights are on, then there exists a set S of vertices of G such that if the light button is pressed at each vertex of S, then all vertex lights of G will be out.*

Exercises for Section 13.3

13.22 Draw the graph that represents all light arrangements when playing Lights Out on the path G_1 of Figure 13.24. What is the fewest number of light buttons that need to be pressed to go from all lights on to all lights out?

13.23 Repeat Exercise 13.22 for the cycle G_2 of Figure 13.24.

13.24 What is the fewest number of light buttons that need to be pressed to go from all lights on to all lights out for the path G_3 of Figure 13.24.

13.25 What is the fewest number of light buttons that need to be pressed to go from all lights on to all lights out for the graph G_4 of Figure 13.24.

13.26 Choose a graph of your own on which to play Lights Out.

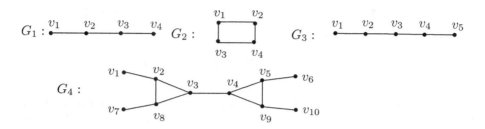

Figure 13.24: Graphs for Exercises 13.22–13.25

13.27 Consider the graph G of Figure 13.25 where all lights of G are on.

   (a) What is the smallest number of vertices whose light buttons need to be pressed to turn off all the lights?

   (b) Let $S = \{v_1, v_2, v_3, v_4\}$. Show that if the light buttons of S are pressed, then all lights of G are out. Is there a proper subset of S that will turn out all lights of G? Does this suggest another question to you?

Figure 13.25: The graph for Exercise 13.27

13.28 Observe that the graph H of Figure 13.26 has one light on, namely the light at v_1. However, we would like all lights to be out. What do we do?

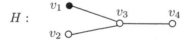

Figure 13.26: The graph for Exercise 13.28

13.29 Prove that in the game of Lights Out, the order in which the buttons are pressed is immaterial.

13.30 (a) Change the rules for Lights Out and play the game under the new rules for a graph that you choose.

   (b) Define a set of rules for playing Lights Out on a digraph and play this game on a digraph of your choice.

13.4   Excursion: And Still It Grows More Colorful

We have seen that graph theory originated with a number of isolated and disconnected results from unexpected sources. Recreational results and truly mathematical theorems alike played major roles in the development of the subject. Authors of the early textbooks on graph theory organized many of the existing theorems and set the stage for what was to follow. Progress in graph theory was greatly aided by numerous attempts to solve a simple-sounding but deceptively difficult problem involving the coloring of maps. Graph theory had the good fortune, however, of attracting a number of talented and dedicated mathematicians to this fascinating subject.

As graph theory progressed further into the 20th century, some well-defined areas of the subject blossomed. Also, the number of mathematicians working in the subject continued to grow. This included researchers who obtained deep results, those who studied graph theory from applied points of view, those who created new and interesting problems to study, those who wrote of the historical perspectives of the subject and its relationships to other more established areas of mathematics and those who wrote and lectured of the many aesthetic aspects of the subject, thereby introducing graph theory to a new generation of mathematicians. That graph theory had grown into a more prominent area of mathematics became increasingly evident during the latter portion of the 20th century and into the 21st century.

In the 1960s a series of conferences that emphasized graph theory came into prominence. One of these was the 1963 Czechoslovak Symposium on Graph Theory held in Smolenice. In 1968 the first of nine Kalamazoo (Michigan) Graph Theory Conferences was held and would continue to take place every fourth year at Western Michigan University throughout the remainder of the 20th century. Yousef Alavi played a leading role in organizing these conferences. In 1969 the first of the Southeastern International Conferences on Combinatorics, Graph Theory and Computing, primarily at Florida Atlantic University and organized by Frederick Hoffman. The British Combinatorial Conferences also began in 1969 and have been held during odd-numbered years since 1973. In more recent times, during even-numbered years, the SIAM (Society for Industrial and Applied Mathematics) Conferences on Discrete Mathematics have taken place.

In 1977 graph theory acquired its own journal when the *Journal of Graph Theory*, founded by Frank Harary, began publication. On the first page of the first issue of the first volume, the publishers (John Wiley & Sons, Inc.) wrote:

> *GRAPH THEORY has definitely emerged as a distinct entity within combinatorial theory .... We are confident that the journal will fill the need for current information dealing with this branch of applicable mathematics.*

The *Journal of Graph Theory* went on to receive an award from the Association of American Publishers for the best new journal published in 1977 in the scientific, medical and technical category.

Another important milestone for graph theory occurred in 1990 when the Institute of Combinatorics and Its Applications was established. The aim of this organization is to promote combinatorics (including graph theory). The mathematician who played the central role in the founding of the Institute was Ralph G. Stanton (1923–2010) who was Distinguished Professor of Computer Science at the University of Manitoba in Winnipeg, Canada.

In 1991 the official journal of the Institute began publication: *The Bulletin of the Institute of Combinatorics and Its Applications*. In its initial volume, the then-president of the Institute and respected mathematician William Tutte, whom we have met often, argued that graph theory (indeed, *any* area of mathematics) is not separate from the rest of mathematics but that mathematics is a single unified subject.

The *Handbook of Graph Theory*, now in its second edition and edited by Jonathan Gross, Jay Yellen and Ping Zhang, is considered by many as the most comprehensive single-source guide to graph theory ever published. In it are described numerous applications of graph theory to diverse areas of study.

And what lies ahead for graph theory? This is difficult to predict, but whatever the future holds is certain to be interesting … and colorful.

Appendix 1: Sets and Logic

1.1 Sets

Many of the sets that are dealt with in graph theory are finite. As for familiar infinite sets, we write $\mathbf{Z}$ for the set of integers, $\mathbf{N}$ for the set of positive integers (natural numbers), $\mathbf{Q}$ for the set of rational numbers and $\mathbf{R}$ for the set of real numbers. Among the infinite sets, we are, by far, most interested in the integers. Even when dealing with a rational number or real number, we are often concerned with a nearby integer. For a real number x, the **floor** $\lfloor x \rfloor$ of x is the greatest integer less than or equal to x. So, for example, $\lfloor 5 \rfloor = 5$, $\lfloor \sqrt{2} \rfloor = 1$, $\lfloor \pi \rfloor = 3$ and

$$\left\lfloor \frac{7 + \sqrt{1 + 96}}{2} \right\rfloor = 8.$$

The **ceiling** $\lceil x \rceil$ of x is the smallest integer greater than or equal to x. For example, $\lceil 5 \rceil = 5$, $\lceil \sqrt{2} \rceil = 2$, $\lceil \pi \rceil = 4$ and

$$\left\lceil \frac{(8 - 3)(8 - 4)}{12} \right\rceil = 2.$$

For a finite set S, we denote its **cardinality** (the number of elements in S) by $|S|$. If $|S| = n$ for some $n \in \mathbf{N}$, then we can write $S = \{s_1, s_2, \ldots, s_n\}$. The set with cardinality 0 is the **empty set**, which is denoted by $\emptyset$. Thus $\emptyset = \{\ \}$.

For two sets A and B, the **Cartesian product** $A \times B$ of A and B is the set

$$A \times B = \{(a, b) : a \in A, b \in B\}.$$

Therefore, $A \times A$ is the set of all ordered pairs of elements of A. For example, if $A = \{a_1, a_2\}$ and $B = \{b_1, b_2, b_3\}$, then

$$A \times B = \{(a_1, b_1), (a_1, b_2), (a_1, b_3), (a_2, b_1), (a_2, b_2), (a_2, b_3)\}$$

and

$$A \times A = \{(a_1, a_1), (a_1, a_2), (a_2, a_1), (a_2, a_2)\}.$$

Two sets S and T are **equal**, written $S = T$, if they consist of the same elements. A set T is a **subset** of a set S if every element of T belongs to S. This is denoted by writing $T \subseteq S$. The number of k-element subsets of an n-element set is given by the binomial coefficient

$$\binom{n}{k} = \frac{n!}{k!(n-k)!},$$

where $n! = n(n-1)(n-2) \cdots 3 \cdot 2 \cdot 1$ if $n \in \mathbf{N}$ and $0! = 1$. In particular,

$$\binom{5}{3} = \frac{5!}{3!(5-3)!} = \frac{120}{6 \cdot 2} = 10.$$

Consequently, the number of 3-element subsets of the set $S = \{1, 2, 3, 4, 5\}$ is 10. These subsets are

$S_1 = \{1, 2, 3\}$, $S_2 = \{1, 2, 4\}$, $S_3 = \{1, 2, 5\}$, $S_4 = \{1, 3, 4\}$, $S_5 = \{1, 3, 5\}$, $S_6 = \{1, 4, 5\}$, $S_7 = \{2, 3, 4\}$, $S_8 = \{2, 3, 5\}$, $S_9 = \{2, 4, 5\}$, $S_{10} = \{3, 4, 5\}$.

The number of 2-element subsets of an n-element set is therefore given by

$$\binom{n}{2} = \frac{n!}{2!(n-2)!} = \frac{n(n-1)}{2}.$$

So $\binom{2}{2} = 1$, $\binom{3}{2} = 3$, $\binom{4}{2} = 6$, $\binom{5}{2} = 10$ and $\binom{6}{2} = 15$. If we were to add the first $n - 1$ positive integers, then the result is $\binom{n}{2}$, that is,

$$1 + 2 + 3 + \ldots + (n-1) = \binom{n}{2}.$$

Another familiar and useful identity is

$$\binom{n}{0} + \binom{n}{1} + \binom{n}{2} + \ldots + \binom{n}{n} = 2^n. \tag{1}$$

Since every subset of an n-element set contains k elements for some k with $0 \leq k \leq n$, (1) states that the total number of subsets of an n-element set is 2^n. For example, there are $2^3 = 8$ subsets of the set $S = \{1, 2, 3\}$, namely,

$$S_1 = \emptyset, \ S_2 = \{1\}, \ S_3 = \{2\}, \ S_4 = \{3\},$$
$$S_5 = \{1, 2\}, \ S_6 = \{1, 3\}, \ S_7 = \{2, 3\}, \ S_8 = \{1, 2, 3\} = S.$$

A **partition** of a nonempty set S is a collection $\mathcal{P}$ of nonempty subsets of S such that every element of S belongs to exactly one of the elements of $\mathcal{P}$. For the set $S = \{1, 2, 3, 4, 5, 6\}$,

$$\mathcal{P} = \{\{1, 5\}, \{2, 3, 6\}, \{4\}\}$$

is a partition of S. A well-known theorem concerning partitions of sets is the following.

The Pigeonhole Principle *If $\mathcal{P}$ is a partition of an n-element set S into k subsets, then some subset of S in $\mathcal{P}$ must contain at least $\lceil \frac{n}{k} \rceil$ elements.*

In particular, if S is a set with 17 elements and $\mathcal{P}$ is a partition of S into five subsets, then some subset of S in $\mathcal{P}$ must contain at least $\lceil 17/5 \rceil = 4$ elements. A related version of the Pigeonhole Principle due to Frank Ramsey is also useful to know.

Ramsey's Theorem *Let $\mathcal{P} = \{S_1, S_2, \ldots, S_k\}$ be a partition of a set S into k subsets and let $n_1, n_2, \ldots, n_k$ be k positive integers such that $|S_i| \geq n_i$ for every integer i with $1 \leq i \leq k$. Then there exists a positive integer n such that every n-element subset of S contains at least n_i elements of S_i for some i $(1 \leq i \leq k)$.*

In particular, the integer

$$n = 1 + \sum_{i=1}^{k}(n_i - 1)$$

has this property. Indeed, it is the smallest integer with this property.

For example, if $\mathcal{P} = \{S_1, S_2, S_3, S_4\}$ is a partition of $S = S_1 \cup S_2 \cup S_3 \cup S_4$, where $|S_1| = 5$, $|S_2| = 6$, $|S_3| = 4$, $|S_4| = 7$ and $n_1 = 4$, $n_2 = 3$, $n_3 = 3$, $n_4 = 5$, then every 12-element subset of S must contain either (1) 4 elements of S_1, (2) 3 elements of S_2, (3) 3 elements of S_3 or (4) 5 elements of S_4.

1.2 Logic

A **statement** P is a declarative sentence that is true or false but not both. If P is a true statement, then its **truth value** is true; otherwise, its truth value is false. The **negation** $\sim P$ (not P) of P has the opposite truth value of P. The **disjunction** $P \vee Q$ (P or Q) of two statements P and Q is true if at least one of P and Q is true and is false otherwise. The **conjunction** $P \wedge Q$ (P and Q) of P and Q is true if both P and Q are true and is false otherwise. Two statements constructed from P and Q and **logical connectives** (such as $\sim$, $\vee$ and $\wedge$) are **logically equivalent** if they have the same truth values for all possible combinations of truth values for P and Q. According to De Morgan's laws, for statements P and Q,

$\sim (P \vee Q)$ is logically equivalent to $(\sim P) \wedge (\sim Q)$ and

$\sim (P \wedge Q)$ is logically equivalent to $(\sim P) \vee (\sim Q)$.

For statements P and Q, the **implication** $P \Rightarrow Q$ often expressed as "If P, then Q." is true for all combinations of truth values P and Q except when P is true and Q is false. Other ways to express $P \Rightarrow Q$ in words are: (1) P implies Q; (2) P only if Q; (3) P is sufficient for Q; (4) Q is necessary for P. In this case, P is a **sufficient condition** for Q and Q is a **necessary condition** for P.

A declarative sentence containing one or more variables is often referred to as an **open sentence**. When the variables are assigned values (from some prescribed set or sets), the open sentence is converted into a statement whose truth value depends on the values assigned to the variables. An open sentence expressed in terms of a real number variable x might be denoted by $P(x)$ or $Q(x)$. Since $P(x)$ and $Q(x)$ are open sentences and not statements, they do not have truth values. Similarly $\sim P(x)$, $P(x) \vee Q(x)$, $P(x) \wedge Q(x)$ and $P(x) \Rightarrow Q(x)$ are then also open sentences, not statements. While

$$P(x): \quad 2x^2 + x - 1 = 0$$

is an open sentence with a real number variable x, assigning x the values -1 and $1/2$ produces the statements

$$P(1): \ 2(-1)^2 + (-1) - 1 = 0 \text{ and } P(\tfrac{1}{2}): \ 2(\tfrac{1}{2})^2 + (\tfrac{1}{2}) - 1 = 0,$$

both of which are true. For every real number $r \neq -1, \tfrac{1}{2}$, however, $P(r)$ is a false statement.

Open sentences can also be converted into statements by means of quantifiers, resulting in a **quantified statement**. For example, suppose that $P(n)$ is an open sentence expressed in terms of an integer variable n. The **universal quantifier**, denoted by $\forall$, represents **for all, for each** or **for every**. Therefore,

$$\forall n \in \mathbf{Z}, \; P(n): \quad \text{For every integer } n, \; P(n).$$

is a quantified statement. This statement is true if $P(n)$ is true for every integer n. The quantified statement $\forall n \in \mathbf{Z}, \; P(n)$ can also be expressed as the implication:

$$\text{If } n \in \mathbf{Z}, \text{ then } P(n).$$

The **existential quantifier**, denoted by $\exists$, represents **there exists, for some** or **for at least one**. Thus

$$\exists n \in \mathbf{Z}, \; P(n): \quad \text{There exists an integer } n \text{ such that } P(n).$$

is a quantified statement. This statement is true if $P(n)$ is true for one or more integers n.

For an integer variable n,

$$P(n): \quad 3n - 5 \text{ is an even integer.}$$

is an open sentence and

$$\forall n \in \mathbf{Z}, \; P(n): \quad \text{For every integer } n, \; 3n - 5 \text{ is even.}$$

is a false statement since $P(2)$ is false. As we saw, this statement can be expressed as the implication:

$$\text{If } n \in \mathbf{Z}, \text{ then } 3n - 5 \text{ is even.}$$

On the other hand,

$$\exists n \in \mathbf{Z}, \; P(n): \quad \text{There exists an integer } n \text{ such that } 3n - 5 \text{ is even.}$$

is true since $P(3)$ is true for example.

If the open sentence $P(n)$ is an implication, say $R(n) \Rightarrow Q(n)$, then the quantified statement

$$\forall n \in \mathbf{Z}, \; R(n) \Rightarrow Q(n).$$

is often expressed as

$$\text{Let } n \in \mathbf{Z}. \text{ If } R(n), \text{ then } Q(n).$$

Then $\forall n \in \mathbf{Z}, \; R(n) \Rightarrow Q(n)$ is true if $R(n) \Rightarrow Q(n)$ is true for every integer n.

For statements P and Q, the **converse** of the implication $P \Rightarrow Q$ is the implication $Q \Rightarrow P$, while the **contrapositive** of $P \Rightarrow Q$ is the implication $(\sim Q) \Rightarrow (\sim P)$. While an implication and its contrapositive are logically equivalent, such is not the case for an implication and its converse.

For open sentences $P(n)$ and $Q(n)$, where n is an integer variable, the **converse** of $\forall n \in \mathbf{Z}, \; P(n) \Rightarrow Q(n)$ is $\forall n \in \mathbf{Z}, \; Q(n) \Rightarrow P(n)$, while the **contrapositive** of $\forall n \in \mathbf{Z}, \; P(n) \Rightarrow Q(n)$ is $\forall n \in \mathbf{Z}, \; (\sim Q(n)) \Rightarrow (\sim P(n))$.

Consider the statement

Let $a, b \in \mathbf{R}$. If $ab = 0$, then $a = 0$ or $b = 0$.

Its converse is

Let $a, b \in \mathbf{R}$. If $a = 0$ or $b = 0$, then $ab = 0$.

and its contrapositive (using one of De Morgan's laws) is

Let $a, b \in \mathbf{R}$. If $a \neq 0$ and $b \neq 0$, then $ab \neq 0$.

For statements P and Q, we write $P \Leftrightarrow Q$ to mean $(P \Rightarrow Q) \wedge (Q \Rightarrow P)$. In words, $P \Leftrightarrow Q$ is expressed as

P if and only if Q

or as

P is necessary and sufficient for Q.

In the case of the quantified statement,

$$\forall n \in \mathbf{Z}, \; P(n) \Leftrightarrow Q(n)$$

can be stated as

Let $n \in \mathbf{Z}$. $P(n)$ if and only if $Q(n)$.

As a specific example, we have

Let $a, b \in \mathbf{R}$. Then $ab = 0$ if and only if $a = 0$ or $b = 0$.

Appendix 2: Equivalence Relations and Functions

2.1: Equivalence Relations

For nonempty sets A and B, a **relation R from** A to B is a subset of the Cartesian product of A and B, that is,

$$R \subseteq A \times B = \{(a, b) : a \in A, b \in B\}.$$

A **relation R on** A is then a relation from A to A, that is, R is a collection of ordered pairs of elements of A. If $(a, b) \in R$, then a and b are said to be **related** by R. This is also expressed by writing $a\ R\ b$. For example, if $A = \{1, 2, 3, 4\}$, then

$$R = \{(1, 2), (1, 3), (2, 2), (2, 3), (2, 4), (4, 1), (4, 2)\}$$

is a relation on A. So $1\ R\ 2$, that is, 1 is related to 2 by R.

A relation R on a nonempty set A is an **equivalence relation** if R has the following three properties:

(1) R is **reflexive**, that is, $x\ R\ x$ for every $x \in A$.

(2) R is **symmetric**, that is, whenever $x\ R\ y$, then $y\ R\ x$ for all $x, y \in A$.

(2) R is **transitive**, that is, whenever $x\ R\ y$ and $y\ R\ z$, then $x\ R\ z$ for all $x, y, z \in A$.

The relation

$$R_1 = \{(1, 1), (2, 2), (3, 3), (4, 4), (1, 3), (3, 1), (1, 4), (4, 1), (3, 4), (4, 3)\} \qquad (2)$$

is an equivalence relation on the set $A = \{1, 2, 3, 4\}$.

The following provides an example of an equivalence relation on an infinite set.

Example 2.1 *A relation R on the set $\mathbf{Z}$ of integers is defined by $x\ R\ y$ if $x + y$ is even. Show that R is an equivalence relation on $\mathbf{Z}$.*

Solution. First, we show that R is reflexive. Let $x \in \mathbf{Z}$. Since $x + x = 2x$ is even, $x\ R\ x$ and so R is reflexive. Next we show that R is symmetric. Assume that $x\ R\ y$, where $x, y \in \mathbf{Z}$. Then $x + y$ is even. Since $y + x = x + y$, it follows that $y + x$ is also even. Therefore, $y\ R\ x$ and R is symmetric.

Finally, we show that R is transitive. Assume that $x\ R\ y$ and $y\ R\ z$, where $x, y, z \in \mathbf{Z}$. Therefore, both $x + y$ and $y + z$ are even. So $x + y = 2a$ and $y + z = 2b$ for integers a and b. Adding $x + y$ and $y + z$, we obtain $(x + y) + (y + z) = 2a + 2b$. Therefore,

$$x + z = 2a + 2b - 2y = 2(a + b - y).$$

Since $a + b - y$ is an integer, $x + z$ is even. Hence $x\ R\ z$ and R is transitive. $\diamond$

For an equivalence relation R defined on a nonempty set A and for an element $a \in A$, the **equivalence class** $[a]$ is defined by

$$[a] = \{x \in A : x \ R \ a\}.$$

Since $a \in [a]$, every equivalence class is nonempty.

For an equivalence relation defined on a nonempty set A, the resulting distinct equivalence classes produce a partition of A, where two elements of A belong to the same equivalence class if and only if they are related. For the relation R_1 defined above in (2) on the set $A = \{1, 2, 3, 4\}$, the distinct equivalence classes are $[1] = \{1, 3, 4\}$ and $[2] = \{2\}$. In this case, $[4] = [3] = [1]$. For the relation R on $\mathbf{Z}$ defined in Example 2.1 by $x \ R \ y$ if $x + y$ is even, the distinct equivalence classes are

$$[0] = \{x \in \mathbf{Z} : x \text{ is even}\} \text{ and } [1] = \{x \in \mathbf{Z} : x \text{ is odd}\}.$$

A common type of equivalence relation is given in the next example.

Example 2.2 *A relation R defined on a nonempty set A of integers by $x \ R \ y$ if $x \equiv y \ (\mathrm{mod}\,3)$ is an equivalence relation.*

For example, if

$$A = \{-4, -2, -1, 0, 3, 4, 5, 8, 11, 12\}$$

is the set of integers in Example 2.2, then the distinct equivalence classes resulting from the equivalence relation R are

$$[-4] = \{-4, -1, 5, 8, 11\}, [-2] = \{-2, 4\}, \text{ and } [0] = \{0, 3, 12\}.$$

Examples of equivalence relations seen in the text are:

(1) Two vertices u and v in a graph are related if u is connected to v. This is discussed in Section 1.2.

(2) Two graphs G and H are related if they are isomorphic. This is discussed in Section 3.2.

(3) Two vertices u and v in a graph G are related if they are similar (that is, if there exists an automorphism α of G for which $\alpha(u) = v$). This is discussed in Section 3.4.

(4) Two edges e and f in a nontrivial connected graph G are related if e and f lie on a common cycle in G. This is discussed in Section 5.2.

(5) Two vertices u and v in a connected graph G are related if they are distance similar (that is, if $d(u, x) = d(v, x)$ for every vertex $x \in V(G) - \{u, v\}$). This is discussed in Section 12.3.

2.2 Functions

For nonempty sets A and B, a **function** f from A to B, written as $f : A \to B$, is a relation from A to B in which each element of A appears as the first coordinate in exactly one ordered pair in f. If the ordered pair (a, b) belongs to f, then we write $b = f(a)$ and b is called the **image** of a. The set of all images of f is called the **range** of f. For example, for $A_1 = \{r, s, t\}$ and $B_1 = \{w, x, y, z\}$,

$$f_1 = \{(r, x), (s, z), (t, z)\} \tag{3}$$

is a function from A_1 to B_1. The range of f_1 is $\{x, z\}$.

A function $f : A \to B$ is **one-to-one** (or **injective**) if distinct elements of A have distinct images in B. Therefore, f is one-to-one if for every two (distinct) elements a_1 and a_2 in A, it follows that $f(a_1) \neq f(a_2)$. Using the contrapositive, we can also say that f is one-to-one if for $a_1, a_2 \in A$, whenever $f(a_1) = f(a_2)$, it follows that $a_1 = a_2$. The function f_1 in (3) is not one-to-one since s and t have the same image, that is, $f_1(s) = f_1(t)$. However, for the sets A_1 and B_1 above, the function

$$g_1 = \{(r, w), (s, z), (t, y)\} \tag{4}$$

is one-to-one.

A function $f : A \to B$ is called **onto** (or **surjective**) if every element of B is the image of some element of A, that is, if the range of f is B. The function f_1 above is not onto since neither w nor y is an image of any element in A_1. The function g_1 in (4) is not onto either since the range of g_1 is not B_1 as x is not in the range of g_1. On the other hand, for $A_2 = \{1, 2, 3\}$ and $B_2 = \{4, 5, 6\}$, the function

$$f_2 = \{(1, 4), (2, 6), (3, 5)\} \tag{5}$$

is both one-to-one and onto.

A function that is both one-to-one and onto is called a **bijective** function or a **one-to-one correspondence**. Therefore, the function f_2 in (5) is bijective. The following gives an example of a bijective function involving an infinite set.

Example 2.3 *Show that the function $f : \mathbf{R} \to \mathbf{R}$ defined by $f(x) = 3x - 8$ for all $x \in \mathbf{R}$ is bijective.*

Solution. First, we show that f is one-to-one. Assume that $f(a) = f(b)$, where $a, b \in \mathbf{R}$. Then $3a - 8 = 3b - 8$. Adding 8 to both sides of this equation and dividing by 3, we obtain $a = b$. Therefore, f is one-to-one.

Next, we show that f is onto. Let r be a real number. Then $x = (r + 8)/3$ is also a real number. Furthermore,

$$f(x) = f\left(\frac{r + 8}{3}\right) = 3\left(\frac{r + 8}{3}\right) - 8 = r,$$

and so f is onto.

Consequently, f is bijective. $\diamond$

For sets A, B and C and functions $f : A \to B$ and $g : B \to C$, the **composition** $g \circ f$ of f and g is the function from A to C defined by

$$(g \circ f)(a) = g(f(a))$$

for all $a \in A$.

For example, let $A = \{1, 2, 3\}, B = \{a, b, c, d\}$ and $C = \{x, y, z\}$ and let $f : A \to B$ and $g : B \to C$ be the functions

$$f = \{(1, c), (2, d), (3, a)\} \text{ and } g = \{(a, y), (b, z), (c, z), (d, y)\}.$$

Then $(g \circ f)(1) = g(f(1)) = g(c) = z$ and, in general,

$$g \circ f = \{(1, z), (2, y), (3, y)\}.$$

An important theorem involving composition of functions is the following.

Theorem 2.4 *If $f : A \to B$ and $g : B \to C$ are bijective functions, then $g \circ f$ is bijective.*

For a function $f : A \to B$, the **inverse relation** f^{-1} of f is defined by

$$f^{-1} = \{(b, a) : (a, b) \in f\}.$$

The most important theorem in this connection is the following.

Theorem 2.5 *For a function $f : A \to B$, the inverse relation f^{-1} is a function from B to A if and only if f is bijective. Furthermore, if f is bijective, then f^{-1} is also bijective.*

For a nonempty set A, a bijective function $f : A \to A$ is a **permutation** of A. The function f considered in Example 2.3 is therefore a permutation of $\mathbf{R}$.

Example 2.6 *Let $A = \{1, 2, 3, 4, 5, 6\}$. The function $f : A \to A$, where*

$$f = \{(1, 3), (2, 6), (3, 4), (4, 1), (5, 5), (6, 2)\}$$

a permutation of A.

The function f in Example 2.6 can also be expressed in terms of permutation cycles, namely, $f = (134)(26)$, which indicates that 1 is mapped into 3, which is mapped into 4 and which is mapped into 1. The integers 2 and 6 map into each other, while 5 is fixed (it maps into itself).

Examples of functions seen in the text are:

(1) isomorphisms (Chapter 3),

(2) automorphisms (Section 3.4),

(3) matchings (Section 8.1),

(4) colorings (Sections 10.2 and 10.3),

(5) the Channel Assignment Problem (Section 12.5).

Appendix 3: Methods of Proof

3.1 Direct Proof

Many, indeed most, theorems in mathematics are (or can be) stated as an implication, typically as a quantified statement $\forall x \in S$, $P(x) \Rightarrow Q(x)$, where $P(x)$ and $Q(x)$ are open sentences involving a variable x whose values are taken from a set S. The most common proof technique is a **direct proof**, where $P(x)$ is assumed to be true for an arbitrary element $x \in S$ and then $Q(x)$ is shown to be true. An example of a direct proof is given next.

Example 3.1 *If n is an even integer, then $5n + 7$ is an odd integer.*

Proof. Assume that n is an even integer. Then $n = 2k$ for some integer k. Therefore,

$$5n + 7 = 5(2k) + 7 = 10k + 7 = 10k + 6 + 1 = 2(5k + 3) + 1.$$

Since $5k + 3$ is an integer, $5n + 7$ is odd. ∎

A few comments about Example 3.1 and its proof might be useful. First, the implication in Example 3.1 can be restated as follows:

For every even integer n, the integer $5n + 7$ is odd.

If we let T denote the set of even integers and define

$$P(n): \quad 5n + 7 \text{ is odd.}$$

then the implication in Example 3.1 can be restated more symbolically as:

$$\forall n \in T, \, P(n): \textit{ For every } n \in T, \, 5n + 7 \textit{ is odd.}$$

When we gave a direct proof of the implication in Example 3.1, we began by assuming that n is an even integer (or letting n be an even integer). Therefore, we began with an arbitrary element in the set T. We then showed that $5n + 7$ is an odd integer.

Two examples in the text in which a direct proof is employed are Theorems 1.11 and 2.1.

Theorem 1.11 *If G is a disconnected graph, then $\overline{G}$ is connected.*

As expected for a direct proof, we began by assuming that G is a disconnected graph and then showed that $\overline{G}$ is connected.

Theorem 2.1 (The First Theorem of Graph Theory) *If G is a graph of size m, then*

$$\sum_{v \in V(G)} \deg v = 2m.$$

Here we started with a graph G of size m and showed that if the degrees of the vertices of G are summed, then $2m$ is obtained.

3.2 Counterexamples

A mathematical statement that can be expressed as an implication can be shown to be false by providing a counterexample. Suppose that a statement we are considering is expressed as $\forall x \in S$, $P(x) \Rightarrow Q(x)$ where $P(x)$ and $Q(x)$ are open sentences concerning a variable x whose values belong to a set S. If some specific element $x \in S$ can be discovered for which $P(x)$ is true and $Q(x)$ is false, then x is a **counterexample** to the statement $\forall x \in S$, $P(x) \Rightarrow Q(x)$. Counterexamples often occur when a **conjecture** is made (that is, a statement that is believed to be true) and an example is found to show that, in fact, the statement is false. This is illustrated below.

Example 3.2 *Determine whether the following statement is true.*

If n is an integer, then $6n + 3$ is not prime.

Solution This statement is false. For $n = 0$, it follows that $6n + 3 = 3$, which is a prime. Therefore, $n = 0$ is a counterexample to this statement. ◇

If the statement in Example 3.2 had read:

If n is a positive integer, then $6n + 3$ is not prime.,

then the statement would be true.

One instance of a counterexample in the text occurs in Section 10.3, where it was mentioned that Peter Guthrie Tait believed that if G is a 3-regular, 3-connected, planar graph, then G is Hamiltonian. However, William Tutte produced an example (a counterexample) of a 3-regular, 3-connected, planar graph (the Tutte graph) that is *not* Hamiltonian.

3.3 Proof by Contrapositive

Recall for statements P and Q that the contrapositive of the implication $P \Rightarrow Q$ is the implication $(\sim Q) \Rightarrow (\sim P)$. Since these two implications are logically equivalent, the statement $\forall x \in S$, $P(x) \Rightarrow Q(x)$ can be established by verifying the statement $\forall x \in S$, $(\sim Q(x)) \Rightarrow (\sim P(x))$ is true. In a **proof by contrapositive** of the statement $\forall x \in S$, $P(x) \Rightarrow Q(x)$, we assume that $Q(x)$ is false for an arbitrary element $x \in S$ and show that $P(x)$ is false. That is, a proof by contrapositive of the statement $\forall x \in S$, $P(x) \Rightarrow Q(x)$ is a direct proof of its contrapositive $\forall x \in S$, $(\sim Q(x)) \Rightarrow (\sim P(x))$. An example of a proof by contrapositive is given next.

Example 3.3 *Let $n \in \mathbf{Z}$. If $11n - 5$ is odd, then n is even.*

Proof. Assume that n is an odd integer. Then $n = 2k + 1$ for some integer k. So

$$11n - 5 = 11(2k + 1) - 5 = 22k + 11 - 5 = 22k + 6 = 2(11k + 3).$$

Since $11k + 3 \in \mathbf{Z}$ is an integer, $11n - 5$ is even. ∎

A theorem in the text where a proof by contrapositive is employed (in fact, *twice*) is Theorem 4.1.

Theorem 4.1 *Let G be a connected graph. An edge e of G is a bridge if and only if e lies on no cycle of G.*

To verify the implication "If an edge e of G is a bridge, then e lies on no cycle of G" using a proof by contrapositive, we assume that e lies on a cycle of G and show that e is not a bridge. To verify the converse "If an edge e lies on no cycle of G, then e is a bridge" using a proof by contrapositive, we assume that e is not a bridge and show that e lies on a cycle of G.

3.4 Proof by Contradiction

In a **proof by contradiction** of some mathematical statement A, we assume that A is false and show that this leads to a contradiction. If A is expressed as $\forall x \in S,\ P(x) \Rightarrow Q(x)$, then assuming that $\forall x \in S,\ P(x) \Rightarrow Q(x)$ is false means assuming that there exists some element $x \in S$ such that $P(x)$ is true and $Q(x)$ is false. An example of a proof by contradiction is given next.

Example 3.4 *Let n be a positive integer. If $n^3 + 1$ is prime, then $n = 1$.*

Proof. Assume, to the contrary, that there is a positive integer n different from 1 such that $n^3 + 1$ be prime. Thus $n \geq 2$. Now $n^3 + 1 = (n+1)(n^2 - n + 1)$. Since $n + 1 > 1$ and $n^2 - n + 1 = n(n - 1) + 1 > 1$, it follows that neither $n + 1$ or $n^2 - n + 1$ is 1 and so $n^3 + 1$ is not prime, producing a contradiction. ∎

Proofs by contradiction are often used to prove negative-sounding results. A theorem in the text that illustrates this is Theorem 5.5.

Theorem 5.5 *Let G be a nontrivial connected graph and $u \in V(G)$. If v is a vertex that is farthest from u in G, then v is not a cut-vertex of G.*

To use a proof by contradiction, we assume that v is a vertex that is farthest from u and that v *is* a cut-vertex of G. We then produce a contradiction.

In the following theorem in the text, a proof by contrapositive and a proof by contradiction are both used.

Theorem 5.1 *Let v be a vertex incident with a bridge in a connected graph G. Then v is a cut-vertex of G if and only if $\deg v \geq 2$.*

A proof by contrapositive is used to verify the implication "If v is a cut-vertex of G, then $\deg v \geq 2$." Thus, we assume that $\deg v = 1$ and show that v is not a cut-vertex of G. A proof by contradiction is used to show the converse "If $\deg v \geq 2$, then v is a cut-vertex of G". So we assume that $\deg v \geq 2$ and that v is not a cut-vertex. We then show that these lead to a contradiction.

3.5 Proof by Minimum Counterexample

A particular type of proof by contradiction is proof by minimum counterexample. This proof technique is often related to the following principle.

The Well-Ordering Principle The set $\mathbf{N}$ of positive integers is well-ordered, that is, every nonempty subset of $\mathbf{N}$ has a smallest element.

Suppose that we have a sequence $S_1, S_2, S_3, \cdots$ of statements, one for each positive integer, that we wish to prove are true. If we assume, to the contrary, that not all of these statements are true, then it follows by the Well-Ordering Principle that there is a smallest positive integer n for which S_n is false. The idea is to use this information to arrive at a contradiction. An example of a proof by minimum counterexample is given next.

Example 3.5 *For every positive integer n, the integer $n^2 - 3n$ is even.*

Proof. Assume that this statement is false. Then among the positive integers n such that $n^2 - 3n$ is odd, let m be the smallest one. If $n = 1$, then $n^2 - 3n = -2$, which is even. Therefore, $m \geq 2$. So we can write $m = k + 1$, where $1 \leq k < m$. Since $1 \leq k < m$, it follows that $k^2 - 3k$ is even. Hence $k^2 - 3k = 2x$ for some integer x. Observe that

$$
\begin{aligned}
m^2 - 3m &= (k+1)^2 - 3(k+1) = (k^2 + 2k + 1) - 3(k+1) \\
&= (k^2 - 3k) + 2k - 2 = 2x + 2k - 2 = 2(x + k - 1).
\end{aligned}
$$

Since $x + k - 1$ is an integer, $m^2 - 3m$ is even, which produces a contradiction. ∎

A theorem in the text that uses a proof by minimum counterexample is Theorem 11.18.

Theorem 11.18 *Every graph of order $n \geq 3$ and size at least $\binom{n-1}{2} + 2$ is Hamiltonian.*

To prove Theorem 11.18 using a proof by minimum counterexample, we assume that the statement is false. Then there is a smallest positive integer $n \geq 3$ for which there exists a graph G of order n and size $\binom{n-1}{2} + 2$ that is not Hamiltonian. We then show that G is, in fact, Hamiltonian, producing a contradiction.

3.6 Proof by Mathematical Induction

Let $S_1, S_2, S_3, \cdots$ be statements, one for each positive integer. To prove that these statements are true using a proof by mathematical induction, we

(1) show that S_1 is true (the basis step) and

(2) verify that the following implication is true:

$$\forall k \in \mathbf{N}, \ S_k \Rightarrow S_{k+1}: \text{Let } k \in \mathbf{N}. \text{ If } S_k, \text{ then } S_{k+1}.$$

An example of a proof by mathematical induction is given next.

Example 3.6 *For every positive integer n, $2^{n+2} \geq 7n + 1$.*

Proof. We proceed by induction. For $n = 1$, we have $2^{1+2} = 2^3 = 8 = 7 \cdot 1 + 1$, which verifies the statement for $n = 1$. Assume that $2^{k+2} \geq 7k + 1$ for some integer $k \geq 1$. We show that $2^{(k+1)+2} \geq 7(k+1) + 1 = 7k + 8$. Now

$$
\begin{aligned}
2^{(k+1)+2} &= 2 \cdot 2^{k+2} \geq 2(7k+1) = 2(7k) + 2 \\
&= 7k + 7k + 2 \geq 7k + 7 + 2 > 7k + 8,
\end{aligned}
$$

as desired. ∎

A theorem in the text that uses a proof by mathematical induction is the following.

Theorem 4.4. *Every tree of order n has size $n - 1$.*

To use a proof by induction, we show first that result is true for $n = 1$. Then we show that if the size of every tree of order k is $k - 1$ for an arbitrary positive integer k, then every tree of order $k + 1$ has size k.

There is a variation of the standard proof by mathematical induction that is used in the text and is useful to know in general. The **Strong Form of Induction** (or the **Strong Principle of Mathematical Induction**) can be used to show that each of the statements $S_1, S_2, S_3, \cdots$ is true. In this case, we need to

(1) show that S_1 is true (the basis step) and

(2) verify the implication:

> For each $k \in \mathbf{N}$, if the statements $S_1, S_2, \ldots, S_k$ are true, then S_{k+1} is true.

An example of a theorem that uses the Strong Form of Induction is the following.

Example 3.7 *Every integer $n \geq 2$ is either prime or can be expressed as a product of primes; that is, $n = p_1 p_2 \cdots p_m$, where $p_1, p_2, \cdots, p_m$ are primes.*

Proof. We employ the Strong Form of Induction. Since 2 is prime, the statement is certainly true for $n = 2$.

For an arbitrary integer $k \geq 2$, assume that every integer i, with $2 \leq i \leq k$, is either prime or can be expressed as a product of primes. We show that $k + 1$ is either prime or can be expressed as a product of primes. Of course, if $k + 1$ is prime, then there is nothing further to prove. We may assume, then, that $k + 1$ is composite. Then there exist integers a and b such that $k + 1 = ab$, where $2 \leq a \leq k$ and $2 \leq b \leq k$. Therefore, by the induction hypothesis, each of a and b is prime or can be expressed as a product of primes. In either case, $k + 1 = ab$ is a product of primes. ∎

A theorem in the text that employs the Strong Form of Induction is Menger's Theorem (Theorem 5.16).

Menger's Theorem *Let u and v be nonadjacent vertices in a graph G. The minimum number of vertices in a $u-v$ separating set equals the maximum number of internally disjoint $u - v$ paths in G.*

To prove Menger's Theorem using the Strong Form of Induction, we first show that the result is true for all empty graphs, that is, graphs of size 0. We then assume that the result is true for all graphs of size *less than m*, where m is a positive integer, and then prove that the result is true for every graph of size m.

3.7 Existence Proofs

There are numerous statements in mathematics that are formed by using an existential quantifier (there exists, there is, for some, there is at least one). To verify such a statement, it suffices to display an appropriate example (or alternatively, to show theoretically that an appropriate example must exist even though a specific example hasn't been found). This is illustrated below.

Example 3.8 *There exist integers a, b and c greater than 1 such that $a^2 + b^3 + c^4$ is prime.*

Proof. Letting $a = 7$ and $b = c = 2$, we have $a^2 + b^3 + c^4 = 49 + 8 + 16 = 73$, which is prime. $\diamond$

Existential quantifiers can occur within an implication. An example of this is stated next.

Example 3.9 *If ϵ is a positive number, then there exists a positive number δ such that if $|x - 2| < \delta$, then $|(2x + 3) - 7| < \epsilon$.*

Proof. Let ϵ be given. Consider $\delta = \epsilon/2$ and suppose that $|x - 2| < \delta$. Then

$$|(2x + 3) - 7| = |2x - 4| = |2(x - 2)| = 2|x - 2| < 2(\epsilon/2) = \epsilon. \qquad \blacksquare$$

For the function $f : \mathbf{R} \to \mathbf{R}$ defined by $f(x) = 2x + 3$ for all $x \in \mathbf{R}$, Example 3.9 would probably be stated more commonly as: For every positive number ϵ, there exists a positive number δ such that if $|x - 2| < \delta$, then $|(2x + 3) - 7| < \epsilon$. What Example 3.9 shows is that f is continuous at 2.

A theorem in the text where an existential quantifier is encountered is the following.

Theorem 2.7 *For every graph G and every integer $r \geq \Delta(G)$, there exists an r-regular graph H containing G as an induced subgraph.*

In the proof of Theorem 2.7, we actually construct an r-regular graph H such that the given graph G is an induced subgraph of H.

Although showing an implication is false can be accomplished by means of an example (a counterexample), showing an existence statement is false requires a proof (to verify that no example of the type being claimed exists).

Example 3.10 *There exists a prime p such that $n^2 - n + p$ is prime for every positive integer n.*

Solution This statement is false. Let p be a prime and let $n = p$. Then $n^2 - n + p = p^2 - p + p = p^2$, which is not prime. $\diamond$

Solutions and Hints for Odd-Numbered Exercises

Chapter 1: Graphs and Graph Models

1.1 Can the seven committees meet during the three time periods? Yes. (An explanation is required.)

1.3 See Figure 1.

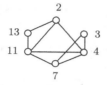

Figure 1: The graph in Exercise 1.3

1.7 (b) $S = \{$CUP, CAP, TAP, PAT, PUT$\}$ (c) $S = \{$RAT, TAR, TAP, CAP, CAT$\}$

1.11 See Figure 2.

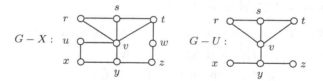

Figure 2: The graph in Exercise 1.11

1.13 (a) See Figure 3.

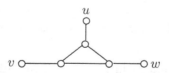

Figure 3: The graph in Exercise 1.13

1.15 Hint: There is only one such graph.

1.17 (a) Hint: Assume, to the contrary, that there exists a connected graph G containing two longest paths P and Q that have no vertex in common. Since G is connected, there exists a $u - v$ path P' where u is on P, v is on Q and no interior vertex (a vertex that is not an end-vertex) of P' belongs to P or Q.

(b) Hint: The statement is false.

1.19 Hint: No.

1.21 See Figure 4.

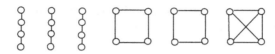

Figure 4: The graph in Exercise 1.21

1.23 (a) Consider the 5-cycle (u, v, x, w, y, u) for $k = 1$ and the 5-cycle (u, x, y, v, z, u) for $k = 2$.

(b) Hint: Note that $k \geq 3$. Consider $P_4 = (u, x, y, v)$.

1.25 **Proof.** If G is not bipartite, then we have the desired result. Thus, we may assume that G is bipartite. Let V_1 and V_2 be two bipartite sets of G. Since the order of G is at least 5, at least one of V_1 and V_2 contains 3 or more vertices, say $|V_1| = k \geq 3$. Since the subgraph of $\overline{G}$ induced by V_1 is the complete graph K_k and $k \geq 3$, it follows that $\overline{G}$ contains a triangle. By Theorem 1.12, $\overline{G}$ is not bipartite. ∎

1.27 (a) $K_5 + K_2 = K_7$ and $K_5 \times K_2$ is shown in Figure 5(a).

(b) $\overline{K}_5 + \overline{K}_3 = K_{5,3}$ and $\overline{K}_5 \times \overline{K}_3 = \overline{K}_{15}$

(c) $C_5 + K_1$ is shown in Figure 5(b) and $C_5 \times K_1 = C_5$.

Figure 5: The graphs in Exercise 1.27

1.29 (a) See the multigraph M in Figure 6.   (b) Add a loop at vertex 2.

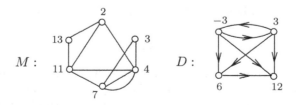

Figure 6: The multigraph and digraph in Exercises 1.29 and 1.33

1.33 See the digraph D in Figure 6.

Chapter 2: Degrees

2.1 (a) No such graph exists. By Corollary 2.3, there is no graph containing an odd number of odd vertices.

   (b) No such graph exists since $\Delta(G) \leq 6$ for every graph G of order 7.

   (c) No such graph exists. Assume, to the contrary, that such a graph G exists. Let $V(G) = \{v_1, v_2, v_3, v_4\}$. We may assume that $\deg v_i = 3$ for $1 \leq i \leq 3$ and $\deg v_4 = 1$. Then each vertex v_i, $1 \leq i \leq 3$, is adjacent to all other vertices of G, including v_4, which is impossible.

2.3 **Solution.** Let x be the number of vertices of degree 4. By the First Theorem of Graph Theory, $(12 - x) \cdot 6 + 4x = 2 \cdot 31$ and so $x = 5$. $\Diamond$

2.5 **Solution.** Let x be a number of vertices of degree 5. By the First Theorem of Graph Theory, $(12 - x) \cdot 3 + 2 \cdot 4 + 5x + 6 \cdot 11 = 2 \cdot 62$ and so $x = 7$. $\Diamond$

2.7 (a) Since every edge of G joins a vertex of U and a vertex of W, both sums $\sum_{u \in U} \deg u$ and $\sum_{w \in W} \deg w$ count every edge of G exactly once, giving the desired result. $\blacksquare$

   (b) The size of G is $3|U| = 3 \cdot 12 = 36$. Let x be the number of vertices of degree 2. Then $2 \cdot x + (10 - x) \cdot 4 = 36$ and so $x = 2$.

2.9 **Proof.** Assume, to the contrary, that these two odd vertices are in different components of G. Then some component of G (which is itself a graph) contains exactly one odd vertex. This contradicts Corollary 2.3. $\blacksquare$

2.11 The bound is sharp. Consider $G = K_5 \cup K_5$. Then $n = 10$ and $\delta(G) = 4 = (n - 2)/2$.

2.13 (a) **Proof.** Assume, to the contrary, that G contains at least three components. Let G_1, G_2 and G_3 be any three components of G. Let $v_i \in V(G_i)$ for $1 \leq i \leq 3$. Since $\deg v_i \geq (n - 2)/3$, each component G_i $(1 \leq i \leq 3)$ contains at least $(n - 2)/3 + 1 = (n + 1)/3$ vertices. Then G contains at least $3 \cdot (n + 1)/3 = n + 1$ vertices, contradicting the fact that G has order n. $\blacksquare$

   Alternative Proof. Assume, to the contrary, that G contains at least three components. Then G contains a component G_1 of order at most $n/3$. Let $v \in V(G_1)$. Then $\deg v \leq (n/3) - 1 = (n - 3)/3$, contradicting that $\delta(G) \geq (n - 2)/3$. $\blacksquare$

   (b) Consider $G = 3K_3$.

2.15 **Proof.** Assume that G is not bipartite. Then G contains an odd cycle C. Let u and v be any two vertices on C. There are two $u - v$ paths on C, one of which has even length and one of odd length. $\blacksquare$

402

2.17 **Proof.** Assume, to the contrary, that G contains a vertex x such that $\deg x \equiv 0 \pmod 3$. Since G is connected, G contains a $w - x$ path, say $P = (w = w_0, w_1, \ldots, w_k = x)$. Let t be the smallest positive integer such that $\deg w_t \equiv 0 \pmod 3$. Then $\deg w_{t-1} \not\equiv 0 \pmod 3$. However, then, $\deg w_{t-1} + \deg w_t \not\equiv 0 \pmod 3$, a contradiction. ∎

2.21 For (a) and (b), see Figure 7.

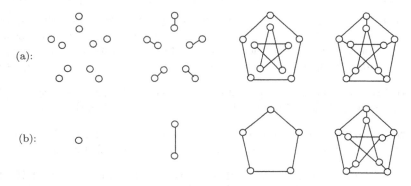

(a):

(b):

Figure 7: The graph in Exercise 2.21

(c) Find four induced subgraphs F_0, F_1, F_2, F_3 *of maximum order* in the Petersen graph, where F_r is r-regular for $0 \le r \le 3$

2.23 See Figure 8.

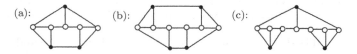

(a): (b): (c):

Figure 8: The graphs in Exercise 2.23

2.25 (a) Since $G - v$ is 3-regular, the order of $G - v$ is even and so the order of G is odd.

2.27 The size of G is $r|U| = r|W|$. Dividing by r, we obtain $|U| = |W|$.

2.29 (a) Let v be any vertex of G and let $\deg_G v = k$. Then $\deg_{\overline{G}} v = n-1-k$. So $\delta(G) + \delta(\overline{G}) \le k + (n-1-k) = n-1$.

(b) **Proof.** Assume that G is regular, say r-regular. Then $\overline{G}$ is $(n-1-r)$-regular. Therefore, $\delta(G) = r$ and $\delta(\overline{G}) = n-1-r$; so $\delta(G) + \delta(\overline{G}) = n-1$.

For the converse, assume that G is not regular. Then G contains two vertices u and v such that $a = \deg_G u < \deg_G v = b$. Therefore, $\delta(G) \le a$. Now $\deg_{\overline{G}} v = n-1-b$ and so $\delta(\overline{G}) \le n-1-b$. Hence $\delta(G) + \delta(\overline{G}) \le a + (n-1-b) = n-1-(b-a) < n-1$. ∎

(c) **Proof.** If G r-regular of order n, then $\overline{G}$ is $(n-1-r)$-regular. For every vertex v of G, $\deg_G v = r = \delta(G)$ and $\deg_{\overline{G}} v = n-1-r = \delta(\overline{G})$. For the converse, suppose that $\deg_G v = r$. Then $\deg_{\overline{G}} v = n-1-r$. Let u be any other vertex of G. Then $\deg_G u \geq r$ and $\deg_{\overline{G}} u \geq n-1-r$. Thus $n-1 = \deg_G u + \deg_{\overline{G}} u \geq r + (n-1-r) = n-1$. This implies that $\deg_G u = r$ and $\deg_{\overline{G}} u = n-1-r$. Thus G r-regular. ∎

(d) **Solution.** Suppose that G has order n. Let $u_1, u_2, \ldots, u_k$ be the vertices of G such that $\deg_G u_i = \delta(G) = r$ for $1 \leq i \leq k$ and let $v_1, v_2, \ldots, v_\ell$ be such that $\deg_{\overline{G}} v_i = \delta(\overline{G}) = s$ for $1 \leq i \leq \ell$. Then $\deg_G v_i = n-1-s > r$. Hence every vertex of G has one of two distinct degrees. ◇

2.31 Hint: For a vertex v of a graph G of order n, we have $\deg_G v + \deg_{\overline{G}} v = n-1$. Thus, if $d_1, d_2, \ldots, d_n$ is a degree sequence of a graph G of order n, then $(n-1) - d_1, (n-1) - d_2, \ldots, (n-1) - d_n$ is a degree sequence of $\overline{G}$.

2.33 Hint: If there exists a graph G with degree sequence $x, 1, 2, 3, 5, 5$, then the order of G is 6. Since there are two vertices of degree 5, it follows that $\delta(G) \geq 2$ and so no vertex of G has degree 1.

2.35 Hint: If there exists a graph G with degree sequence $x, 7, 7, 5, 5, 4, 3, 2$, then $\delta(G) \geq 2$. Since every graph has an even number of odd vertices, x must be odd and so the only possible values of x are 3, 5, 7. Apply the Havel-Hakimi Theorem to show that $x = 5$ or $x = 3$.

Chapter 3: Isomorphic Graphs

3.1 See Figure 9.

Figure 9: The graphs in Exercise 3.1

3.3 (a) $G_1 \not\cong G_2$,　(b) $G_1 \cong G_2$.

3.5 We cannot conclude that $G_1 \not\cong G_2$ from this information. For example, perhaps some vertex of G_2 other than v_2 has degree 3 and is adjacent to a vertex of degree 2. See Figure 10.

3.7 The solution is not correct. In fact, $G_1 \cong G_2$. That there is no path of length 2 lying inside a 5-cycle of G_2 is only a feature of the way G_2 is drawn.

3.9 The graphs G_1 and G_2 are not isomorphic. For example, the two vertices of degree 2 in G_1 are mutually adjacent to two vertices, while the two vertices of degree 2 in G_2 are not.

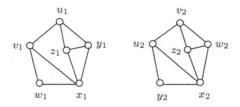

Figure 10: The graphs in Exercise 3.5

3.11 Proof. Suppose that $|W| = a$. Then $|U| = a$. Consider a vertex $v \in W$. Then $\deg_G v \geq n/2$. Therefore, $\deg_{\overline{G}} v = n - 1 - \deg_G v \leq n/2 - 1$. Since there are a vertices v in G with $\deg_G v \geq n/2$, there are a vertices v in G with $\deg_G v \leq n/2 - 1$. Because there are a vertices v in G with $\deg_G v \leq n/2$, it follows that there are no vertices v in G with $\deg_G v = n/2$. ∎

3.13 The graphs G and H are isomorphic.

Proof. The function ϕ is one-to-one and onto. Consider any two vertices u and v of G. Then $uv \in E(G)$ or $uv \notin E(G)$. Assume first that $uv \in E(G)$ and so $d_G(u, v) = 1$. Since $d_H(\phi(u), \phi(v)) = 1$, it follows that $\phi(u)$ and $\phi(v)$ are adjacent in H. Next, assume that $uv \notin E(G)$. Since $d_G(u, v) \geq 2$, we have $d_H(\phi(u), \phi(v)) \geq 2$. That is, if u and v are nonadjacent in G, then $\phi(u)$ and $\phi(v)$ are nonadjacent in H. So ϕ is an isomorphism. ∎

3.15 The statement is false. Let $G_1 = P_3 = (u_1, v_1, w_1)$ and $G_2 = P_3 = (u_2, v_2, w_2)$. Then $G_1 \cong G_2$. Define a one-to-one correspondence $\phi : V(G_1) \rightarrow V(G_2)$ by $\phi(u_1) = u_2$, $\phi(v_1) = w_2$ and $\phi(w_1) = v_2$. Then $d_{G_1}(u_1, v_1) = 1$, while $d_{G_2}(\phi(u_1), \phi(v_1)) = d_{G_2}(u_2, w_2) = 2$.

3.17 (a) Yes, (b) No, (c) Yes.

3.19 Construct a graph G with $V(G) = \{v_1, v_2, \ldots, v_n\}$ and where $v_i v_j \in E(G)$ if and only if G_i is isomorphic to G_j. Since G is a graph, it contains an even number of odd vertices.

Chapter 4: Trees

4.1 Consider the graph G obtained from the graph K_3 with $V(K_3) = \{v_1, v_2, v_3\}$ by adding two new vertices x and y and the two edges $v_1 x$ and $v_2 y$.

4.3 Proof. Since $e = uv$ is an edge, G contains the $u - v$ path $P = (u, v)$. We show that P is the only $u - v$ path in G. Assume, to the contrary, that G contains another $u - v$ path P'. Observe that $P' \neq P$ and so the path P' contains at least three vertices. Since u and v are the end-vertices of P' and P' is a path, it follows that $u, v \notin V(P') - \{u, v\}$ and so $e \notin E(P')$. Thus P' together with $e = uv$ form a cycle containing e. Since e is a bridge in G, a contradiction is produced. ∎

4.5 (a) The size of G is $n-1$. Hint: If e_1 is an edge of G, then $G_1 = G - e_1$ has two components. If e_2 is an edge of G_1, then $G_2 = G_1 - e_2$ contains three components.

    (b) The size of G is $n - k$.

4.7 (a) There are three trees of order 5.

    (b) There are six forests of order 6 with one component, six forests of order 6 with two components, four forests of order 6 with three components, two forests of order 6 with four components, one forest of order 6 with five components and one forest of order 6 with six components. So there are 20 forests of order 6.

4.9 Let $G = C_{n-1} \cup K_1$ for $n \geq 4$.

4.11 Hint: $T_2 = P_4$.

4.13 Let x be the number of vertices of degree 5. Thus $21 - 15 - 1 - x = 5 - x$ vertices of T have degree 3. Since the size of T is $21 - 1 = 20$, it follows by the First Theorem of Graph Theory that $\sum_{v \in V(T)} \deg v = 15 + 6 + 5x + 3(5 - x) = 2 \cdot 20$. So $x = 2$.

4.15 Let x be the number of vertices of degree 2. Then the order of T is $n = 50 + 4x$. Thus $m = 49 + 4x$. Summing the degrees, we obtain $50 \cdot 1 + 2x + 3x + 4x + 5x = 2(49 + 4x)$ and so $x = 8$. Thus $n = 50 + 4 \cdot 8 = 82$.

4.17 (a) The graph T is a double star with two vertices of degree 4.

    (b) Let T be a tree of order n where 75% of the vertices have degree 1 and the remaining 25% vertices have degree 4. Then $(3n/4) \cdot 1 + (n/4) \cdot 4 = 2(n - 1)$. Solving for n, we have $n = 8$. Therefore, T is the double star in (a).

    (c) Let T be a tree of order n where 75% of the vertices have degree 1 and the remaining 25% vertices have a fixed degree x. Then $(3n/4) \cdot 1 + (n/4) \cdot x = 2(n - 1)$. Then $n(5 - x) = 8$, implying that $x \leq 4$. So $x = 2, 3, 4$. Since n is an integer, $x \neq 2$. Thus $x = 3$ or $x = 4$. If $x = 3$, then $n = 4$ and $T = K_{1,3}$; while if $x = 4$, then $n = 8$ and T is the double star in (a).

4.19 (a) Hint: Since $n = \sum_i n_i$ and $2(n - 1) = 2m = \sum_i i n_i$, it follows that $2 \left(\sum_i n_i - 1 \right) = \sum_i i n_i$ and so $2 \sum_i n_i - 2 = \sum_i i n_i$. Thus, $2 \sum_i n_i - \sum_i i n_i = \sum_i (2 - i) n_i = 2$. Simplifying, we have $n_1 = \sum_{i \neq 1} (i - 2) n_i + 2 = n_3 + 2n_4 + \cdots + 2$.

    (b) $n_1 = 5 + 2 \cdot 2 + 2 = 11$.

4.21 The graph $\overline{C}_{n+2}$ is $(n - 1)$-regular. Since $\delta(\overline{C}_{n+2}) = n - 1$, it follows by Theorem 4.9 that T is isomorphic to a subgraph of $\overline{C}_{n+2}$.

4.23 **Solution.** $T = K_1$ or $T = P_4$.

    Proof. Let T be a tree of order n. Since T and $\overline{T}$ are both trees of order n, it follows that the sizes of T and $\overline{T}$ are $n-1$. Thus $n-1+n-1 = 2(n-1) = \binom{n}{2} = \frac{n(n-1)}{2}$. Hence $4(n-1) = n(n-1)$ and so $(n-1)(n-4) = 0$, implying that $n = 1$ or $n = 4$. If $n = 1$, then $T = K_1$. If $n = 4$, then $T = P_4$ or $T = K_{1,3}$. Since $\overline{P_4} = P_4$ and $\overline{K}_{1,3}$ is not a tree, it follows that $T = P_4$. $\blacksquare$

4.25 (a) There are 8 spanning trees of G and two nonisomorphic spanning trees.

      (b) There are 9 spanning trees of G and three nonisomorphic spanning trees.

4.27 See Figure 11 for one example.

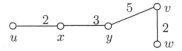

Figure 11: The graph in Exercise 4.27

4.29 Hint: By Kruskal's algorithm, there is only one choice at each step for the edge selected.

4.31 For $k = 2$, let $G = C_3$ with weights $1, 2, 2$. For $k \geq 3$, let $G = C_k$ such that each edge of G has the same weight.

Chapter 5: Connectivity

5.1 (a) Every nontrivial tree has this property.

    (b) See the three graphs in Figure 12, where v is a cut-vertex in each graph.

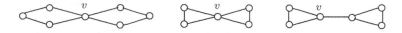

Figure 12: The graphs in Exercise 5.1(b)

5.3 Each of the statements (a)-(d) is false.

    (a) Let G be any graph in Figure 12.

    (b) Let v be an end-vertex of G.

    (c) Let $G = K_{1,n-1}$ $(n \geq 3)$.

    (d) Every tree of order n has $n-1$ bridges and at most $n-2$ cut-vertices.

5.5 (a) Let $G = K_{1,12}$.

(b) Let G be either of the first two graphs in Figure 12.

(c) Let G be either of the first two graphs in Figure 12.

(d) Let $G = K_2$.

5.7 **Proof.** Let x and y be two vertices of T such that $d(x,y) = \text{diam}(T) \geq 2$. Then x and y are not cut-vertices by Theorem 5.5 and so x and y are end-vertices of T. Let v be a vertex on the $x - y$ geodesic P that is adjacent to y. Then v is a cut-vertex. Assume, to the contrary, that v is adjacent to two cut-vertices. Then v is adjacent to a cut-vertex w that is not on P. Then $G - w$ has a component containing a vertex z that does not lie on the $x - v$ subpath of P. Since T has a unique $x - z$ path, the length of this path is $d(x,z) = d(x,y) + 1 = \text{diam}(T) + 1$, which is impossible. ∎

5.9 The cut-vertices of G are r, t, w and the bridges of G are qr, tw. The blocks of G are shown in Figure 13.

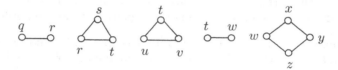

Figure 13: The graphs in Exercise 5.9

5.11 **Proof.** By Corollary 2.5, G is connected. Thus it remains only to show that no vertex of G is a cut-vertex. Let $v \in V(G)$. Let u and w be any two vertices of G that are distinct from v. We show that G contains a $u - w$ path that does not contain v. This is obvious if $uw \in E(G)$, so we can assume that u and w are nonadjacent vertices. The set $N(u)$ of vertices adjacent to u contains at least $n/2$ elements, as does $N(w)$. Since G contains n vertices, $N(u) \cap N(w)$ must contain at least two vertices, at least one of which, say x, is not v. Then (u, x, w) is a $u - w$ path not containing v. ∎

5.13 The statement is false. See Figure 14.

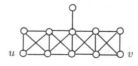

Figure 14: The graph in Exercise 5.13

5.15 Hint: Let B be a nonseparable subgraph of G that is not a proper subgraph of any other nonseparable subgraph of G. Let e and f be two edges of B. We show that either $e = f$ or e and f lie on a common cycle of G. Suppose that $e \neq f$. Let C be a cycle that contains e. If f is on C, then the proof is complete. Otherwise, there is a $u - v$ path $P = (u = u_0, u_1, \ldots, u_k = v)$ in B with $f = u_0 u_1$ such that u_i is not on C for $0 \leq i \leq k - 1$. Since u_k is not a cut-vertex in B, there is a $u_{k-1} - w$ path P' not containing u_k such that w is the only vertex of P' on C. (Show that there is a cycle of B containing both e and $u_{k-1} u_k$. Then complete the proof in this direction.)

For the converse, let B be the subgraph of a nontrivial connected graph G induced by the edges in an equivalence class resulting from the equivalence relation defined in Theorem 5.8. We show that B is a nonseparable subgraph of G that is not a proper subgraph of any other nonseparable subgraph of G. Since this is true if B consists of a single edge, we assume that the order of B is 3 or more. Since every two edges of B lie on a common cycle, B is connected. Let w be a vertex of B. (Show that if w is a cut-vertex of B, then there exist vertices u and v, both adjacent to w, such that every $u - v$ path contains w. Show that this is impossible.) Thus w is not a cut-vertex and B is nonseparable. (It now remains to show that B is not a proper subgraph of any other nonseparable subgraph of G.)

5.17 A vertex-cut U of a connected graph G is **minimal** if no proper subset of U is a vertex-cut of G. Then every minimum vertex-cut is minimal, but the converse is not true. One question to ask is what is the maximum cardinality of a minimal vertex-cut in a graph G.

5.19 The statement is false. For example, let $G = K_1 + (K_1 \cup K_2)$.

5.21 (a) Let $G = C_5$.

(b) No such example exists.

(c) Let $G = C_5$.

(d) No such example exists

5.23 (a) **Proof.** Let $H = G + K_1$, where v is a vertex of H that is not in G. We show that $\kappa(H) \geq k+1$. Let S be a set of vertices H with $|S| = k$. There are two cases.

Case 1. $v \notin S$. Since every vertex in G is adjacent to v in H, every vertex in $H - S$ is adjacent to v in $H - S$ and so $H - S$ is connected.

Case 2. $v \in S$. Then $H - S = G - (S - \{v\})$. Since $\kappa(G) \geq k$ and $|S - \{v\}| = k - 1$, it follows that $G - (S - \{v\})$ connected.

In either case, S is not a vertex-cut of H. Thus the removal of k or fewer vertices from H does not disconnect H and so $\kappa(H) \geq k + 1$. Therefore, H is $(k + 1)$-connected. ∎

(b) Hint: Use an argument similar to the one in (a).

5.25 (a) See Figure 15(a).

Figure 15: The graphs in Exercise 5.25(a) and (d)

    (b) No such example exists.

    (c) No such example exists.

    (d) See Figure 15(d).

5.27 (a) Hint: Let G be a (connected) graph with connectivity $k \geq 1$. Then there exists a vertex v_1 of G that is not a cut-vertex. Thus $G_1 = G - v_1$ is connected. Let v_2 be a vertex of G_1 that is not a cut-vertex.

    (b) Hint: The answer depends on k and G.

    (c) The statement is false. Let $G = K_{1,3}$.

    (d) The statement is true.

       Proof. Assume, to the contrary, that there is a vertex-cut W such that $v \notin W$. Since v is adjacent to every vertex in $G - W$, it follows that $G - W$ is connected, a contradiction. ∎

    (e) Hint: The statement is true.

5.29 (a) Let $G = P_3 \times K_2$. (b) Let $G = K_{2,3}$.

5.31 (a) $k = 2 = \lambda(G) - 1$. (b) $k = 1 = \kappa(G) - 1$.

5.33 **Proof.** Since G is 5-connected, $G - w$ is 4-connected. Therefore, there are four internally disjoint $u - v$ paths in $G - w$, each pair of which produces a cycle. Let P_1, P_2, P_3, P_4 denote these four paths and let C be the cycle produced by P_1 and P_2 and C' be the cycle produced by P_3 and P_4. These cycles have only u and v in common and neither contains w since the paths occur in $G - w$. ∎

5.35 **Proof.** Construct a new graph H by adding a new vertex w and joining w to v_i for $1 \leq i \leq k$. Since G is k-connected, it follows by Corollary 5.18 that H is k-connected. By Theorem 5.17, there are k internally disjoint $u - w$ paths in H. The restriction of these paths to G yields the desired internally disjoint $u - v_i$ paths ($1 \leq i \leq k$). ∎

5.37 Hint: $\kappa(Q_n) = \lambda(Q_n) = n$.

Chapter 6: Traversability

6.1 **Solution.** A multigraph M can be constructed that models this situation, where $V(M)$ is the set of rooms and two vertices of M are joined by the number of edges equal to the number of doorways between the rooms in

Figure 16. Since M is connected and contains exactly two odd vertices (R3 and R6), it follows that M contains an Eulerian trail. So such a walk *is* possible, either starting at R3 and ending at R6 or starting at R6 and ending at R3. $\Diamond$

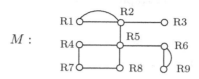

M :

Figure 16: The multigraph M in Exercise 6.1

6.3 **Proof.** Let G_i be r_i-regular of order n_i ($i = 1, 2, 3$). Since G_1 is Eulerian, r_1 is even. Since $\overline{G}_1$ is Eulerian, $n_1 - r_1 - 1$ is even. Thus n_1 is odd. Since G_2 is not Eulerian, r_2 is odd and so n_2 is even. Similarly, r_3 is odd and n_3 is even. Observe that

(1) every vertex of G_1 in G has degree $r_1 + n_2 + n_3$, which is even,

(2) every vertex of G_2 in G has degree $r_2 + n_1 + n_3$, which is even,

(3) every vertex of G_3 in G has degree $r_3 + n_1 + n_2$, which is even.

Since G is connected and every vertex has even degree, G is Eulerian. ∎

6.5 Let $G = K_5 - e$.

6.7 **Solution.** (b) is true. Since n is odd, r is even and $n - 1 - r$ is even. In H, every vertex of G has degree $r + 2$ and every vertex of $\overline{G}$ has degree $(n - 1 - r) + 2$, both of which are even. Both u and v have degree $2n + 1$. Thus H is a connected graph having exactly two vertices of odd degree. $\Diamond$

6.9 **Solution.** Assume, to the contrary, that there is a nonempty subset S of $V(G)$ such that $k(G - S) > |S|$. Since G contains no cut-vertices, $|S| \geq 2$. Moreover, since z is adjacent to all other vertices of G, we must have $z \in S$; otherwise, $G - S$ is connected. Also, since the order of G is 7, removing four or more vertices from G results in a graph with three or less components. So $|S| = 2$ or $|S| = 3$. The graph $G - z$ is in Figure 17. If $|S| = 2$, then $G - S$ is disconnected only if the remaining vertex of S is u, w or y but then, $k(G - S) = |S| = 2$. If $|S| = 3$, then $G - S$ can only have more than two components if the remaining vertices of S are selected from $\{u, w, y\}$ in which case $k(G - S) = |S| = 3$. This says that the condition for G to be Hamiltonian is only necessary and not sufficient, that is, the converse of this theorem does not hold. $\Diamond$

6.11 Hint: Note that $\overline{C}_n$ is a $(n - 3)$-regular graph. If $n = 5$, then $\overline{C}_5 = C_5$, which is Hamiltonian. If $n \geq 6$, then $n - 3 \geq n/2$.

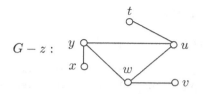

$$G - z : \quad$$

Figure 17: The graph $G - z$ in Exercise 6.9

6.13 (a) Let $G = K_{2,4}$. Removing the vertices in the partite set having cardinality 2 produces a graph with four components.

(b) Let $G = K_4 - e$. This graph contains two odd vertices.

(c) Let $G = K_4 - e$.

(d) Let $G = P_3$.

6.15 **Solution.** Yes, it is true.

Proof. Let F be the subdivision graph of the graph G. First, observe that if F is Hamiltonian, then G must be connected. If G contains an end-vertex, then F does as well. So $\delta(G) \geq 2$. If G contains a vertex of degree 3 or more, then F contains a vertex adjacent to at least three vertices of degree 2. Since no Hamiltonian graph has such a vertex, $\Delta(G) \leq 2$. So G is a connected, 2-regular graph; that is, $G = C_n$ for some integer $n \geq 3$. Hence G is Eulerian. ∎

6.17 **Solution.** All graphs of order 4 together with C_3 and P_3.

Proof. First, if G has order 3 and contains an isolated vertex, then $G(3)$ is not Hamiltonian; otherwise, G is connected and $G(3)$ is Hamiltonian. If $G = \overline{K}_4$, then $G(3) = Q_3$, which is Hamiltonian. So if G has order 4, then $G(3)$ is Hamiltonian. So let $n \geq 5$ and let G be a graph of order n. Let $V(G(3)) = V(G) \cup W$, where $W = \{v_S : S \subseteq V(G), |S| = 3\}$. Then $|W| = \binom{n}{3}$ and the number of components in $G(3) - V(G)$ is $k(G(3) - V(G)) = |W|$. It is known that if $k(G(3) - V(G)) > |V(G)|$ (that is, if $\binom{n}{3} > n$), then $G(3)$ is not Hamiltonian. For a positive integer n, the inequality $\binom{n}{3} > n$ is equivalent to $n > 4$. Since $n \geq 5$, the graph $G(3)$ is not Hamiltonian. ∎

6.19 **Proof.** First, observe that the order of G is $2n$. Since $\delta(G_1) \geq n/2$ and $\delta(G_2) \geq n/2$, the degree of every vertex of G (in G_1 or G_2) is at least $n/2 + n/2 = n$. So G is Hamiltonian by Dirac's Theorem. ∎

6.21 Hint: Consider $G' = G + K_1$, where $V(K_1) = \{x\}$. Then $\deg_{G'} u + \deg_{G'} v \geq n + 1$ for every two nonadjacent vertices u and v in G'. Thus G' contains a Hamiltonian cycle C and so G contains the Hamiltonian path $C - x$.

6.23 Hint: (a) Yes   (b) No.

Chapter 7: Digraphs

7.1 (a) **Proof.** Let v be a vertex of D. By hypothesis, $D - v$ is a strong
oriented graph. Hence $D - v$ is a directed cycle (see Figure 18(a)).
On the other hand, $D - u$ is strong, so it is a directed cycle as well.
Since (w, x) is a directed edge, $D - u$ is the directed cycle shown
in Figure 18(b). Thus D contains the digraph of Figure 18(c) as a
subdigraph and so D is strong. ∎

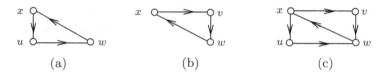

(a)   (b)   (c)

Figure 18: A directed cycle in the proof of Exercise 7.1(a)

(b) From (a), D contains the digraph of Figure 18(c) as a subdigraph.
Regardless of whether u and v are adjacent or not, $D - w$ is not
strong.

7.3 See Figure 19.

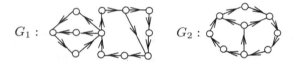

Figure 19: The graphs of Exercise 7.3

7.5 **Proof.** Let D be a strong digraph. Assume, to the contrary, that there
exists an edge-cut S of the underlying graph G of D separating $V(G - S)$
into two sets A and B such that there are no arcs directed from A to B.
Let $u \in A$ and $v \in B$. Then there is no $u - v$ path in D, contradicting the
fact that D is strong. For the converse, assume that D is not strong. Then
D contains two vertices u and v such that there is no $u - v$ path in D. Let
$A = \{x \in V(D) : D$ contains a $u - x$ path$\}$ and $B = V(D) - A$. Since
$u \in A$ and $v \in B$, it follows that $A \neq \emptyset$ and $B \neq \emptyset$. The set S of edges
of G joining a vertex of A and a vertex of B is an edge-cut of G. Since D
contains a $u - x$ path for all $x \in A$ and no $u - x$ path for all $x \in B$, there
is no arc in D directed from A to B. ∎

7.7 **Proof.** First observe that since T is strong, T contains no vertex x with
od $x = 0$ or id $x = 0$. Also, since $T - (u, v) + (v, u)$ is strong for every arc
(u, v) of T, it follows that T contains no vertex x with od $x = 1$ or id $x = 1$.

Consequently, $\operatorname{od} x \geq 2$ and $\operatorname{id} x \geq 2$ for every vertex x of T. Since the order of T is n, where $3 \leq n \leq 5$, it follows that $n = 5$ and every vertex x of T has $\operatorname{od} x = \operatorname{id} x = 2$. There is a unique tournament of order 5 with this property. ∎

7.9 **Proof.** First, assume that T is a transitive tournament. Let u and v be two vertices of T. Assume, without loss of generality, that (u, v) is an arc of T. Let U be the set of all vertices to which v is adjacent. Then $\operatorname{od} v = |U|$. Therefore, if $x \in U$, then (v, x) is an arc of T. Because T is transitive and (u, v) and (v, x) are arcs, it follows that (u, x) is an arc of T. Therefore, u is adjacent to every vertex of U and so $\operatorname{od} u \geq 1 + |U|$, implying that $\operatorname{od} u \neq \operatorname{od} v$.

For the converse, assume that T is a tournament of order n whose vertices have distinct outdegrees. Then these outdegrees are $0, 1, 2, \ldots, n-1$. Hence we may assume that $V(T) = \{v_1, v_2, \ldots, v_n\}$, where $\operatorname{od} v_i = n - i$ for $1 \leq i \leq n$. We claim that each vertex v_i ($1 \leq i \leq n - 1$) is adjacent to $v_{i+1}, v_{i+2}, \ldots, v_n$, which we verify by induction. Since $\operatorname{od} v_1 = n - 1$, this is certainly true for the vertex v_1. Assume that v_i is adjacent to $v_{i+1}, v_{i+2}, \ldots, v_n$ for all vertices v_i, where $1 \leq i \leq k$ and $1 \leq k < n$. Consider v_{k+1}. By the induction hypothesis, all of the vertices $v_1, v_2, \ldots, v_k$ are adjacent to v_{k+1}. Since $\operatorname{od} v_{k+1} = n - k - 1$, it follows that v_{k+1} is adjacent to $v_{k+2}, v_{k+3}, \ldots, v_n$. We now show that T is transitive. Let (u, v) and (v, w) be arcs of T. Then $u = v_r, v = v_s$ and $w = v_t$, where $r < s < t$. Since $r < t$, it follows that (u, w) is an arc of T and so T is transitive. ∎

7.11 **Proof.** Recall that $\sum_{i=1}^n \operatorname{od} v_i = \sum_{i=1}^n \operatorname{id} v_i$; so $\sum_{i=1}^n (\operatorname{od} v_i - \operatorname{id} v_i) = 0$. Now

$$
\begin{aligned}
0 &= \sum_{i=1}^n (\operatorname{od} v_i - \operatorname{id} v_i) = \sum_{i=1}^{n-1} (\operatorname{od} v_i - \operatorname{id} v_i) + (\operatorname{od} v_n - \operatorname{id} v_n) \\
&\geq (n - 1) + \operatorname{od} v_n - \operatorname{id} v_n.
\end{aligned}
$$

Therefore, $\operatorname{id} v_n - \operatorname{od} v_n \geq n - 1$, which implies that $\operatorname{id} v_n = n - 1$ and $\operatorname{od} v_n = 0$. Since $\operatorname{od} v_n = 0$, it follows that T is not strong. ∎

7.13 **Solution.** Let u and v be two distinct vertices in a tournament T. We may assume that $(u, v) \in E(T)$ and so $(v, u) \notin E(T)$. Then $\vec{d}(u, v) = 1$ and $\vec{d}(v, u) \neq 1$. Therefore, $\vec{d}(u, v) \neq \vec{d}(v, u)$. ◇

7.15 **Proof.** We proceed by induction. Let T be a strong tournament of order $n \geq 3$ and let v be a vertex of T. We first show that T contains a cycle of length 3. Since T is strong, $\operatorname{od} v > 0$ and $\operatorname{id} v > 0$. Let $N^+(v)$ be the set of all vertices to which v is adjacent and let $N^-(v)$ be the set of all vertices from which v is adjacent. Thus $N^+(v)$ and $N^-(v)$ are nonempty. For the same reason, there is a vertex $u \in N^+(v)$ that is adjacent to some vertex w in $N^-(v)$. Hence (v, u, w, v) is a cycle of length 3.

Suppose that T contains a cycle of length k, where $3 \leq k < n$. We show that T contains a cycle of length $k + 1$.

Let $C = (v = v_1, v_2, \ldots, v_k, v_1)$ be a cycle of length k. Suppose that there exists a vertex u of T not on C such that u is adjacent from some vertex of C and is adjacent to some other vertex of C. Then there exists a pair of adjacent vertices of C, say v_i and v_{i+1} (where the subscripts are expressed modulo k) such that (v_i, u) and (u, v_{i+1}) are both arcs of T. In this case, $(v_1, v_2, \ldots, v_i, u, v_{i+1}, \ldots, v_k, v_1)$ is a cycle of length $k + 1$ containing v.

Assume now that every vertex of T not on C is either adjacent from all vertices of C or is adjacent to all vertices of C. Let U be the set of vertices of $V(T) - V(C)$ that are adjacent from all vertices of C and let W be the set of vertices of $V(T) - V(C)$ that are adjacent to all vertices of C. Since T is strong, there is some vertex $u \in U$ that is adjacent to some vertex $w \in W$. However then, $(v_1, v_2, \ldots, v_{k-1}, u, w, v_1)$ is a cycle of length $k + 1$ containing v. ∎

Chapter 8: Matchings and Factorization

8.1 (a) See Figure 20.

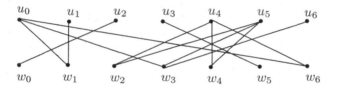

Figure 20: The graph G in Exercise 8.1

(b) A perfect matching of G is $M = \{u_0 w_6,\ u_1 w_1,\ u_2 w_0,\ u_3 w_5,\ u_4 w_2,\ u_5 w_4,\ u_6 w_3\}$.

8.3 For the graph G_1, the set U can be matched to W as $M = \{av, bw, cy, dz, ex\}$ is a matching. The set U in G_2 cannot be matched to W since U is not neighborly. For example, let $X = \{v, x, y\}$. Then $N(X) = \{a, c\}$.

8.5 **Proof.** Assume, to the contrary, that there exists a tree T having two distinct perfect matchings M_1 and M_2. Hence there exists a vertex v incident with distinct edges e_1 and e_2 such that $e_1 \in M_1$ and $e_2 \in M_2$, where, say, $e_1 = uv$ and $e_2 = vw$ and $u \neq w$. Therefore, T contains a path of length 2 whose edges are alternately in $M_1 - M_2$ and $M_2 - M_1$. Let P be a path of greatest length whose edges are alternately in $M_1 - M_2$ and $M_2 - M_1$. Suppose that P is an $x - y$ path and x is incident with an edge of $M_1 - M_2$ on P. Since M_2 is a perfect matching, there is an edge e of $M_2 - M_1$ not on P incident with x, say $e = xz$. If z is not on P, then z, P is a $z - y$ path whose edges are alternately in $M_1 - M_2$ and $M_2 - M_1$ and whose length is greater than that of P. This is impossible. If z is on P (which can only occur if $z = y$ and y is incident with an edge of $M_1 - M_2$), then a cycle is formed in T, which is also impossible. ∎

8.7 Consider the graph (see Figure 21). Let $M_1 = \{uv, ws, tx, yz\}$ and $M_2 = \{us, vw, xy, tz\}$.

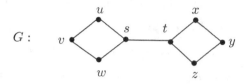

$G:$

Figure 21: A graph for Exercise 8.7

8.9 Consider $G_1 = K_8$, $G_2 = K_6 - e$, $G_3 = P_5$, $G_4 = K_{1,4}$.

8.11 If G is a complete bipartite graph, then $\alpha(G) = \beta'(G)$.

8.13 Observe that $\beta(H) = \alpha'(H) = 2$ and $\alpha(H) = \beta'(H) = 5$. The set $\{t, u\}$ is a minimum vertex cover, $\{u, w, x, y, z\}$ is a maximum independent set of vertices, $\{uv, tw, tx, ty, tz\}$ is a minimum edge cover and $\{uv, tw\}$ is a maximum independent set of edges.

8.15 Observe that $\alpha'(G) = (n_1 + n_2)/2$. By Theorem 8.7, $\beta'(G) = (n_1 + n_2)/2$.

8.17 The graph G_1 has a 1-factor but is not 1-factorable, G_2 does not have a 1-factor (and therefore is not 1-factorable either) and G_3 has a 1-factor but is not 1-factorable.

8.19 Hint: Construct a Hamiltonian factorization of K_9.

8.21 Hint: Let S be the partite set of $K_{3,5}$ with $|S| = 3$. Then $k_o(G - S) = 5$.

8.23 Hint: By Theorem 8.11, G has a 1-factor if G has no bridges; otherwise, let $P = (u = u_0, u_1, \ldots, u_k = v)$ be a $u - v$ path containing all bridges of G. Without loss of generality, we may assume that uu_1 and $u_{k-1}v$ are bridges. Let G_1 be the component of $G - uu_1$ containing u and let G_2 be the component of $G - u_{k-1}v$ containing v. For each $i = 1, 2$, let $e_i = x_i y_i$ be an edge of G_i. Furthermore, let G' be the graph obtained from G by deleting the edges e_i and adding a new vertex w_i and the edges $x_i w_i$ and $w_i y_i$ for $i = 1, 2$. Thus w_1 and w_2 are the only vertices of degree 2 in G'. Let F_1, F_2 and F_3 be three copies of G'. For each j with $1 \leq j \leq 3$, let w_{1j} be the vertex in F_j corresponding to w_1 in G' and $w_{2,j}$ be the vertex in F_j corresponding to w_2 in G'. Construct a graph F from F_1, F_2 and F_3 by (1) adding two new vertices z_1 and z_2 and (2) joining z_1 to w_{1j} for $1 \leq j \leq 3$ and joining z_2 to w_{2j} for $1 \leq j \leq 3$. Then F is 3-regular and bridgeless and so F has a 1-factor by Theorem 8.11. Complete the proof by showing that at least one of F_1, F_2 and F_3 has a 1-factor.

8.25 Hint: If n is even, then C_n is 1-factorable. If n is odd, consider two n-cycles $C = (v_1, v_2, \ldots, v_n, v_1)$ and $C' = (v_1', v_2', \ldots, v_n', v_1')$ in $C_n \times K_2$, where $v_i v_i'$ is an edge for $1 \le i \le n$. Consider three subgraphs F_1, F_2 and F_3 of $C_n \times K_2$, where $v_1 v_1' \in E(F_1)$, $v_2 v_2' \in E(F_2)$ and $v_i v_i' \in E(F_3)$ for $3 \le i \le n$.

8.27 Hint: Consider $C^* = (v_1, v_2, v_3, v_5, v_4, v_6, v_1, v_4, v_2, v_5, v_6, v_3, v_1)$.

8.29 Hint: Observe that $K_{n+1} = K_n + K_1$.

8.31 Hint: The graph K_7 can be decomposed into three copies of $C_3 \cup C_4$.

8.33 Hint: If $K_{2,2,2}$ were $K_{1,4}$-decomposable, then $K_{2,2,2}$ could be decomposed into three copies of $K_{1,4}$ and so there are vertices of $K_{2,2,2}$ that are not the center of any star $K_{1,4}$.

8.35 Hint: The graph C_6 is not graceful. Consider the parity of the labels as one moves cyclically about C_6. On the other hand, C_8 is graceful. Let $C_8 = (v_1, v_2, \ldots, v_8, v_1)$. Consider the labeling f defined by $f(v_1) = 0$, $f(v_2) = 8$, $f(v_3) = 1$, $f(v_4) = 4$, $f(v_5) = 5$, $f(v_6) = 7$, $f(v_7) = 2$ and $f(v_8) = 6$.

8.37 Hint: Show that T is a graceful tree of size 5 and then use Theorem 8.24.

Chapter 9: Planarity

9.1 See Figure 22. For G_1, $n = 6, m = 10, r = 6$. For G_2, $n = 10, m = 17, r = 9$. For G_3, $n = 6, m = 12, r = 8$. In each case, we have $n - m + r = 2$.

G_1 :  G_2 : 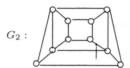 G_3 :

Figure 22: Graphs for Exercise 9.1

9.3 (a) **Solution.** The graph G has order $n = 7$ and size $m = 16$. Thus $m = 16 > 3 \cdot 7 - 6 = 15$. By Theorem 9.2, the graph G is nonplanar. ◇

    (b) **Solution.** The graph G has order $n = 12$ and size $m = 33$. Thus $m = 33 > 3 \cdot 12 - 6 = 30$. By Theorem 9.2, the graph G is nonplanar. ◇

9.5 (a) The graph of the octahedron is a 4-regular planar graph and the complete bipartite graph $K_{4,4}$ is a 4-regular nonplanar graph.

    (b) The graph of the icosahedron is a 5-regular planar graph and the complete bipartite graph $K_{5,5}$ is a 5-regular nonplanar graph.

    (c) By Corollary 9.3, every planar graph contains a vertex of degree 5 or less.

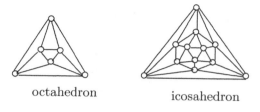

octahedron   icosahedron

Figure 23: Graphs for Exercise 9.5(a) and (b)

9.7 (a) C_4.

(b) No such graph exists since a nonplanar graph must have at least five vertices to contain K_5 or $K_{3,3}$ (or a subdivision of either) as a subgraph.

(c) A graph obtained by subdividing a single edge of K_5 exactly once.

(d) No such graph exists. If G has 5 vertices and 10 edges, then $G = K_5$, which is nonplanar. (Note that the Euler Identity may appear to hold since $n - m + r = 5 - 10 + 7 = 2$, but there is no such planar graph.)

(e) Let $G = K_3$.

(f) Let $G = K_6 \cup K_1$.

9.9 The graph G is nonplanar since G contains a subdivision of $K_{3,3}$ as shown in Figure 24.

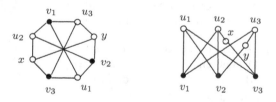

Figure 24: The graphs in Exercise 9.9

9.11 The graph is planar as shown in Figure 25.

9.13 (a) **Proof.** First, suppose that G is connected. Note that the inequality holds if $m = 2, 3$. Thus, we may assume that $m \geq 4$. We draw the graph G as a plane graph and denote the number of regions of G by r. For each region R of G, we determine the number of edges lying on the boundary of R and then sum these numbers over all regions of G. We denote this number by M. Since G has no triangle, there are at least 4 edges belonging to the boundary of each region. Thus $M \geq 4r$. On the other hand, the number M counts every edge of G

418

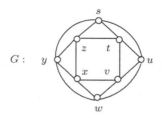

Figure 25: The graph in Exercise 9.11

once or twice, that is $M \leq 2m$. Hence $4r \leq M \leq 2m$ or $2r \leq m$. Since $n - m + r = 2$, it follows that

$$4 = 2n - 2m + 2r \leq 2n - 2m + m.$$

Therefore, $m \leq 2n - 4$.

If G is disconnected, then edges can be added to G to produce a connected plane graph of order n and size m' without triangles, where $m' > m$. Then $m' \leq 2n - 4$ and so $m < 2n - 4$. ∎

(b) Since $K_{3,3}$ is bipartite, $K_{3,3}$ has no triangles. The order of $K_{3,3}$ is $n = 6$ and the size is $m = 9$. Since $9 > 8 = 2 \cdot 6 - 4 = 2n - 4$, it follows by (a) that $K_{3,3}$ is nonplanar.

(c) **Proof.** Assume, to the contrary, that there exists a planar bipartite G of order n and size m such that $\delta(G) \geq 4$. Then $2m \geq 4n$ and so $m \geq 2n$. On the other hand, G is bipartite and so G has no triangles. By (a), $m \leq 2n - 4$, which is impossible. ∎

9.15 **Proof.** Assume, to the contrary, that there exists a planar graph G of order of $n \leq 11$ such that $\delta(G) \geq 5$. Thus $2m \geq 5n$. On the other hand, G is planar and so $m \leq 3n - 6$. Therefore, $5n \leq 2m \leq 6n - 12$, implying that $n \geq 12$, which is a contradiction. ∎

9.17 Hint: The graph $\overline{C_6}$ is planar. The graph $\overline{C_7}$ contains a subdivision of $K_{3,3}$ and is nonplanar. The graph $\overline{C_8}$ is nonplanar since $m = \binom{8}{2} - 8 = 20$, $3n - 6 = 3 \cdot 8 - 6 = 18$ and $m > 3n - 6$.

9.19 Hint: Since G is maximal planar, $m = 3n - 6$. Then use the Euler Identity.

9.21 **Proof.** Assume, to the contrary, that G contains a vertex v with $\deg v \leq 2$. Suppose that G has order n and size m. Since G is maximal planar, $m = 3n - 6$. Then $G - v$ is planar. Hence $G - v$ has order n and size $m' \geq m - 2$. Thus $m' \leq 3(n - 1) - 6$ and so $m - 2 \leq 3n - 3 - 6$ and $m \leq 3n - 7$, a contradiction. ∎

9.23 See Figure 26.

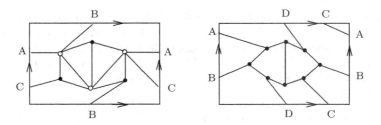

Figure 26: Embedding the graph in the torus in Exercise 9.23

9.25 Solution. Since $K_{3,3}$ is a subgraph of $K_{4,4}$, it follows that $K_{4,4}$ is non-planar and so $\gamma(K_{4,4}) \geq 1$. Since $K_{4,4}$ can be embedded in the torus, $\gamma(K_{4,4}) = 1$. ◇

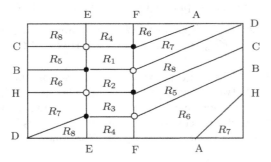

Figure 27: Embedding $K_{4,4}$ in the torus in Exercise 9.25

9.27 (a) False. Every planar graph can be embedded on the sphere and therefore on the torus as well.

   (b) True. The complete graph K_8 has genus 2 (by Theorem 9.12).

   (c) True. Draw G on the sphere (possibly with edges crossing). For each edge, insert a handle and draw that edge only on the handle.

   (d) False. Every planar graph can be embedded on the torus.

Chapter 10: Colorings

10.1 $\chi(G_1) = \chi(G_4) = 3$, $\chi(G_2) = \chi(G_3) = \chi(G_5) = 4$. Since G_1 contains a triangle, $\chi(G_1) \geq 3$. Because there is a 3-coloring of G_1, $\chi(G_1) \leq 3$. Since G_2 contains K_4 as a subgraph, $\chi(G_2) \geq 4$. Since G_2 is planar, $\chi(G_2) \leq 4$. Since G_3 contains K_4 as a subgraph, $\chi(G_3) \geq 4$. Because there is a 4-coloring of G_3, $\chi(G_3) \leq 4$. Since G_4 contains a triangle, $\chi(G_4) \geq 3$ Because there is a 3-coloring of G_4, $\chi(G_4) \leq 3$. Since G_5 contains K_4 as a subgraph, $\chi(G_5) \geq 4$. Since there is a 4-coloring of G_5, $\chi(G_5) \leq 4$.

10.3 The chromatic number of a tree of order at least 2 is 2 since a tree is a bipartite graph.

10.5 Proof. Let G be a graph of order 6 and chromatic number 3 and let there be a 3-coloring of G. If there are three (or more) vertices that are assigned the same color, then G has an independent set of three or more vertices. Thus the size of G is at most $\binom{6}{2} - 3 = 12$. Otherwise, $V(G)$ can be partitioned into three independent sets of two vertices each and the size of G is at most $\binom{6}{2} - 1 - 1 - 1 = 12$. ∎

The result cannot be improved since $G = K_{2,2,2}$ is a graph of order 6 and size 12 having chromatic number 3.

10.7 Since $\Delta(G) = 5$ and G is neither complete nor an odd cycle, $\chi(G) \leq \Delta(G) = 5$.

10.9 (a) **Proof.** Let $G = K_5$. Then G is nonplanar and $\chi(G) = 5$. Observe that $G - v = K_4$ is planar for every vertex v of G and $\chi(K_4) = 4$. ∎

(b) **Proof.** Let $G = K_{3,3}$. Then G is nonplanar and $\chi(G) = 2$. Observe that $G - v = K_{2,3}$ is planar for every vertex v of G and $\chi(K_{2,3}) = 2$. ∎

10.11 Solution. Let G be a graph whose vertex set is the set of chemicals. Place an edge between two vertices (chemicals) in G if it is risky to ship them in the same container. The graph G is shown in Figure 28. Since $\chi(G) = 4$, the minimum cost of shipping the chemicals is $125 + 3 \cdot 85 = 380$. Container 1: c_1, c_7; Container 2: c_4, c_5; Container 3: c_2, c_8; Container 4; c_3, c_6. ◇

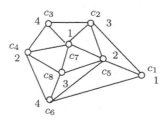

Figure 28: The graph G in Exercise 10.11

10.13 Proof. Let $V_1, V_2, \ldots, V_k$ be the k color classes resulting from a k-coloring of G, where the vertices of V_i are colored i $(1 \leq i \leq k)$. Then every two distinct color classes have different cardinalities. Hence we may assume that $1 \leq |V_1| < |V_2| < \cdots < |V_k|$. Since $\alpha(G) = k$, it follows that $|V_k| \leq k$. This implies that $|V_i| = i$ for every i $(1 \leq i \leq k)$. Hence V_1 consists of only one vertex v, which is colored 1. If there is a vertex $u \in V_i$ $(i \neq 1)$ that is not adjacent to v, then u could be colored 1, which would produce a contradiction. Thus v is adjacent to all other vertices of G and so $\deg v = n - 1$. ∎

10.15 Let G be the graph with vertex set A and two vertices a_i and a_j are adjacent in G if $\{a_i, a_j\} \in S$. Then f is a coloring of G. The cardinality of the range of f is $\chi(G)$.

10.17 $\chi'(G_1) = \chi'(G_2) = \chi'(G_5) = 4$ and $\chi'(G_3) = \chi'(G_4) = 5$,

10.19 **Solution** Draw an edge between two vertices (teams) if the teams have to play each other. The graph G has odd order $n = 7$, size $m = 13$, and $\Delta(G) = 4$. Since $m > \frac{(n-1)\Delta(G)}{2}$, it follows that $\chi'(G) = 1 + \Delta(G) = 5$, which is the minimum number of days to schedule all 13 games. A coloring of the edges of G is given in Figure 29.

Day 1 : A - B, N - M, D - L
Day 2 : B - N, A - C, L - M
Day 3 : C - L, D - M
Day 4 : A - N, C - D
Day 5 : A - M, D - N, B - C ◇

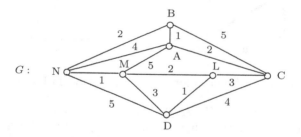

Figure 29: The graph in Exercise 10.19

Chapter 11: The Ramsey Number of Graphs

11.1 Hint: F_1 is a subgraph of K_s and F_2 is a subgraph of K_t.

11.3 **Proof.** Let there be given a red-blue coloring of $G = K_{18}$. For $v \in V(G)$, there are 9 edges incident with v that are colored the same color. Suppose that v is joined to each vertex of $S = \{v_1, v_2, \ldots, v_9\}$ by a red edge. Since $r(K_3, K_4) = 9$, there is a red K_3 or a blue K_4 in $H = G[S] = K_9$. If there is a red K_3 in H, then G has a red K_4; if there is a blue K_4 in H, then G has a blue K_4. Therefore, $r(K_4, K_4) \le 18$. ∎

11.5 $r(2K_2, P_3) = 4$. **Proof.** Since the order of $2K_2$ is 4, it follows that $r(2K_2, P_3) \ge 4$. Next we show that $r(2K_2, P_3) \le 4$. Let there be given a red-blue coloring of K_4. If all edges of K_4 are colored red, then we have a red $2K_2$. Thus we may assume that at least one edge is colored blue, say uv is colored blue as shown in Figure 30(a). If either ux or vy is colored blue, then we have a blue P_3 as shown in Figure 30(b). Otherwise, both ux and vy are colored red and we have a red $2K_2$ as shown in Figure 30(c). ∎

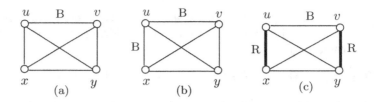

Figure 30: A red-blue coloring of K_4 in Exercise 11.5

11.7 $r(2K_2, 2K_2) = 5$. **Proof.** In the coloring of K_4 as shown in Figure 31(a) where each red edge of K_4 is drawn as a bold edge, there is no red $2K_2$ and no blue $2K_2$. Thus $r(2K_2, 2K_2) \geq 5$.

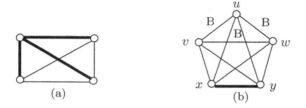

Figure 31: Red-blue colorings of K_4 and K_5 in Exercise 11.7

Next we show that $r(2K_2, 2K_2) \leq 5$. Let there be given a red-blue coloring of K_5 and suppose that there is no red $2K_2$ and no blue $2K_2$. If all edges of K_5 are colored blue, then we have a blue $2K_2$. Thus we may assume that at least one edge is colored red, say xy is colored red as shown in Figure 31(b). Since there is no red $2K_2$, all edges uv, uw, vw are colored blue. Since uw is blue, vx must be red. Since uv is blue, wy is red. But vx and wy produce a red $2K_2$, which is a contradiction. ∎

11.9 $r(K_{1,3}, K_{1,3}) = 6$. **Proof.** There is a red-blue coloring of K_5 such that the red subgraph and blue subgraph are both C_5 and so contains no red $K_{1,3}$ and no blue $K_{1,3}$. Thus $r(K_{1,3}, K_{1,3}) \geq 6$. Let there be given a red-blue coloring of K_6. For a vertex v of K_6, there are 5 edges incident with v. Thus at least three of these 5 edges are colored same, say red, and so it contains a red $K_{1,3}$. Therefore, $r(K_{1,3}, K_{1,3}) = 6$. ∎

11.11 $r(K_{1,4}, K_{1,4}) = 7$. **Proof.** Consider the red-blue coloring of K_6 in which the red subgraph is C_6. Since the blue subgraph is 3-regular, there is no red $K_{1,4}$ and no blue $K_{1,4}$. Thus $r(K_{1,4}, K_{1,4}) \geq 7$.

Next we show $r(K_{1,4}, K_{1,4}) \leq 7$. Assume, to the contrary, that there is a red-blue coloring of K_7 with no red $K_{1,4}$ and no blue $K_{1,4}$. Then every vertex of K_7 must be incident with exactly three red edges and three blue edges, for otherwise, K_7 contains a red $K_{1,4}$ or a blue $K_{1,4}$. However then, the red subgraph of K_7 is 3-regular of order 7, which is impossible. ∎

11.13 $r(C_4, C_4) = 6$. **Proof.** The coloring of K_5 of Figure 32(a) contains no red C_4 and no blue C_4. Thus $r(C_4, C_4) \geq 6$.

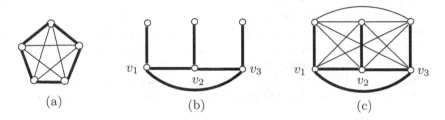

Figure 32: A red-blue coloring of K_5 and a subgraph of K_6 in Exercise 11.13

Next we show that $r(C_4, C_4) \leq 6$. Assume, to the contrary, that there is a red-blue coloring of K_6 with no red C_4 and no blue C_4, where $V(K_6) = \{v_1, v_2, \ldots, v_6\}$. Since $r(K_3, K_3) = 6$, there is either a red K_3 or a blue K_3, say the former. Let the subgraph induced by $\{v_1, v_2, v_3\}$ be a red K_3. If there is a vertex in $\{v_4, v_5, v_6\}$ that is joined to two vertices in $\{v_1, v_2, v_3\}$ by red edges, then K_6 contains a red C_4. Thus each vertex in $\{v_4, v_5, v_6\}$ is joined to at least two vertices in $\{v_1, v_2, v_3\}$ by blue edges. Furthermore, no two vertices in $\{v_4, v_5, v_6\}$ are joined by blue edges to the same two vertices in $\{v_1, v_2, v_3\}$, for otherwise, a blue C_4 is produced, giving a contradiction. This implies that every vertex in $\{v_4, v_5, v_6\}$ is joined to exactly two vertices in $\{v_1, v_2, v_3\}$ by blue edges and joined to one vertex in $\{v_1, v_2, v_3\}$ by a red edge. Thus K_6 contains the red subgraph shown in Figure 32(b). If there is a red edge joining two vertices in $\{v_4, v_5, v_6\}$, then K_6 contains a red C_4, a contradiction. Thus the subgraph induced by $\{v_4, v_5, v_6\}$ is a blue K_3 as shown in Figure 32(c). However, then this produces a blue C_4, a contradiction. ∎

11.15 **Proof.** Let $p = r(K_s, K_t)$ and let $G = K_{p-1}$. Now let there be given red-blue coloring of G. Furthermore, let H be the graph obtained from G by adding a new vertex v and joining v to every vertex of G by a red edge. Then $H = K_p$. Since $p = r(K_s, K_t)$, it follows that H has a red K_s or a blue K_t. If H has a blue K_t, then G contains a blue K_t and therefore a blue K_{t-1}. Otherwise, H contains a red K_s. If the red K_s does not contain v, then G contains a red K_s and therefore a red K_{s-1}. If the red K_s in H contain v, then G contains a red K_{s-1}. ∎

11.19 (a) The graph $G = K_{3,3,4}$, which has size 33.

(b) The smallest such positive integer m is $33 + 1 = 34$.

11.21 $T_{5,1} = \overline{K}_5$, $T_{7,2} = K_{3,4}$, $T_{6,3} = K_{2,2,2}$, $T_{6,4} = K_{1,1,2,2}$, $T_{5,5} = K_5$.

11.23 Hint: $1 + \lfloor k(n-1)/2 \rfloor$.

11.25 The smallest positive integer m for which every graph of order $n \geq 2$ and size m contains a Hamiltonian path is $\binom{n-1}{2} + 1$.

Proof. First observe that the disconnected graph $K_{n-1} \cup K_1$ has order n and size $\binom{n-1}{2}$ but contains no Hamiltonian path. Next we show that every graph of order n and size $\binom{n-1}{2} + 1$ has a Hamiltonian path. Let G be such a graph and let $H = G + K_1$, where H is obtained by adding a vertex v to G and joining v to every vertex of G. Then H has order $n + 1$ and size $n + \binom{n-1}{2} + 1 = \binom{n}{2} + 2$. By Theorem 11.18, H has a Hamiltonian cycle C. Deleting v from C produces a Hamiltonian path in G. ∎

Chapter 12: Distance

12.1 $\operatorname{rad}(G) = \operatorname{diam}(G) = 5$ and $\operatorname{Cen}(G) = G$.

12.3 Note that $\operatorname{rad}(K_{s,t}) = 1$ if $s = 1$ and $\operatorname{rad}(K_{s,t}) = 2$ if $s \geq 2$, while $\operatorname{diam}(K_{s,t}) = 1$ if $t = 1$ and $\operatorname{diam}(K_{s,t}) = 2$ if $t \geq 2$. Thus $\operatorname{Cen}(K_{s,t}) = K_1$ if $s = 1$ and $t \geq 2$ and $\operatorname{Cen}(K_{s,t}) = K_{s,t}$ otherwise.

12.5 Consider the graph G_4 in Figure 12.3.

12.7 (a) Hint: Let $v \in V(G)$ such that $e(v) = \operatorname{diam}(G)$. Fir each i with $0 \leq i \leq 2$, let $S_i = \{u \in V(G) : d(v,u) = i\}$ and let $S_3 = \{u \in V(G) : d(v,u) \geq 3\}$. Then $\{S_0, S_1, S_2, S_3\}$ is a partition of $V(G)$. Complete the proof by showing that $d_{\overline{G}}(x,y) \leq 3$ for all $x,y \in V(G)$.

(b) Let $G = P_4$.

12.9 **Proof.** Assume, without loss of generality, that $e(u) \geq e(v)$. Let x be a vertex that is farthest from u. So $d(u,x) = e(u)$. By the triangle inequality,

$$e(u) = d(u,x) \leq d(u,v) + d(v,x) \leq d(u,v) + e(v).$$

Hence $e(u) \leq d(u,v) + e(v)$, which implies that $0 \leq e(u) - e(v) \leq d(u,v)$. Therefore, $|e(u) - e(v)| \leq d(u,v)$. ∎

12.11 **Proof.** Let $x, y \in V(G)$ such that $e(x) = \operatorname{rad} G$ and $e(y) = \operatorname{diam} G$ and let $P = (x = u_0, u_1, u_2, \ldots, u_a = y)$ be an $x - y$ path. By Theorem 12.2, $|e(u_j) - e(u_{j+1})| \leq 1$ for $0 \leq j \leq a - 1$. Let i be the greatest integer with $0 \leq i \leq a - 1$ such that $e(u_i) < k$. Therefore, $e(u_i) \leq k - 1$ and $e(u_{i+1}) \geq k$. Thus $e(u_{i+1}) - e(u_i) \geq 1$. Since $e(u_{i+1}) - e(u_i) \leq 1$ by Theorem 12.2, it follows that $e(u_{i+1}) - e(u_i) = 1$ and so $e(u_{i+1}) = 1 + e(u_i) = k$. ∎

12.13 (a) **Proof.** Let v be a vertex of T' and suppose that $e_T(v) = k$. Let u be a vertex of T such that $d(v,u) = k$. Necessarily, u is an end-vertex of T, for otherwise the $v - u$ path in T could be extended to a longer path in T, contradicting the fact that $e_T(v) = k$. So $e_{T'}(v) = k - 1$. Therefore, a vertex of minimum eccentricity in T' is a vertex of minimum eccentricity in T and so $\operatorname{Cen}(T) = \operatorname{Cen}(T')$. ∎

(b) Hint: Note that the center of a graph G lies in a block of G. For a tree T, the only blocks are K_1 and K_2.

(c) Hint: Observe that if the end-vertices of a tree T are removed, producing a tree T', then $\operatorname{diam} T' = \operatorname{diam} T - 2$.

12.15 $\operatorname{Per}(G) = G$.

12.17 $\operatorname{Per}(K_{s,t}) = \overline{K}_t$ if $s = 1$ and $t \geq 2$, while $\operatorname{Per}(K_{s,t}) = K_{s,t}$ otherwise.

12.19 Consider $P_9 = (v_1, v_2, \ldots, v_9)$ and let $v = v_3$.

12.21 Hint: No.

12.23 Hint: If G is self-centered, then $\operatorname{Per}(G) = G$. For the converse, use Theorem 12.7 to show that $e(v) = 2$ for every vertex v of G.

12.25 Hint: No.

12.27 See Figure 33.

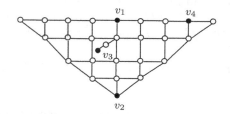

Figure 33: The graph G in Exercise 12.27

12.31 Hint: Let $V(G) = \{u_1, u_2, \ldots, u_n\}$ and let G' be the graph obtained from G by adding n vertices $v_1, v_2, \ldots, v_n$ and joining each v_i to u_i for $1 \leq i \leq n$. Let P and Q be two copies of the path P_5, where $P = (x_1, x_2, \ldots, x_5)$ and $Q = (y_1, y_2, \ldots, y_5)$. Let H be the graph obtained from G', P and Q by joining each end-vertex of P and Q to every vertex of G. Show that $\operatorname{Cen}(H) = \operatorname{Int}(H) = G$.

Chapter 13: Domination

13.1 (a) $\gamma(G) = 3$. (b) $\gamma_o(G) = 4$.

13.3 For $n \geq 2$, $\gamma(K_n) = 1$, $\gamma(P_n) = \lceil n/3 \rceil$, $\gamma(K_{s,t}) = 1$ if $\min\{s, t\} = 1$ and $\gamma(K_{s,t}) = 2$ if $\min\{s, t\} \geq 2$, $\gamma(Q_3) = 2$, $\gamma(PG) = 3$. For $n \geq 2$, $\gamma_o(K_n) = 2$ and

$$\gamma_o(P_n) = \begin{cases} \frac{n+1}{2} & \text{if } n \text{ is odd} \\ \frac{n}{2} & \text{if } n \equiv 0 \pmod 4 \\ \frac{n}{2} + 1 & \text{if } n \equiv 2 \pmod 4. \end{cases}$$

$\gamma_o(K_{r,s}) = 2$, $\gamma_o(Q_3) = 3$ and $\gamma_o(PG) = 4$.

13.5 (a) Let $G = C_4 = (u, x, v, y, u)$. Then $S = \{u, v\}$ is a minimum dominating set for G, where $N(x) \cap S = S$ and $N(y) \cap S = S$.

  (b) Hint: Let $n = 2k + 1$, where $k \geq 1$ and let G be the graph obtained by subdividing each edge of $K_{1,k}$. Then $\gamma(G) = k$.

13.7 Hint: Let G be the graph obtained from K_{n-k} with $V(K_{n-k}) = \{v_1, v_2, \cdots, v_{n-k}\}$ by adding k new vertices $u_1, u_2, \cdots, u_k$ and k new edges $u_i v_i$ for $1 \leq i \leq k$. Then G is a connected graph of order n with $\gamma(G) = k$.

13.9 Hint: For each $v \in S$, the vertex v is not dominated by any vertex in $S - v$ and so $S - v$ is not a dominating set for G. Thus S is necessarily a minimal dominating set. But S is not necessarily a minimum dominating set. For example, let $G = C_6 = (v_1, v_2, \ldots, v_6, v_1)$ and let $S = \{v_1, v_3, v_5\}$.

13.11 (a) **Proof.** The lower bound follows immediately from the observation that if $\gamma(G) = 1$, then $\gamma(\overline{G}) \geq 2$. It remains to verify the upper bound. If G has an isolated vertex, then $\gamma(G) \leq n$ and $\gamma(\overline{G}) = 1$; while if $\overline{G}$ has an isolated vertex, then $\gamma(\overline{G}) \leq n$ and $\gamma(G) = 1$. So in these cases, $\gamma(G) + \gamma(\overline{G}) \leq n + 1$. If neither G nor $\overline{G}$ has an isolated vertex, then $\gamma(G) \leq n/2$ and $\gamma(\overline{G}) \leq n/2$ by Corollary 13.5 and so $\gamma(G) + \gamma(\overline{G}) \leq n$. ∎

  (b) Let $G = K_{1,n-1}$, where $n \geq 2$.

  (c) Let $G = \overline{K}_n$.

13.13 For $n > k \geq 2$, let G be the graph obtained from $P_{k+2} = (v_0, v_1, v_2, \ldots, v_{k+1})$ by adding k new vertices $u_1, u_2, \ldots, u_k$ and k new edges $u_i v_i$ for $1 \leq i \leq k$. Then G is a connected graph without isolated vertices for which $\gamma(G) = \gamma_o(G) = k$.

13.15 Let G be the graph obtained from the 4-cycle $(v_1, v_2, v_3, v_4, v_1)$ by adding the vertices u_1 and u_3 and joining u_i and v_i for $i = 1, 3$. Then $\gamma(G) = 2$ and $\gamma_o(G) = 3$.

REFERENCES

Chapter 2

1. Chartrand G., Erdős P. and Oellermann O. R., How to define an irregular graph. *College Math. J.* **19** (1988) 36-42.

2. Chartrand G., Holbert K. S., Oellermann O. R. and Swart H. C., *F*-Degrees in graphs. *Ars Combin.* **24** (1987) 133-148.

3. Chartrand G., Jacobson M. S., Lehel J., Oellermann O. R., Ruiz S. and Saba F., Irregular networks. *Congress. Numeran.* **64** (1988) 197-210.

4. Erdős P. and Kelly P. J., The minimal regular graph containing a given graph. *Amer. Math. Monthly* **70** (1963) 1074-1075.

5. Hakimi S. L., On the realizability of a set of integers as degrees of the vertices of a graph. *SIAM J. Appl. Math.* **10** (1962) 496-506.

6. Havel V., A remark on the existence of finite graphs (Czech.) *Časopis Pěst. Mat.* **80** (1955) 477-480.

7. König D., *Theorie der endlichen und unendliehen Graphen.* Teubner, Leipzig (1936).

8. Wells D., Are these the most beautiful? *Math. Intelligencer.* **12** (1990) 37-41.

Chapter 3

1. Baker G., Bruckner A., Michael E. and Yaqub A., *Paul J. Kelly*, Mathematics: Santa Barbara. 1995, University of California: In Memoriam.

2. Bondy J. A. and Hemminger R. L., Graph reconstruction - a survey. *J. Graph Theory* **1** (1977) 227-268.

3. Frucht R. W., How I became interested in graphs and groups. *J. Graph Theory* **6** (1982) 101-104.

4. König D., *Theorie der endlichen und unendliehen Graphen.* Teubner, Leipzig (1936).

5. O'Connor J. J. and Robertson E. F., *Ferdinand Georg Frobenius.*
   http://www-history.mcs.st-andrews.ac.uk/Biographies/Frobenius.html

6. O'Connor J. J. and Robertson E. F., *Issai Schur.*
   http://www-history.mcs.st-andrews.ac.uk/Biographies/Schur.html

7. O'Connor J. J. and Robertson E. F., *Stanislaw Marcin Ulam.*
   http://www-history.mcs.st-andrews.ac.uk/Biographies/Ulam.html

428

Chapter 4

1. Borůvka O., O jistém Problému minimálním. *Práce Mor. Přírodověd. Spol. v Brně* (*Acta Societ. Scient. Natur. Moravicae*) **3** (1926) 37-58.

2. Cayley A., A theorem on trees. *Quart. J. Math.* **23** (1889) 276-378.

3. Graham R. L. and Hell P., On the history of the minimum spanning tree problem. *Annals of the History of Computing* **7** (1985) 43-57.

4. Kirchhoff G., Über die Auflösung der Gleichungen, auf welche man bei der Untersuchung der linearen Verteilung galvanischer Ströme gefürht wird. *Ann. Phys. Chem.* **72** (1847) 497-508.

5. Kruskal J. B., On the shortest spanning tree of a graph and the traveling salesmen problem. *Proc. Amer. Math. Soc.* **7** (1956) 48-50.

6. Prim R. C., Shortest connection networks and some generalizations. *Bell Syst. Tech. J.* **36** (1957) 1389-1401.

7. Moon J. W. (personal communication).

8. O'Connor J. J. and Robertson E. F., *Arthur Cayley.*
http://www-history.mcs.st-andrews.ac.uk/Biographies/Cayley.html

9. O'Connor J. J. and Robertson E. F., *Gustaf Robert Kirchhoff.*
http://www-history.mcs.st-and.ac.uk/Biographies/ Kirchhoff.html

Chapter 5

1. Chartrand G., Harary F. and Zhang P., On the geodetic number of a graph. *Networks* **39** (2002) 1-6.

2. Dirac G. A., In abstrakten Graphen vorhande vollständigene 4-Graphen und ihre Unterteilungen. *Math. Nachr.* **22** (1960) 61-85.

3. Harary F., The maximum connectivity of a graph. *Proc. Nat. Acad. Sci. U.S.A.* **48** (1962) 1142-1146.

4. Menger K., Zur allgemeinen Kurventheorie. *Fund. Math.* **10** (1927) 95-115.

5. Menger K., On the origin of the n-arc theorem. *J. Graph Theory* **5** (1981) 341-350.

6. O'Connor J. J. and Robertson E. F., *Karl Menger*
http://www-history.mcs.st-and.ac.uk/Biographies/Menger.html

7. O'Connor J. J. and Robertson E. F., *Hassler Whitney*
http://www-history.mcs.st-and.ac.uk/Biographies/ Whitney.html

8. Whitney H., Congruent graphs and the connectivity of graphs. *Amer. J. Math.* **54** (1932) 150-168.

Chapter 6

1. Berge C., *Théorie des Graphes et Ses Applications.* Dunod, Paris (1958).

2. Biggs N. L., Lloyd E. K. and Wilson R. J., *Graph Theory* 1736-1936. Oxford University Press, London (1976).

3. Bondy J. A. and Chvátal V., A method in graph theory. *Discrete Math.* **15** (1976) 111-136.

4. Chartrand G., Thomas T., Zhang P. and Saenpholphat V., A new look at Hamiltonian walks. *Bull. Inst. Combin. Appl.* **42** (2004) 37-52.

5. Chvátal V., *Claude Berge*: 5.6.1926-30.6.2002.

   http://users.encs.concordia.ca/~chvatal/perfect/claude2.pdf

6. Descartes B., The expanding unicurse. In *Proof Techniques in Graph Theory* (edited by F. Harary), Academic Press, New York (1969) 25.

7. Dirac G. A., Some theorems on abstract graphs. *Proc. London Math. Soc.* **2** (1952) 69-81.

8. Euler L., Solutio problematis ad geometriam situs pertinentis. *Comment. Academiae Sci. I. Petropolitanae* **8** (1736) 128-140.

9. Gallai T., Dénes König: A biographical sketch. In *Theory of Finite and Infinite Graphs by* D. König (translated into English by R. McCoart), Birkäuser, Boston (1990).

10. Goodman S. E. and Hedetniemi S. T., On Hamiltonian walks in graphs. *SIAM J. Comput.* **3** (1974) 214-221.

11. Harary F., *Graph Theory.* Addison-Wesley, Reading, PA (1969).

12. Harary F. (personal communication).

13. Hersh R. and John-Steiner V., A visit to Hungarian mathematics. *Mathematical Intelligencer* **15** (1993) 13-26.

14. Hierholzer C., Über die Möglichkeit, einen Linienzug ohne Wiederholung und ohne Unterbrechung zu umfahren. *Math. Ann.* **6** (1873) 30-32.

15. Januszewski M., *Kaliningrad.*

   http://castlesofpoland.com/prusy/krol_hist_en.htm

16. O'Connor J. J. and Robertson E. F., *Paul Adrien Maurice Dirac*

   http://www-history.mcs.st-and.ac.uk/Biographies/Dirac.html

17. O'Connor J. J. and Robertson E. F., *Leonhard Euler*
    http://www-history.mcs.st-and.ac.uk/Biographies/Euler.html

18. O'Connor J. J. and Robertson E. F., *Sir William Rowan Hamilton*
    http://www-history.mcs.st-and.ac.uk/Biographies/ Hamilton.html

19. Ore O., Note on Hamilton circuits. *Amer. Math. Monthly* **67** (1960) 55.

20. Ore O., *Theory of Graphs.* Amer. Math. Soc. Providence, RI (1962).

21. Ore O., *Graphs and Their Uses.* Random House, New York (1963).

22. Posa L., A theorem concerning Hamilton lines. *Magyar Tud. Akad. Mat. Kutató Int. Közl.* **7** (1962) 225-226.

23. Russian News Network, *Kaliningrad.*
    http://www.russiannewsnetwork.com/Kaliningrad.html

24. Zykov A. A., Teoriia Konechnykh Grafov. Nauka, Novosibirsk (1969).

25. Zykov A. A. (personal communication).

Chapter 7

1. Camion P., Chemins et circuits hamiltoniens des graphes complets. *C. R. Acad. Sci. Pairs* **249** (1959) 2151-2152.

2. Page W., An interview with Herbert Robbins. *College Mathematics Journal* **15** (1984) 2-24.

3. Rédei L., Ein Kombinatorischer Satz. *Acta Litt. Szeged* **7** (1934) 39-43.

4. Robbins H. E., A theorem on graphs, with an application to a problem in traffic control. *Amer. Math. Monthly* **46** (1939) 281-283.

5. Siegmund D., Herbert Robbins and sequential analysis. *Annals of Statistics* **31** (2003) 349-365.

Chapter 8

1. Armbruster F. O.,
   http://www.coloradobootstrap.com/cas/about_ frank_armbruster.html

2. Armbruster F. O. (personal communication).

3. Biggs N., *Professor W. T. Tutte: Mathematician and Bletchley code-breaker.*
   http://www-history.mcs.st-and.ac.uk/Obits2/ Tutte_Independent.html

4. Chartrand G., Erwin D., VanderJagt D. W. and Zhang P., γ-Labelings of graphs. *Bull. Inst. Combin. Appl.* **44** (2005) 51-68.

5. Erdős P., Personal reminiscences and remarks on the mathematical work of Tibor Gallai. *Combinatorica* **2** (1982) 207-212.

6. Gallai T., Über extreme Punkt- und Kantenmengen. *Ann. Univ. Sci. Budapest, Eötvöst Sect. Math.* **2** (1959) 133-138.

7. Gallian J. A., A dynamic survey of graph labeling. *Electron. J. Combin.* **17** (2010) #DS6 (November 2010 Version).

8. Hall P., On representation of subsets. *J. London Math. Soc.* **10** (1935) 26-30.

9. Hobbs A. M. and Oxley J. G., William T. Tutte (1917 - 2002). *Notices Amer. Math. Soc.* **51** (2004) 320-332.

10. Kirkman T. P., On a problem in combinations. *Cambridge and Dublin Math. J.* **2** (1847) 191-204.

11. König D., Über Graphen und ihre Anwendung auf Determinantentheorie und Mengenlehre. *Math. Ann.* **77** (1916) 453-465.

12. Lovász L., Tibor Gallai is seventy years old. *Combinatorica* **2** (1982) 203-205.

13. Lützen T., Sabidussi G. and Toft B., Julius Petersen 1839-1910: A biography. *Discrete Math.* **100** (1992) 9-82.

14. Mulder H. M., Julius Petersen's theory of regular graphs. *Discrete Math.* **100** (1992) 157-175.

15. O'Connor J. J. and Robertson E. F., *August Leopold Crelle*
http://www-history.mcs.st-and.ac.uk/Biographies/Crelle.html

16. O'Connor J. J. and Robertson E. F., *Philip Hall*
http://www-history.mcs.st-and.ac.uk/Biographies/Hall.html

17. O'Connor J. J. and Robertson E. F., *Thomas Penyngton Kirkman*
http://www-history.mcs.st-and.ac.uk/Biographies/Kirkman.html

18. O'Connor J. J. and Robertson E. F., *Jakob Steiner*
http://www-history.mcs.st-and.ac.uk/Biographies/Steiner.html

19. Petersen J., Die Theorie der regulären Graphen. *Acta Math.* **15** (1891) 193-220.

20. Ray-Chaudhuri D. K. and Wilson R. M., Solution of Kirkman's schoolgirl problem. *Combinatorics (Proc. Sympos. Pure Math.,* Vol. XIX, Univ. California, Los Angeles, Calif., 1968), Amer. Math. Soc., Providence, RI. (1971) 187-203.

432

21. Rosa A., On certain valuations of the vertices of a graph. In *Theory of Graphs, Proc. Internat. Sympos. Rome* 1966. Gordon and Breach, New York (1967) 349-355.

22. Sabidussi G., Correspondence between Sylvester Petersen, Hilbert and Klein on invariants and the factorisation of graphs 1889-1891. *Discrete Math.* **100** (1992) 99-155.

23. Saxon W., William Tutte, 84, Mathematician and code-breaker, dies. *New York Times*, October 30, 2002.

24. Tutte W.T., A short proof of the factor theorem for finite graphs. *Canad. J. Math.* **6** (1954) 347-352.

Chapter 9

1. Chartrand G., Gavlas. H. and Schultz M., Framed! A graph embedding problem. *Bull. Inst. Combin. Appl.* **4** (1992) 35-50.

2. Erdős P. and Kelly P. J., The minimal regular graph containing a given graph. *Amer. Math. Monthly* **70** (1963) 1074-1075.

3. Euler L., Demonstratio nonnullarum insignium proprietatum quibus solida hedris planis inclusa sunt praedita. *Novi Comm. Acad. Sci. Imp. Petropol.* **4** (1758) 140-160.

4. Krasinkiewicz J., A note on the work and life of Kazimierz Kuratowski. *J. Graph Theory* **5** (1981) 221-223.

5. Kuratowski K., Sur la problème des courbes gauches en topologie. *Fund. Math.* **15** (1930) 271-283.

6. O'Connor J. J. and Robertson E. F., *Kazimierz Kuratowski*

   http://www-history.mcs.st-and.ac.uk/Biographies/Kuratowski.html

7. Ringel G. and Youngs J. W. T., Solution of the Heawood map-coloring problem. *Proc. Nat. Acad. Sci. USA* **60** (1968) 438-445.

8. Robertson N. and Seymour P. D., Generalizing Kuratowski's theorem. *Congr. Numeran.* **45** (1984) 129-138.

9. Robertson N. and Seymour P. D., Graph minors. VIII A Kuratowski theorem for general surfaces. *J. Combin. Theory Ser. B* **48** (1990) 255-288.

10. Robertson N. (personal communication).

11. Wagner K., Üer eine Eigenschaft der ebene Komplexe. *Math. Ann.* **114** (1937) 570-590.

Chapter 10

1. Appel K., Haken W. and Koch J., Every planar map is four-colorable. *Illinois J. Math.* **21** (1977) 429-567.

2. Beineke L. W. and Wilson R. J., On the edge-chromatic number of a graph. *Discrete Math.* **5** (1973) 15-20.

3. Brooks R. L., On coloring the nodes of a network. *Proc. Cambridge Philos. Soc.* **37** (1941) 194-197.

4. Chartrand G., Saba F., Salehi E. and Zhang P., Local colorings of graphs *Util. Math.* **67** (2005) 107-120.

5. Descartes B., On some recent progress in combinatorics. *J. Graph Theory* **1** (1977) 192.

6. Gutin G. and Toft B., Interview with Vadim G. Vizing, *European Mathematical Society Newsletter* No. 38, (December 2000) 22-23.

7. Harary F., Conditional colorability in graphs. *Graphs and Applications.* (edited by F. Harary and J. Maybee) Wiley, New York (1985) 127-136

8. König D., Über Graphen und ihre Anwendung auf Determinantentheorie und Mengenlehre. *Math. Ann.* **77** (1916) 453-465.

9. Mycielski J., Sur le coloriage des graphes. *Colloq. Math.* **3** (1955) 161-162.

10. O'Connor J. J. and Robertson E. F., *Garrett Birkhoff*
    http://www-history.mcs.st-and.ac.uk/Biographies/Birkhoff_Garrett.html

11. O'Connor J. J. and Robertson E. F., *George David Birkhoff*
    http://www-history.mcs.st-and.ac.uk/Biographies/Birkhoff.html

12. O'Connor J. J. and Robertson E. F., *Percy John Heawood*
    http://www-history.mcs.st-and.ac.uk/Biographies/ Heawood.html

13. O'Connor J. J. and Robertson E. F., *Alfred Bray Kempe*
    http://www-history.mcs.st-and.ac.uk/Biographies/Kempe.html

14. O'Connor J. J. and Robertson E. F., *August Ferdinand Möbius*
    http://www-history.mcs.st-and.ac.uk/Biographies/Mobius.html

15. O'Connor J. J. and Robertson E. F., *James Joseph Sylvester*
    http://www-history.mcs.st-and.ac.uk/Biographies/ Sylvester.html

16. O'Connor J. J. and Robertson E. F., *Peter Guthrie Tait*
    http://www-history.mcs.st-and.ac.uk/Biographies/Tait.html

434

17. O'Connor J. J. and Robertson E. F., *Heinrich Franz Friedrich Tietze*
    http://www-history.mcs.st-and.ac.uk/Biographies/Tietze.html

18. Ringel G. and Youngs J. W. T., Solution of the Heawood map-coloring
    problem. *Proc. Nat. Acad. Sci. USA* **60** (1968) 438-445.

19. E. Salehi, F. Okamoto and P. Zhang A checkerboard problem and modular
    colorings of graphs *Bull. Inst. Combin Appl.* **58** (2010) 29-47.

20. Robertson N., Sanders D., Seymour P. D. and Thomas R., The four-color
    theorem. *J. Combin. Theory Ser.* B **70** (1997) 2-44.

21. Szekeres G. and Wilf H. S., An inequality for the chromatic number of a
    graph. *J. Combin. Theory* **4** (1968) 1-3.

22. Tutte W. T., On hamiltonian circuits. *J. London Math. Soc.* **2** (1946)
    98-101.

23. Vizing V. G., On an estimate of the chromatic class of a p-graph. *Diskret.
    Analiz.* **3** (1964) 25-30.

24. Vizing V. G. (personal communication).

25. Wilson R., *Four Colors Suffice: How the Map Problem was Solved.* Prince-
    ton University Press, Princeton, NJ (2002).

Chapter 11

1. Bialostocki A. and Voxman W., Generalizations of some Ramsey-type theorems for matchings. *Discrete Math.* **239** (2001) 101-107.

2. Chvátal V., Tree-complete ramsey numbers. *J. Graph Theory* **1** (1977) 93.

3. Erdős P. and Rado R., A combinatorial theorem. *J. London Math. Soc.*
   25 (1950) 249-255.

4. Eroh L., *Rainbow Ramsey Numbers.* Ph.D. Dissertation, Western Michigan
   University (2000).

5. Eroh L., Rainbow Ramsey numbers of stars and matchings. *Bull. Inst.
   Combin. Appl.* **40** (2004) 91-99.

6. Grossman J., *The Erdős Number Project.*
   http://www.oakland.edu/enp/

7. Hoffman P., *The Man Who Loved Only Numbers: The Story of Paul Erdős
   and the Search for Mathematical Truth.* Hyperion, New York (1998).

8. Mellor D. H., The eponymous F. P. Ramsey. *J. Graph Theory* **7** (1983)
   9-13.

9. O'Connor J. J. and Robertson E. F., *Paul Erdős*

http://www-history.mcs.st-and.ac.uk/Biographies/Erdos.html

10. Odda T., On properties of a well-known graph or what is your Ramsey number? In *Topics in Graph Theory* (edited by F. Harary) New York Academy of Sciences, New York (1977) 166-172.

11. Ramsey F., On a problem of formal logic. *Proc. London Math. Soc.* **30** (1930) 264-286.

12. Simonovits M., On Paul Turán's influence on graph theory. *J. Graph Theory* **1** (1977) 102-116.

13. Spencer J., Ramsey Theory and Ramsey theoreticians. *J. Graph Theory* **7** (1983) 15-23.

14. Szemerédi E., On sets of integers containing no four elements in arithmetic progression. *Acta Math. Acad. Sci. Hung.* **20** (1969) 89-104.

15. Szemerédi E., On sets of integers containing no k elements in arithmetic progression. *Acta Arith.* **27** (1975) 199-245.

16. Turán P., Eine Extremalaufgabe aus der Graphentheorie. *Mat. Fiz. Lapok* **48** (1941) 436-452.

17. Turán P., A note of welcome. *J. Graph Theory* **1** (1977) 7-9.

18. van der Waerden B., Beweis einer Baudetschen vermutung. *Nieuw Arch. Wisk* **15** (1927) 212-216.

19. *Erdős-Bacon Numbers.*

http://www.simonsingh.net/Erdos-Bacon_Numbers.html

20. *N is a Number*, video, Mathematical Association of America. (youtube)

21. *UVA Computer Science: The Oracle of Bacon at Virginia.*

http://www.cs.virginia.edu/misc/news-bacom.html

Chapter 12

1. Bielak H. and Syslo M.M., Peripheral vertices in graphs. *Studia Sci. Math. Hungar.* **18** (1983) 269-275

2. Buckley F., Miller Z. and Slater P. J., On graphs containing a given graph as center. *J. Graph Theory* **5** (1981) 427-434.

3. Chartrand G., Erwin D., Johns G. L. and Zhang P., Boundary vertices in graphs. *Discrete Math.* **263** (2003) 25-34.

4. Chartrand G., Erwin D. and Zhang P., A graph labeling problem suggested by FM channel restrictions. *Bull. Inst. Combin. Appl.* **43** (2005) 43-57.

5. Chartrand G., Escuadro H. and Zhang P., Detour distance in graphs. *J. Combin. Math. Combin. Comput.* **53** (2005) 75-94.

6. Chartrand G., Erwin D., Harary F. and Zhang P., Radio labelings of graphs. *Bull. Inst. Combin. Appl.* **33** (2001) 77-85.

7. Chartrand G., Gu W., Schultz M. and Winters S. J., Eccentric graphs. *Networks.* **34** (1999) 115-121.

8. Chartrand G., Saba F. and Zou H. B., Edge rotations and distance between graphs. *Časopis Pěst. Mat.* **113** (1985) 87-91.

9. Chartrand G., Schultz M. and Winters S. J., On eccentric vertices in graphs. *Networks.* **28** (1996) 181-186.

10. Hale W., Frequency assignment: theory and applications. *Proc. IEEE* **68** (1980) 1497-1514.

11. Harary F. and Norman R. Z., The dissimilarity characteristic of Husimi trees. *Ann. of Math.* **58** (1953) 134-141.

12. Hedetniemi S.T. (personal communication).

13. Georges J. P., and Mauro D. W., On the size of graphs labeled with a condition at distance two. *J. Graph Theory* **22** (1996) 47-57.

14. Georges J. P., and Mauro D. W., On the criticality of graphs labelled with a condition at distance two. *Congr. Numer.* **101** (1994) 33-49.

15. Georges J. P., and Yeh R. K., Labelling graphs with a condition at distance two. *Siam J. Discrete Math* **5** (1992) 586-595.

16. Ostrand P. A., Graphs with specified radius and diameter. *Discrete Math.* **4** (1973) 71-75.

17. Slater P.J., Leaves of trees. *Congress. Numer.* **14** (1975) 549-559.

18. Slater P.J., Dominating and reference sets in graphs. *J. Math. Phys. Sci.* **22** (1988) 445-455.

19. *Edwin Armstrong: The creator of FM radio.*

    http://securehosts.com/techa/armstrong.htm

20. Minimum distance separation between stations. *Code of Federal Regulations*, Title 47, sec. 73.207.

Chapter 13

1. Berge C., *Théorie des Graphes et Ses Applications*. Dunod, Paris (1958).

2. Chartrand G., Haynes T. W., Henning M. A. and Zhang P., Stratification and domination in graphs. *Discrete Math.* **272** (2003) 171-185.

3. Cockayne E.J. and Hedetniemi S.T., Towards a theory of domination in graphs. *Networks* **7** (1977) 247-261.

4. Cockayne E.J., Dawes R.M. and Hedetniemi S.T., Total domination in graphs. *Networks* **10** (1980) 211-219.

5. Coonce H. B. (personal communication).

6. Haynes T.W., Hedetniemi S.T. and Slater P.J., *Fundamentals of Domination in Graphs*, Marcel Dekker, New York, 1998.

7. Ore O., *Theory of Graphs*. Amer. Math. Soc. Providence, RI (1962).

8. Rashidi R., *The Theory and Applications of Stratified Graphs*. Ph.D. Dissertation, Western Michigan University (1994).

9. Sutner K., Linear cellular automata and the Garden-of-Eden. *Math. Intelligencer* **11** (1989) 49-53.

10. Walikar H.B., Acharya B. D. and Sampathkumar E., Recent developments in the theory of domination in graphs.. Mehta Research Institute Allahabad, MRI *Lecture Notes in Math.* **1** (1979)

11. *Lights Out*. http://mathworld.wolfram.com/LightsOutPuzzle.html

12. John Wiley & Sons, Inc. *Publsher's Note. J. Graph Theory* **1** (1977) 1.

13. *The Mathematics Genealogy Project*. (H.B. Coonce, managing director) http://genealogy.math.ndsu.nodak.edu/

Index

Names

Mathematical Terms

442

Symmetric group 69
Symmetric property of distance 328
System of distinct representatives 187-
    188

Tait coloring 287
Three Houses and Three Utilities Prob-
    lem 229
Torus 241
Total
    deficiency 253
    dominating set 367
    domination 367
Total domination number 368
Tournament 169
    of paired comparisons 176
Trail
    in a digraph 162
    in a graph 12-13
Transformed by an edge rotation 357
Transitive tournament 170
Tree 87
Triangle 13
Triangle inequality for distance 328
Triangle-free graph 276
Triple 206, 209
Trivial
    graph 3
    walk 11
Turán graph 311
Turán's Theorem 312
Tutte-Coxeter graph 224

Unavoidable set of reducible configura-
    tions 264
Underlying graph
    of a digraph 163
    of a multigraph 53
Uniformly embedded 255
Union of graphs 14
Unlabeled graph 3
Upper Hamiltonian number 154

Value
    of a radio k-coloring 353
    of a radio labeling 355
Vertex (vertices) 2, 27, 162
    set 2
Vertex-connectivity 115
Vertex cover 191
Vertex covering number 191
Vertex-cut 115
Vertex independence number 191, 269
Vertex-transitive graph 68
Vizing's Theorem 282

Wagner's Theorem 250
Walk
    in a digraph 162
    in a graph 11
Weakly connected digraph 163
Weight
    of an edge 96
    of a subgraph 96
Weighted graph 53
Word graph 5

List of Symbols

448